Freshwater Fishes of the Eastern Himalayas

Freshwater Fishes of the Eastern Himalayas

Waikhom Vishwanath

Department of Life Sciences, Manipur University, Imphal, India

ACADEMIC PRESS

An imprint of Elsevier

ELSEVIER

Academic Press is an imprint of Elsevier
125 London Wall, London EC2Y 5AS, United Kingdom
525 B Street, Suite 1650, San Diego, CA 92101, United States
50 Hampshire Street, 5th Floor, Cambridge, MA 02139, United States
The Boulevard, Langford Lane, Kidlington, Oxford OX5 1GB, United Kingdom

Notices

Knowledge and best practice in this field are constantly changing. As new research and experience broaden our understanding, changes in research methods, professional practices, or medical treatment may become necessary.

Practitioners and researchers must always rely on their own experience and knowledge in evaluating and using any information, methods, compounds, or experiments described herein. In using such information or methods they should be mindful of their own safety and the safety of others, including parties for whom they have a professional responsibility.

To the fullest extent of the law, neither the Publisher nor the authors, contributors, or editors, assume any liability for any injury and/or damage to persons or property as a matter of products liability, negligence or otherwise, or from any use or operation of any methods, products, instructions, or ideas contained in the material herein.

British Library Cataloguing-in-Publication Data
A catalogue record for this book is available from the British Library

Library of Congress Cataloging-in-Publication Data
A catalog record for this book is available from the Library of Congress

ISBN: 978-0-12-823391-7

For Information on all Academic Press publications
visit our website at https://www.elsevier.com/books-and-journals

Publisher: Charlotte Cockle
Acquisitions Editor: Anna Valutkevich
Editorial Project Manager: Billie Jean Fernandez
Production Project Manager: Niranjan Bhaskaran
Cover Designer: Christian J. Bilbow

Typeset by MPS Limited, Chennai, India

Contents

About the author ix

Foreword *by Maurice Kottelat* xi

Foreword *by Kailash Chandra* xiii

Preface xv

Acknowledgments xvii

CHAPTER 1   **Introduction** 1

    Eastern Himalaya 1

    Fish biogeography 3

        Center of origin of freshwater fishes 3

        Plate tectonics 5

        Drainage basin evolution 5

        Evolution by vicariance 7

    Ichthyofaunal diversity 11

    Purpose of the book 12

    Taxonomy and nomenclature 12

CHAPTER 2   **Methods Adopted** 15

    Fish collection 15

    Fish preservation 15

    Fish photography 16

    Morphological features 16

        Measurements 16

        Fins 18

    Fins ray counts 19

        Simple rays 19

        Branched rays 20

        Principal rays of caudal fin 20

        Rays in paired fins 20

        Last (most posterior) ray of dorsal and anal fins 20

    Scale counts 21

        Lateral line scales 21

        Transverse scales 22

CHAPTER 3   **Systematic index** 27

CHAPTER 4   **Systematic Account** 43

    Family Notopteridae 43

        Featherfin knifefishes 43

Family Anguillidae 45
    Freshwater eels 45
Family Engraulididae 46
    Anchovies 46
Family Clupeidae 47
    Herrings 47
Family Cyprinidae 50
    Carps and minnows 50
Family Psilorhynchidae 147
    Mountain carps 147
Family Botiidae 157
    Botiid loaches 157
Family Cobitidae 161
    Loaches 161
Family Balitoridae 170
    Stream loaches 170
Family Nemacheilidae 174
    River loaches 174
Family Amblycipitidae 205
    Torrent catfishes 205
Family Akysidae 210
    Stream catfishes 210
Family Sisoridae 212
    Sisorid catfishes 212
Family Siluridae 273
    Sheat fishes 273
Family Chacidae 278
    Angler or frogmouth fishes 278
Family Clariidae 280
    Air breathing catfishes 280
Family Ariidae 283
    Sea catfishes 283
Family Ailiidae 285
    Asian schilbeids 285
Family Horobagridae 289
    Imperial or sun catfishes 289
Family Pangasiidae 291
    Shark catfishes 291
Family Bagridae 292
    Bagrid catfishes 292

Family Salmonidae 315
   Trouts 315
Family Mugilidae 316
   Mullets 316
Family Belonidae 317
   Needle fishes 317
Family Aplocheilidae 318
   Rivulines 318
Family Poeciliidae 320
   Livebearers 320
Family Syngnathidae 321
   Pipe fishes and sea horses 321
Family Synbranchidae 322
   Swamp eels 322
Family Chaudhuridae 324
   Earthworm eels 324
Family Mastacembelidae 325
   Spiny eels 325
Family Ambassidae 329
   Asiatic glassfishes 329
Family Sciaenidae 334
   Drums, croakers 334
Family Nandidae 335
   Asian leaffishes 335
Family Badidae 337
   Chameleon fishes 337
Family Cichlidae 344
   Cichlids 344
Family Gobiidae 345
   Gobies 345
Family Anabantidae 348
   Climbing perches 348
Family Osphronemidae 349
   Gouramies 349
Family Channidae 352
   Snakeheads 352
Family Tetraodontidae 362
   Puffers, globefishes 362

CHAPTER 5   **Miscellaneous Notes** **365**

    Oromandibular structures of *Bangana* 365

       Bangana dero 365

       Bangana devdevi 366

    Barilius versus *Opsarius* 366

       Danionin notch and mandibular symphyseal knob 368

    Devario horai 369

    Devario aequipinnatus 369

    Snout morphology of *Garra* 369

       Snout with smooth dorsal surface group 369

       Snout with transverse lobe group 371

       Snout with proboscis group 371

    Garra nasuta 371

    Poropuntius shanensis 372

    Semiplotus 372

    Tor yingjiangensis Chen and Yang, 2004 372

    Type locality of *Lepidocephalichthys irrorata* 372

    Aborichthys cataracta and *A. verticauda* 373

    Aborichthys kempi 373

    Schistura chindwinica 373

    Paracanthocobitis marmorata 373

    Rhyacoschistura manipurensis 374

    Glyptothorax burmanicus and *G. cavia* 374

    Rama and *Chandramara* 374

    Mystus carcio 375

    Olyra kempi 375

    Channa aurolineata and *C. marulius* 376

Bibliography 379

Index 401

About the author

Waikhom Vishwanath is a renowned fish taxonomist of northeastern India and the adjoining water bodies. He is a PhD of Manipur University and served the university in different capacities till 2019. After superannuation as a professor of Higher Academic Grade, he is now availing BSR-Faculty Fellowship of the University Grants Commission, India. During his research career of about 40 years, he has discovered and described 100 fish species new to science, redescribed some species, and clarified taxonomic ambiguity of several species. He has guided more than 30 PhD students and published more than 100 research papers and 5 books. He has also conducted Fish Taxonomy workshops in Manipur and in other institutes of India. He has trained a dozen of ICAR (Indian Council of Agricultural Research) scientist probationers. He was a member of Research Advisory Committees (RAC) of ICAR-DCFR (Directorate of Coldwater Fisheries Research) and ICAR-CIFRI (Central Inland Fisheries Research Institute) and was the Chairman, RAC of ICAR-NBFGR (National Bureau of Fish Genetic Resources). He took part in the assessment and evaluation of freshwater fishes of the Eastern Himalaya conducted by IUCN and had an important role in writing the IUCN-Red list: The Status and Distribution of Freshwater Biodiversity in the Eastern Himalaya, 2010. He was Cochair of the South Asia, IUCN-FFSG (Freshwater Specialist Group). Vishwanath was awarded EK Janaki Ammal National Award in Animal Taxonomy-2015 by the Ministry of Forests, Environment and Climate Change, Government of India, for his contributions in fish taxonomy. He is now a member of the Research Advisory and Monitoring Committee of the Zoological Survey of India.

Foreword *by Maurice Kottelat*

The Eastern Himalaya is drained by the Ganges, Brahmaputra, Meghna, Kaladan, and Irrawaddy Rivers, and it is no surprise that its waters are inhabited by a very diverse fish fauna. Until recent times, this fauna was considered quite uniform, with many species inhabiting all drainages. It then appeared that the different populations of many of widespread " species" in different drainages are in fact several different species, which, despite the superficial similarity, differ in details of morphology, anatomy, ecology, and genetics. Even within a single drainage, the species diversity is high, especially in the small tributaries and headwaters, where a great number of species new to science have been discovered.

Many of the newly discovered species had remained unnoticed because of their small size, because they have specialized environmental requirements (like fast current and rapids), because they live in remote areas difficult to access or in habitats difficult to explore (like caves and waterfalls), or because their distribution range is limited. Sometimes most of these conditions are met. It is often hard to discover the existence of these fishes, observe them, and study their biology.

While the small size, small distribution, and specialized habitat allowed these fishes to escape from science for a long time, they now are also a threat to their survival. Their diversity is underestimated in environmental impact assesments, and their presence is (willingly or not) often ignored. The mountainous landscapes that characterize the Eastern Himalaya are targets for hydropower development, mining, forestry, and so on, which all irremediably impact aquatic habitats; road constructions often destroy kilometers of streams in hilly areas; introduction of exotic animals decimates or extirpates the original fauna. While a number of protected areas target on large mammals and birds, they usually ignore the native, endemic, stenotopic fishes, except maybe for their value as food for water birds.

Although the discovery phase is not yet finished, the challenge for biology now is also the study of the biology of these fishes and the ecology of their diverse communities, in difficult to access and survey habitats; the challenges for conservation is to understand life under the surface, in an habitat foreign to human senses, and to fight for its survival; the challenge for environmental management is take these habitats into account. Any human impact, also on land and air, ends in the water and has a significance for the survival of aquatic biodiversity. And at the end, fishes have great significance for the survival of humans, primarily as food, but also to stimulate our curiosity and need for intellectual melioration.

With 40 years of experience in research on the taxonomy and biology of the fishes of northeastern India, Waikhom Vishwanath has compiled the available information in the present *Freshwater Fishes of the Eastern Himalayas*.

Freshwater Fishes of the Eastern Himalayas provides a comprehensible review of the 512 species of fishes presently recognized in the Eastern Himalaya in India and adjacent waters. For each, it provides a diagnosis or a short description and information on the distribution. Most species are illustrated with color photographs, most of which are published for the first time.

I wish that this work will help those working on the biology of the fishes in Eastern Himalaya. I hope it will attract attention to their great diversity and that it will be an important tool for their conservation and management. It should also be of interest for aquarists, anglers, or more generally anybody interested in fishes.

Freshwater Fishes of the Eastern Himalayas is certainly not the last word on the topic. Numerous new discoveries are still awaiting naturalists, it is for them to go out, in unsurveyed areas or insufficiently searched habitats, and open their eyes.

Maurice Kottelat
Independant Aquatic Biodiversity Consultant, Delemont,
Switzerland Honorary Research Associate,
Lee Kong Chian Natural History Museum,
National University of Singapore, Singapore

Foreword *by Kailash Chandra*

Eastern Himalayas is one of the biodiversity hotspots in India, covering an area from Sikkim to Arunachal Pradesh, adjoining the northeastern states of India and northern Bengal in the Himalayan foothills, as well as the adjacent areas sharing similar drainages. The region is very rich in freshwater fauna especially the fish diversity. The diversity is attributed to its physiography, variety of eco-climatic conditions, drainage pattern, and tectonic history. The region is very rich in endemic fish elements. The paleo drainage evolution contributing to the diversity and evolution of fishes in the region have become a matter of great interest. Recognizing the importance of the freshwater fauna in the region, research on the fish diversity of the Eastern Himalayas have been given great importance in the last few decades and hundreds of new taxa have been published in the recent past from the region.

Since the literature on the subject is scattered in the form of research papers and not yet comprehended, the students and researchers find difficulty in gathering all the available information for their studies. Hence the up-to-date information on the freshwater fish diversity of the Eastern Himalayas has been brought out by the author in the form of a book. This book provides diagnostic characters of about 500 species of fishes under 128 genera and 40 families known so far. Key to genera of the family is also provided. Color images of almost all the fishes, except a few, have been included. Images of parts of fishes showing the diagnostic characters have also been incorporated wherever necessary. This book provides a basic foundation of the freshwater fish diversity of the Eastern Himalayas. I congratulate the author, who has the authority on freshwater fish diversity in India, to bring out the book titled *"Freshwater Fishes of the Eastern Himalayas."* This book will be useful not only for the assessment of fishery resources of eastern Himalayas but also for the conservation of the biodiversity for the future generation.

Kailash Chandra
Director, Zoological Survey of India, Ministry of Environment,
Forest and Climate Change, Government of India, India

Preface

The students of *Fish and Fisheries* special paper (I was one of them) were taken by our teacher, Dr. A.K. Mittal, to a fishing trip to the River Varuna, a tributary of the Ganges at Varanasi in 1974. We all boarded a fishing boat; the fisherman went on catching fishes using a dip net. The teacher told us to identify the fishes immediately after the catch, as fast as we could, using some books and class notes we took to the site. Since then, fish identification has become a fascination for me.

After the completion of a master's degree course, I used to go to different water bodies of Manipur and of the adjoining areas of Assam and Myanmar. Sufficient books and literature were not available, and fishes could not be identified without proper guidance. I often visited Zoological Survey of India, where I met Dr. K.C. Jayaram, Dr. A.G.K. Menon, and Dr. P.K. Talwar, renowned ichthyologists of India. They encouraged me to perform field surveys in northeastern India which they described as the gold mine of freshwater fishes. It turned out to be true, and I could describe a few interesting species new to science. My interest on fish taxonomy was enhanced after I attended European Ichthyological Congress at Trieste, Italy, in 1996 and had a chance to interact with some actively working ichthyologists of the world. Many of them had an interest in my work, and some named me a new-generation Indian ichthyologist. I had also a chance to discuss on the manuscripts of their unpublished research papers and books.

Fish identification was not taken seriously in the past. Students often compared specimens with the published figures in books like that of Day's (1879) Vol. 2 of *Fishes of India* or followed the artificial keys provided by some authors in their books. Systematic examination of specimens was seldom done, and people followed shortcut ways. This led to misidentifications and created several problems. In fact the science of taxonomy was regarded as outdated. However, identifying correct species before proceeding to further biological studies is very important.

Fishes are aquatic and thus water is their boundary. Majority of them are confined to a particular drainage basin, except for those which are air breathing and those which are prolific breeders and easily propagated intentionally or otherwise. Some fishes particularly those inhabiting hill streams are stenotopic and have habitat preference. Species previously thought to be widely distributed are now, after careful examination, proved to be different species. Drainage basin and habitat preference of hill stream fishes are important aspects in studying fish taxa.

With the gradual understanding of the fact, a number of new species have been described from northeastern India and the adjoining water bodies. Taxonomic problems of ambiguous species have been resolved. Osteological studies have been carried out, and it formed the basis for other workers to establish taxa with additional molecular characters. In fact, a change in the status of fish taxonomy studies has been seen in the past few decades. Many workshops and hands-on training for fishery scientists have been conducted, and probational training for scientist probationers has been given.

The need for a book on the fishes of the Eastern Himalaya was strongly felt since the author attended the IUCN workshop on the assessment and evaluation of freshwater fishes of the region in 2009 at Kathmandu. Eastern Himalaya has been recognized as a freshwater biodiversity hotspot. At the workshop the list of fishes for assessment was not readily available. A list of 296 fishes included in a book entitled "*Fishes of Northeast India*" which was published in 2007 was available with the author. Other resource persons who attended the workshop added on further to the list and formed the basis for the assessment. The rich diversity of freshwater fishes of the region in view of the presence of different drainages and their evolution, ecological conditions, etc. have attracted ichthyologists in the recent past. This has resulted in several surveys and contributed much to the literature of the fauna of the region. However, the information is scattered, and researchers find difficulty in finding the relevant papers for their work.

Molecular characterization and DNA barcoding have added another dimension in the traditional morphological technique of fish identification now and have proved to be very useful in solving taxonomic problems. Scientists involved in the molecular characterization may not have basic ideas of fish taxonomy. Barcodes of misidentified species are sometimes submitted to gene banks, and this often leads to confusions to future workers. Thus correct identification of species is an essential step before proceeding further works.

Fish identification thus involves many considerations: careful examination of characters, habit and habitat, stages of life history, behavior, drainage basin and its evolution, etc. The book is a modest attempt to compile the diagnostic characters, original references, maximum size, and up-to-date systematic status of all the available fish species in the Eastern Himalaya. It is hoped that the book will be helpful to all concerned related to fish and fisheries for the identification of species, pursue further research, and fisheries planning.

Acknowledgments

It would not have been possible to write this book without the generous help and contribution of several persons. I am indebted to Maurice Kottelat for briefly going through the manuscript and furnishing valuable comments to improve the book, especially on the boundary of the Eastern Himalaya and also agreeing to write the Foreword. I am thankful to Kailash Chandra for his support in the study and also for writing a Foreword.

I extend gratefulness to the following persons for giving access to their collections: Maurice Kottelat (Switzerland), Heok Hee Ng [(Singapore), Fisheries Research Division, Nepal Agricultural Research Council (Kathmandu), Director, Zoological Survey of India, Kolkata, Officer in charges of ZSI-SRS (Chennai), ZSI-ERS (Shillong), ZSI-APRC (Itanagar), ZSI-NRC (Dehra Dun)], D.N. Das (Rajiv Gandhi University, Itanagar), Dandadhar Sharma and Hrishikesh Choudhury (Department of Zoology, Gauhati University), Rupak Nath (St Anthony College, Shillong), Lalramliana (Pachunga College, Mizoram), Sewali Pathak (Chirang District, Assam), Mrigendra Mohan Goswami (Gauhati University), and Shyama Prasad Biswas (Dibrugarh University). The collections of Niti Sharma, Satish Kumar Koushlesh, Jyotish Barman, N. Pitambari, and all ICAR Scientist probationers attached to the Department of Life Sciences, Manipur University, were also useful for the present work.

Valuable reprints and PDF files of research publications were made available by E. Zhang, Heok Hee Ng, K. Nebeshwar, Lalramliana, Lukas Ruber, Maurice Kottelat, Ralf Britz, R.J. Thoni, Rohan Pethiyagoda, Sven Kullander, Tyson Roberts, Wei Zhou, Carl Ferraris, Petru M. Bănărescu, Larry M. Page, and Kevin K. Conway.

Complimentary copies of valuable books and monographs given to me by some authors were of great help in writing the present book, to name a few: Achom Darshan (Biodiverity of Fishes in Arunachal Himalaya), Maurice Kottelat (Indochinese Nemacheilines, Handbook of European Freshwater Fishes, Fishes of Laos, Conspectus Cobitidum), Promož Zupančič (Rijetke I Ugrožene Slatkovodne Ribe Jadranskog Slijeva Hrvatske, Lovenije I Bosne I Hercegovine), Ralf Britz (Francis Hamilton's Gangetic Fishes in Colour), Sven Kullander (The Fishes of the Kashmir Valley IN River Jhelum, Kashmir Valley, impacts on the aquatic environment), and Tej Kumar Shreshtha (Ichthyology of Nepal).

I owe my thanks to Achom Darshan, Bungdon Shangningam, Chinglemba Yengkhom, Chungkham Sarojnalini, Irengbam Linthoingambi, Juliana Laishram, Keisham Shanta, Khangjarakpam Geetakumari, Kongbrailatpam Nebeshwar, Laishram Kosygin, Laishram Shakuntala, Manohermayum Shantakumar, Mayanglambam Dishma, Narengbam Roni, Sapam Anganthoibi, Selim Keishing, Wahengbam Manojkumar, Yumnam Bedajit, Yumnam Lokeshwor, Yumnam Ramananda, and Yumnam Rameshori, who conducted several field trips to different water bodies of the Eastern Himalaya and also managed to obtain fish collections from the remote headwaters of several river drainages and working on the species accounts. I am grateful to Mutua Bahadur for taking the trouble to collect fishes from Kalemeu, Myanmar to Churamani alias Lalcharliana, Chawntalaipui, Lunglei District, Mizoram, and his family members for their help in the collections of specimens from various parts of Mizoram; B.D. Shangningam and her brothers and sisters for arranging many fishing trips in the Ukhrul and Chandel districts of Manipur; Sarbojit Thaosen for specimens in the north flowing tributaries of the Brahmaputra in Dima Hasao;

Sewali Pathak for collections of fishes in the streams of Assam bordering Bhutan; J. Laishram for arranging fishing trips in the Rangeet and Teesta rivers, Sikkim; R.K. Sinha of Patna for deputing Rajesh Sinha to Manipur with the collections of fishes from the Ganges; S.P. Biswas and Santosh Abujam for arranging materials from upper Assam; scholars of the College of Fisheries, Tripura for collections from Tripura and the adjoining water bodies in Bangladesh; Kento Kadu and Kenjum Bagra for specimens from Arunachal Pradesh.

The following persons permitted to use their unpublished photographs of fishes. They are Achom Darshan, Arpita Dey, Bungdon Shangningam, Gurumayum Shantabala, Hrishiokesh Chowdhury, Laith A. Jawad, Lalramliana, Beta Mahatvaraj, and Maurice Kottelat.

Narengbam Roni helped in inserting the maximum size of the fishes; Achom Darshan helped in checking the references and in arranging the figures and copyright forms; and Yengkhom Chinglemba and A. Darshan helped in sorting out the fishes in the museum for study and photography.

I shall be failing on my part if I do not record the huge moral support given by my father W. Tomchou Singh who left for heavenly abode and could not see the book released. I am indebted to my wife Achom Umabati for her support and for caring my children and also to my daughters: Varuna, Gangotri, and Triveni and my son Gajanand for their support in the work.

BSR (Basic Scientific Research) Faculty Fellowship awarded by the University Grants Commission, India No. F. 18-1/2011(BSR), dated 21 December, 2018, is gratefully acknowledged.

Waikhom Vishwanath

Introduction

Freshwater ecosystems, which hold about 0.01% of world's water are the most endangered ecosystems in the world and the declines in biodiversity in it are far greater than in the most affected terrestrial ecosystems (Sala et al., 2000). The reasons are the environmental change and human activities. Freshwaters house several plants and animals of which 40% is global fish diversity and 25% is global vertebrate diversity.

While there is much attention paid to the threats to charismatic terrestrial organisms such as mammals, birds, and orchids, or to disappearing habitats such as rain forests, the same is not so to many freshwater habitats that are under serious threat and to the very large number of aquatic organisms within them that face imminent extinction (Kottelat and Whitten, 1996). Wetlands are often given importance for conservation of migratory birds, and so on, although they are rich in freshwater biodiversity.

Freshwater biodiversity concerns will be addressed by paying attention to freshwater biodiversity issues, collecting good and current data, executing studies to understand the whole economic values of freshwater biodiversity, maintaining the ecological balance, preparing field guides and manuals, nurturing partnerships between engineers and biodiversity specialists to achieve better project designs and more effective mitigation measures, and releasing of scientific data collected by project proponents as part of environmental assessments.

The threats to global freshwater biodiversity can be grouped under five interacting categories: overexploitation, water pollution, flow modification, destruction or degradation of habitat, and invasion by exotic species (Dudgeon et al., 2006). Additional threat now, more seriously for cold-blooded organisms is the climate change. Knowledge of the total diversity of freshwaters is woefully incomplete, particularly among invertebrates and microbes, and especially in tropical latitudes that support most of the world's species. Even vertebrates are incompletely known, including well-studied taxa such as fishes (Stiassny, 2002).

Eastern Himalaya

Eastern Himalaya comprises the lowlands of western Nepal and the montane regions of central and eastern Nepal; Bhutan, the northern extent of West Bengal, and all the northeastern states of India, namely, Arunachal Pradesh, Assam, Manipur, Meghalaya, Mizoram, Nagaland, Sikkim, and Tripura (CEPF, 2005). The region has a much more sophisticated geomorphic history and pervasive topographic features than the Central Himalayas. The region's topography, in part, has facilitated the region's rich biological diversity and ecosystem structure. In fact, the region is a part of two biodiversity hotspots: the Himalaya and Indo-Burma (CEPF, 2005). The region is dominated by a monsoon climate from June to September and westerly disturbances in the remaining months. The Himalaya region acts as a

Freshwater Fishes of the Eastern Himalayas. DOI: https://doi.org/10.1016/B978-0-12-823391-7.00004-7

climatic barrier between lower and midlatitudes in the global atmospheric circulation systems and is responsible for the moist summers and mild winters in south Asia (Shrestha and Devkota, 2010).

Allen et al. (2010) presented a map of the Eastern Himalaya Freshwater Biodiversity Hotspot (Fig. 2.1) in black boundary, which is reproduced here as Fig. 1.1, overlaid with drainages and freshwater ecoregions. The region in the black boundary is treated herein as the Eastern Himalaya. Freshwater fishes within the boundary and those in the adjacent areas sharing similar drainages have been included in the present work.

Eastern Himalaya is drained by several river drainages, namely, the Ganga, Teesta, Brahmaputra, Barak−Surma−Meghna, Kaladan, and the Chindwin. Kottelat and Whitten (1996) also presented a map of freshwater biodiversity hotspot based mainly on fishes, in which large parts of eastern Himalaya is included. The region covers five ecoregions on the basis of the map of the freshwater ecoregions presented by Abell et al. (2008), namely, the Ganga Delta and Plain, Ganga Himalayan Foothills, Middle Brahmaputra, Chin Hills-Arakan Coast, and the Sittaung-Irrawaddy (Fig. 1.1).

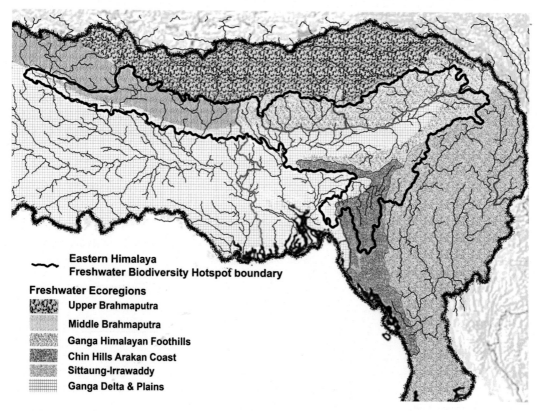

FIGURE 1.1

Eastern Himalaya (EH) boundary in diffused black and EH Freshwater Biodiversity Hotspot boundary in black, overlayered with the freshwater ecoregions of the region as per Abell et al. (2008) and the drainage system.

Redrawn after Allen DJ, Molur S, Daniel BA: The status and distribution of freshwater biodiversity in the Eastern Himalaya, Cambridge, UK and Gland, Switzerland: IUCN, and Coimbatore, India, 2010: Zoo Outreach Organisation.

Eastern Himalaya is a biodiversity-rich region. The diversity is attributed to the recent geological history (the collision of Indian, Chinese, and Burmese plates) and the Himalayan orogeny played an important role in the speciation and evolution of groups inhabiting mountain streams (Kottelat, 1989). The evolution of the river drainages in this part of the world has been the subject of several studies that utilize geological evidence to reconstruct the palaeodrainage patterns during much of the Cenozoic era (65.6 million years ago to the present) (e.g., Brookfield, 1998; Clark et al., 2004). Molecular phylogenetic studies of the fishes of this region (e.g., Guo et al., 2005; Rüber et al., 2004) have indicated that vicariance events in the Miocene (23.0−5.3 million years ago) may have played a substantial role in shaping the current distribution pattern of the freshwater fishes of the region.

Kottelat (1989) divided the fish fauna of Asia into three subregions, namely, (1) South Asia, (2) South-East Asia, and (3) East Asia. The South Asia subregion included Indus, Ganga, Brahmaputra and Irrawaddy basins, and Peninsular India. Thus Eastern Himalaya is the home of a part of the South Asian fish fauna.

The fish fauna of the Eastern Himalaya region may be subdivided into three drainage-based geographic units:

1. The Ganga−Brahmaputra drainage that flows in the Ganga Himalayan Foothills, Ganga Delta and Plain ecoregions, and the Upper and Middle Brahmaputra;
2. The Chindwin−Irrawaddy drainage in the Sittaung-Irrawaddy freshwater ecoregion;
3. The Kaladan/Kolodyne drainage and a number of short drainages along the western face of the Rakhine-Yoma of Myanmar in the Chin Hills-Arakan freshwater ecoregion.

Freshwater ecosystems have become the most endangered ecosystems in the world (Dudgeon et al., 2006). Freshwater habitats and the biodiversity that they support are especially vulnerable to human activities and environmental change. The reason is the enormous richness of inland waters as a habitat for plants and animals. More than 10,000 fish species which is approximately 40% of global fish diversity and one-fourth of the global vertebrate diversity live in freshwater (Lundberg et al., 2000). If aquatic vertebrates, namely, amphibians, reptiles (crocodiles, turtles), and mammals (otters, river dolphins, and platypus) are taken together in addition to the freshwater fish population, it becomes clear that as much as one-third of all vertebrate species are confined to freshwater. However, freshwater habitats contain only around 0.01% of the world's water and cover only about 0.8% of the Earth's surface (Gleick, 1996). About 100,000 species out of approximately 1.75 million species described inhabit freshwater (Hawksworth and Kalin-Arroyo, 1995) and an additional 50−100 thousand species inhabit the groundwater (Gibert and Deharveng, 2002). There is a serious and immediate danger to the freshwater because of several threats and human activities (Stiassny, 1999).

Fish biogeography

Center of origin of freshwater fishes

"Center of dispersal" is the area within a larger area inhabited by a genus, the most progressive species being at the center and the most primitive or generalized species at the remote end (Matthews, 1915). Based on the theory of Matthews (1915), Oriental region is considered as the "center of origin" of ostariophysans (Briggs, 1989) and South-East Asia as that of the freshwater fish fauna of India (Bǎnǎrescu and Nalbant, 1982). It was believed that species from the center of origin distributed centrifugally resulting in the more conservative species inhabiting farthest from it (Fig. 1.2).

The discontinuous distribution of some hill stream fishes in North-East India and Peninsular India was also explained by Hora's (1944, 1949) Satpura Hypothesis. The hypothesis assumed the dispersal of the South-East Asian fauna through the North-East India corridor, via a mountain range through the Garo Hills and Rajmahal Hills (Garo-Rajmahal gap) by imagining a complex geological history.

Most past workers including Darlington (1957) were active supporters of continental stability. Later, continental drift has become a generally accepted fact and an increasingly greater number of biogeographers explained distributions of species in the light of continental drift and plate tectonics (Bǎnǎrescu, 1985). Croizat et al. (1974) rejected the concept of "center of origin" of species and Kottelat (1989) pointed out the Satpura Hypothesis to have no geological basis, referring to important works.

He et al. (2001) while studying the distribution of glyptosternoid fishes of Tibet found the fishes to be not having a center of origin, their ancestors occurring in a wide area and not possible to trace a concept of dispersal.

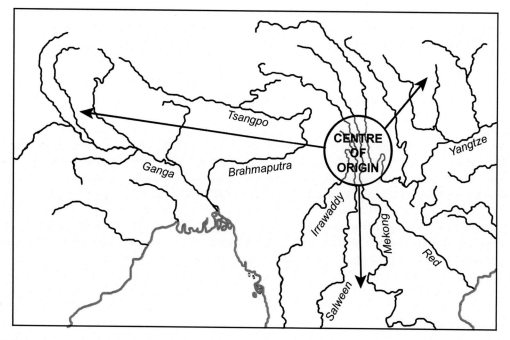

FIGURE 1.2

South-East Asia as the center of origin of freshwater fishes.

Redrawn after He S, Wenxuan C, Chen Y: The uplift of Qinghai-Ziang (Tibet) plateau and the vicarian speciation of glyptosternoid fishes (Siluriformes: Sisoridae), Sci China (Ser C); 44(6):644–651, 2001.

Plate tectonics

Earlier theories of the distribution of species were mostly based on the concept of continental stability (including Darlington, 1957) and the occurrence of land bridges (Bering, Panama, Suez, Indo-Australian archipelago, etc.) in the past for dispersal of species. Now it is accepted that continents do drift and thus the earlier theories of distribution have been rejected.

Continental drift is the theory that the Earth's continents have moved over geologic time relative to each other, thus appearing to have drifted across the ocean bed. The idea of the drift has been supplemented by the theory of plate tectonics, which explains that the continents move by riding on plates of the lithosphere.

Wegener (1912) was the first to propose the concept of continental drift, a large-scale horizontal movements of continents relative to one another and to the ocean basins during one or more episodes of geologic time. He assembled several evidences to show that Earth's continents were once connected in a single supercontinent, now known as Pangaea. The theory was based on his observations that: (1) fossils of animals and plants of similar kind occur in different continents, the dispersal of which is not possible through the seawater; (2) the shorelines of the continents have interlocking fits; and (3) the mountain ranges in the adjoining continents have continuity and the composition of rocks are similar. The heat of the inner "core" of the earth heats up the middle layer, the "mentle" in which the crust is partially melted causing a convection current. The current cause the uppermost layer, the "crust" to move in the form of plates. The concept of the drifting of continents was an important precursor to the development of the theory of plate tectonics. The theory is now widely accepted after the discovery of plate edges through magnetic surveys of the ocean floor and through the seismic listening.

Kumar et al. (2007) predicted that the mental plume heating from below resulted in the breakup of the Gondwana into Africa, Antarctica, Australia, and India about 140 million years ago. The Indian plate attained a very high speed of about 18–20 cm/year during the late Cretaceous period, due to ridge push or slab pull, unique among the other fragments which was attributed to the melting of the lithospheric roots by the plume, making it the thinnest plate of about 100 km thickness. The other plates maintained a thickness of about 180–300 km. The continental collision with Asia some 50 million years ago resulted in the highest mountain chain, the west-east extending Himalaya in the world and subsequent formation of the north-south extending Indo-Myanmar (Indo-Burman) range in the east (Figs. 1.3, 1.4, 1.5).

Drainage basin evolution

The present existing river system of the Eastern Himalaya and Southeast Asia are very much related to the plate tectonics, collision of the Indian Plate with the Eurasian Plate and the Himalayan orogeny. Kottelat (1989) discussed the existing river system of South-East Asia with the post-Himalayan river system hypothesized by Gregory (1925). The hypothesis included Upper Yangtze connecting to Red River, Upper Mekong to the Chao Phraya, the Upper Salween to the Chao Phraya, the Upper Irrawaddy to the Sittang, and the Tsangpo to the Chindwin and lower Irrawaddy. Based on the fish fauna, and their affinities as observed by Kottelat (1989), some of the conclusions contradict others. It was found that the Brahmaputra and Irrawaddy faunae had strong affinities with that of the Salween.

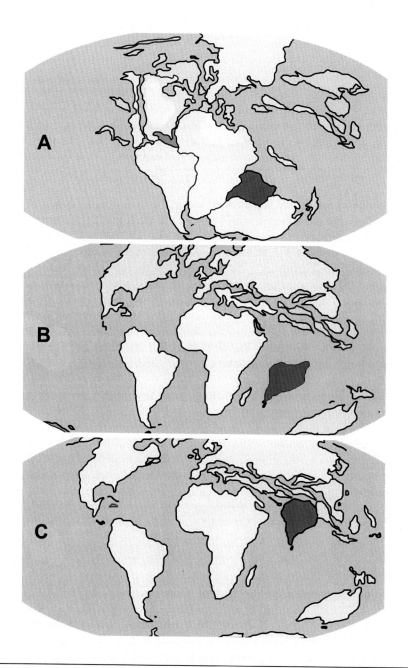

FIGURE 1.3

Drift of Indian plate (dark gray): (A) 200 MY, (B) 70 MY, and (C) 40 MY.

Roughly reconstructed on the basis of Van der Voo R: A plate-tectonic model for the Paleozoic assembly of Pangea based on paleomagnetic data. In Hatcher RD Jr, Williams H, Zietz I, editors: Contributions to the tectonics and geophysics of mountain chains, *1983, pp 19–23 and Chatterjee S, Scotese CR, Bajpai S: The restless Indian plate and its epic voyage from Gondwana to Asia: its tectonic, paleoclimatic and paleobiogeographic evolution.* Geol Soc Am Spl Pap 529:1–149, 2017.

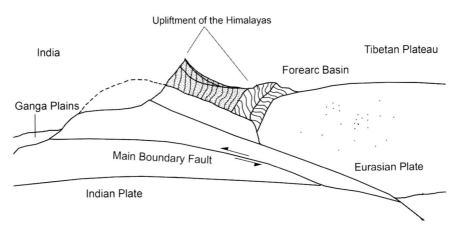

FIGURE 1.4

Himalayan upheaval in the north.

Redrawn after Williams MW: Mountain geography, University of Colarado at Boulder Week 11: Class notes. (Accessed 31 December 2019). Snobear.colarado.edu. > Marckw > Mountains > week11.

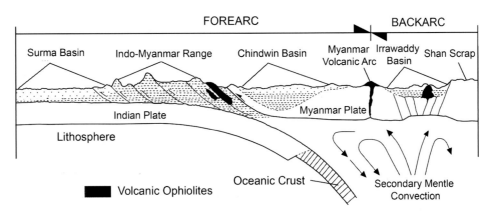

FIGURE 1.5

Formation of Indo-Myanmar range in the east.

Redrawn after Soibam I, Khuman MCH, Subhamenon SS: Ophiolitic rockes of the Indo-Myanmar ranges, NE India: relicts of an inverted and tectonically imbricated hyperextended continental margin basin?. In Gibson GM, Roure F, Manatschal, editors: Sedimentary basins and crustal processes at continental margins: from modern hyperextended margins of deformed ancient analogues, The Geological Society, London, 2015, pp 301–331.

Evolution by vicariance

Bănărescu (1985) reviewed Croizat's (1952, 1958, 1962, 1982) principles of biogeography and freshwater zoogeography which have the main components: (1) species originate through vicarism, that is, geographical isolation, ancestral species splitting in two or more daughter species, each with a smaller range as a result of geographical and climatic changes; (2) splitting of wide ancestral ranges into smaller ones being the main phenomenon in the speciation process, with the rejection of the concept of "center of dispersal"; (3) the present-day occurrence of a monophyletic lineage in distant continental areas separated by seas is, at least usually, not the result of long-distance colonization across the barrier, but the consequence of the presence of the ancestors on a continuous area that later spilt. This necessarily implies that past geography was different from the present one.

Kottelat (1989) subscribes to the concept of the disjunct pattern of distribution of fishes including North-East India on the one hand and South India on the hand as the remnants of an older wider distribution and not recent migration. The occurrence of torrential fishes in these parts of the continent has been considered to be due to convergent evolution. He was of the view that high diversity of fish fauna in South-East Asia is because the area consists of three very distinct units with almost nothing in common (grossly corresponding to the Salween, Mekong, and Red River basins), suggesting that three faunae evolved without contact for a long time. Considering this trichotomy, the whole area cannot be accepted as a potential "center of origin." Taken separately, these units no longer give this impression of great taxonomic diversity. In fact, the diversity observed is better explained as three distinct faunae; an additional factor, recent geological history, especially the Himalayan orogeny which played an important role in the speciation and evolution of groups inhabiting mountain streams.

He et al. (2001) and Peng et al. (2006) on the basis of phylogenetic studies of glyptosternoids recognized that the Tibetan plateau uplift was likely in three main steps. He et al. (2001), on the basis of the phylogenetic studies of the glyptosternoid fishes in the Tibet area, proposed the following hypothesis: the speciation of the group has a direct relationship with the three uplifts of the Tibet (Fig. 1.6). The process was explained by the theory of vicariance of biogeography.

1. Tethys sea closed and the Indian plate collided with the Eurasian Plate in the Cenozoic era, this area gradually uplifted. This made the Tibetan plateau appear, and the plain uplift higher and higher. The *glyptothorax*-like ancestor that adapted to the environment of large rivers was obligated to live in a very different environment and modified gradually as the primitive glyptosternoid fishes. This period lasted a very long time, allowing this ancestor to become distributed over a very wide area: Kabul and Amu in Afganistan and Indus in Pakistan in the west and in the upper Yangtze and Tsangpo in the east. In this phase, the plateau was not very high, the climate was moist and warm, which facilitated the wide distribution of glyptosternoid fishes.

2. After the Pleistocene, with the enforced colliding, the gradual uplift of the Qinghai-Tibet Plateau brought about the current water environment, and the Glyptosternoids were generated from *Glyptothorax*-like fish under this environment. The present *Glyptosternum*, distributed across the Himalayas, is the ancestor of Glyptosternoids. The second uplift allows speciation resulting in the appearance of the genus *Euchiloglanis*, allopatrically distributed in small tributaries of Brahmaputra Basin and Amu River. The second uplift of

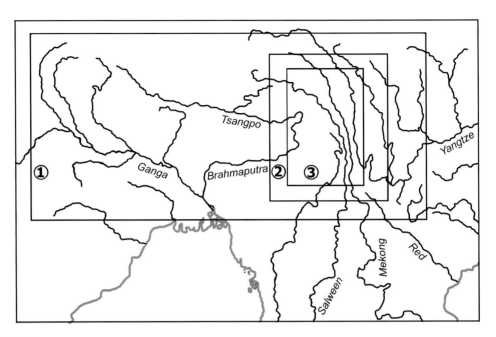

FIGURE 1.6

Evolution of fish species by vicariance due to uplift of Tibetan Plateau.

Redrawn after He S, Wenxuan C, Chen Y: The uplift of Qinghai-Ziang (Tibet) plateau and the vicarian speciation of glyptosternoid fishes (Siluriformes: Sisoridae), Sci China (Ser C) *44(6):644–651, 2001.*

the plateau produced a harsher environment. This phase may have been very short as not much speciation occurred.

3. The third uplift, the so-called post-Himalayas orogenic movement, by its largest scale and altitude, enormously affected speciation within the glyptosternoid fishes. In two billion years since the beginning of the Quaternary, this region rose over 3000 m, implying an average uplift rate of 1−1.5 mm per year. As the land uplifted, the rivers cut deeply and separated, an *Euchiloglanis*-like ancestor was assigned to these rivers. After isolating for a long time, they were specialized into different species.

In the three uplift intervals of the plateau, the water system of this region was separated gradually and *Glyptosternum*-like ancestor was isolated in different rivers and evolved into various species. All this resulted in the speciation and formation of the biogeographical pattern of glyptosternoids.

Rüber et al. (2004) reconstructed phylogenetic relationships of the family Badidae using both mitochondrial and nuclear nucleotide sequence data to address badid systematics and to evaluate the role of vicariant speciation on their evolution and current distribution. They also used their molecular phylogeny to test a vicariant speciation hypothesis derived from geological evidence of large-scale changes in drainage patterns in the Miocene affecting the Irrawaddy— and Tsangpo—Brahmaputra drainages, in the southeastern Himalaya. Within both genera, *Badis* and *Dario*, they observed a divergence into Irrawaddy— and Tsangpo—Brahmaputra clades which were tentatively dated the vicariant event at the Oligocene—Miocene boundary (19—24 Myr), which seems to agree with geological evidence for the separation of these drainages caused by tectonic uplifts in the Eastern Tibet. These geological events appear to have played a major role in the diversification of badids thus supporting the hypothesis of primarily vicariant-based speciation in this group (Fig. 1.7).

However, given the current uncertainty in the paleobiogeographic reconstruction of North-East India and Myanmar, establishing temporal congruence between potential vicariant and cladogenetic events remains provisional. It is expected that new paleobiogeographic insights may come out from broader comparative phylogenetic studies of freshwater fishes, from the Indo-Burma region (North-East India and Myanmar), one of the world's leading biodiversity hotspots.

Rüber et al. (2004) also mentioned that their data were concordant with a hypothesized paleo connection of the Tsangpo River with the Irrawaddy drainage that was most likely interrupted during Miocene orogenic events through tectonic uplifts in eastern Tibet. Their data indicated a substantial role of vicariant-based speciation shaping the current distribution patterns of badids.

The Indo-Myanmar (also known as Indo-Burman) Range is on the convergent boundary of the Indian and Burma microplates in Myanmar. The subduction between the two plates resulted in the development of accretionary wedges and then thrusting, folding, and uplifting formed the Indo-Myanmar ranges (Mohiuddin and Mustafa, 2003). The mountain belt comprises the Arakan-Yoma (also called Rakhine-Yoma) mountains and the Chin-, Naga-, Manipur-, Lushai-, and Patkai hills (Wang et al., 2014). The Indo-Myanmar range merged with Eastern Himalayan Syntaxis further north, submerged into the Andaman Sea and resurfaced as Andaman Islands further south (Mukhopadhyay and Sujit, 1988). Central plain of Manipur is in between the Siruhi Hill range in the east which further continues southward merging with the Chin Hills and Koubru Hill range in

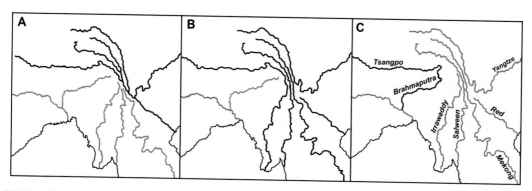

FIGURE 1.7

River basin evolution in South and Southeast Asia.

Redrawn after Rüber L, Britz R, Kullander SO, Zardoya R: Evolutionary and biogeographic patterns of the Badidae (Teleostei: Perciformes) inferred from mitochondrial and nuclear DNA sequence data, Mol Phyl Evol 32:1010—1022, 2004.

the west which continues southward with Mizoram Hills. The central plain and the eastern hills of Manipur are drained by the headwaters of the Chindwin and the western hills of the state, by the headwaters of the Barak, a part of the Surma-Meghna drainage. Soibam and Hemanta Singh (2007) and Soibam et al. (2015) reported that the rifting and stretching of the crustal layer (lithosphere) of this region possibly initiated sometimes toward the close of the Mesozoic era (upper Cretaceous), and the Manipur central plain was evolved because of the passive rifting of the continental margins, that is, sinking of a former plateau. There is a probability of the stream captures between the Barak and Chindwin headwaters before the rifting and stretching or during the complex geomorphologic changes, which might have contributed to the distribution of some species, sharing of genera, and so on.

Rüber et al. (2020) opined that collision of the Indian and Eurasian landmasses in the Cenozoic was a decisive factor in shaping biodiversity patterns in Southern and Southeast Asia. They also reported that most studies have been focused on the biotic interchange between India and Eurasia and evolutionary diversification on or around the Tibetan plateau. However, little attention has been paid to the biodiversity buildup in the Eastern Himalaya biodiversity hotspot which harbors over 540 freshwater fishes with high degree of endemicity. Many species of *Channa*, particularly those of the *gachua* group are found in the Eastern Himalaya making this area an outstanding hotspot for channid diversity. The endemics are restricted to the southern foothills of the Eastern Himalaya and the Shillong-Mikir Hills Plateau, areas west of Indo-Myanmar ranges. Their results reveal complex biogeographic patterns indicating that both vicariance and dispersal events have potentially been responsible for shaping current distribution patterns in Asian channids.

Ichthyofaunal diversity

Estimates of fish diversity across the Eastern Himalaya vary widely. Kottelat and Whitten (1996) estimated the Ganga River drainage to contain 350 and Tsangpo and Brahmaputra river drainages to contain 200 species of fishes, respectively. Inferred from the Fig. 1.2 of Abell et al. (2008), the region is the home of about 500 fish species, of which the Irrawaddy ecoregion contains more endemic species of freshwater fish (between 119 and 195) than any other Eastern Himalayan freshwater ecoregions: 28−40 in Ganga Delta and Plains Chin Hills-Arakan, 12−19 in Ganga Himalayan Foothills, and 1−11 in the Middle and Upper Brahmaputra.

Vishwanath et al. (2010) presented a map showing portions of the Brahmaputra drainage in Arunachal Pradesh, parts of Assam, Meghalaya, northern Bengal, the Himalayan foothills between Nepal and Bihar to exhibit the most diverse fish fauna (107−154). Species richness is highest in the Teesta, Kameng, Dikrong, Subansiri, and Siang basins. They commented the richness due to the diversity of habitats and environments existing between the plains of the Brahmaputra at a low altitude (120−200 m) to the upland cold water regions (1500−3500 m) in the hill ranges in Arunachal Pradesh and also in Meghalaya and Assam within a short aerial distance of 200−500 km. Similar levels of richness are expected in other parts of Assam, Arunachal Pradesh, and Bhutan having connections with the Brahmaputra drainage and also in the Barak−Surma−Meghna basins in northeastern India and Bangladesh and also in the Chindwin and Kaladan drainages. The Kaladan River is a drainage that flows between the Barak−Surma−Meghna and the Chindwin−Irrawaddy drainages.

The river is separated from the Barak–Surma–Meghna drainage by the Chittagong hill tract in the west and from the Chindwin–Irrawaddy by the Rakhine Yoma hill range in the east. The ichthyofauna of the Kaladan is poorly explored and expected to contain many endemic species (Anganthoibi and Vishwanath, 2010a).

Ichthyofaunal survey in the Surma-Meghna drainage in Bangladesh, Kaladan, and Chindwin drainages have gained momentum in the past about 10 years. There have been contributions to the fish fauna of Bhutan. More than 100 species of freshwater fishes have been added to the list of fishes of the eastern Himalaya recently.

Purpose of the book

Considering the high species richness in the Eastern Himalaya and the region identified as a freshwater biodiversity hotspot, great attention is now being paid to the faunae of the region. Freshwater fishes are of considerable economic value and are an important source of food for many people. There has been an increasing demand for new fishes in the ornamental fish trade. This diversity can be adversely affected by projects related to dams, flood control, water supply, mining, fisheries management, introduction of exotic species, irrigation, bridges, industrial effluent, domestic waste, waterway modification for navigation and other purposes, and forest clearance (Kottelat and Whitten, 1996). Since correct species identification is the basic starting point for any type of biological study, it is important that each name applies to only a single species, and each species is known by a single name (Rainboth, 1996a, 1996b). In order to encourage awareness and attention to freshwater fishes, make appropriate environmental management, categorize species under threat and identify conservation dependent ones, it is important that all fish species available in the region be correctly identified and documented with easy to identification diagnostic characters and photographs. The book is a sincere approach to achieve this goal.

Due to limited space, descriptive parts have been minimized as much as possible and attempt has been made to incorporate photographs and diagrams as much as possible for easy guide to the reader. Conservation status of the fishes have not been included here since the reports of the IUCN assessment and evaluation (IUCN, 2010) of freshwater fishes of the region are available. Moreover, hundreds of species described thereafter have not been assessed.

It is not possible to include vernacular/local names of species since there are several ethnic groups who call a species with different names. Many fishes do not have a name and are commonly named for a group having a similar look. Many species are collected from the areas where there is no human inhabitation.

Taxonomy and nomenclature

All names used in the headings are valid scientific names according to the Code. Only those species names which comply with the ICZN (1999) and amendment to expand and refine methods of publication ICZN (2012) are included. Only references of original description of species/genus are used after the headings. Synonyms are excluded as far as possible and references of revisions are included wherever possible. Fish species described can be easily diagnosed from its congeners, but the availability of the name is doubtful because of the compliance of the publication with the Article 8.5.3 regarding criteria of publication of the International Code of Zoological Nomenclature is inserted with "?" mark in the book.

Classification of fishes, particularly of the families follow Nelson (2006) and of the families, genera, and species Kottelat (2013) except for the newly revised genera and species. In view of the comments of Britz (2017), the classification of Nelson et al. (2016) is not fully followed here.

The present work includes 512 fish species under 128 genera and 40 families.

Methods Adopted

Fish collection

Fishes were collected by local and modified techniques. Some fishermen use boats equipped with gill/seine nets for fishing regularly in the rivers. We wait for the fishermen to return and select fishes from the catch. Various types of traps and nets are used. Sometimes we take part in community fishing, like drive-in by using 5−10 canoes toward previously set gill nets. Local fishermen often dam a diversion of a hill stream and dry-up to harvest fishes. Plant poisons are sometimes used. Dynamite-fishing and electro-fishing are also the techniques used. However, these are discouraged by village authorities as the destructive impacts have been gradually understood. Market sampling is one easy way to obtain fish specimens. Ujjan Bazar in Guwahati, Assam on the bank of the Brahmaputra, Agartala market for fishes from Tripura and Bangladesh, and Tamu market in Myanmar are some of them.

However, all fish species are not targeted by fishermen. Small fishes like *Balitora, Erethistes, Glyptothorax, Psilorhynchus*, and so on are neglected. Electro-fishers (sling type) with 12 V lithium battery for localized shock are also used when required to obtain species of interest.

Fish preservation

Fishes are tranquilized and then fixed in 10% formalin (formalin contains 40% formaldehyde), that is, 4% formaldehyde in a plastic bucket. Fixing and preserving specimens in congestion and tightly in containers should be avoided. This results in the distortion of fishes and loss of coloration at the points of contact with other specimens. After fixation, the specimens are tagged with a piece of paper lined with aluminum foil (alternatively fruit juice box may be used). The paper is cut into 4×1.5 cm with a hole at the corner (Fig. 2.1). The tag is marked with a number using a needle on the aluminum foil side and then tied at the caudal peduncle of the fish. The details about the collection site, date, coordinates, and collector's name are noted down on a field notebook with the corresponding number written on the tag.

FIGURE 2.1

Aluminum foil fish tag.

Freshwater Fishes of the Eastern Himalayas. DOI: https://doi.org/10.1016/B978-0-12-823391-7.00009-6

The fish specimens are transferred to freshly prepared 10% formalin or 70% ethanol and stored in wide-mouthed translucent plastic bottles with the head facing downward. The specimens are transferred in 70% ethanol after 1 week.

Fish photography

Photography is done usually before the color of the fishes fades away. A board with a needle fixed vertically is designed. For the lateral view, the specimen is pinned with the head facing left and the photograph is taken with a Nikon 300S DSLR camera with a macro 60 mm lens. Good photographs obtained from other sources or taken with other cameras are also taken when it is not possible. Natural and diffused light and soft reflectors are used to obtain uniform lighting. Flash guns are avoided as far as possible. Photography in a specially designed aquarium is also done for some fishes. For close-up images of tubercles, lip structures, taste buds, unculi, and so on of *Garra*, nemacheilines, balitorids, and so on, a stereo-zoom binocular microscope, Leica-S8APO with transmitted light was used and photographed with Leica DFC 425 fitted on it. The image was processed using Leica Application Suit ver. 3.6. In a few cases where detailed topography is needed, Environmental Scanning electron microscope Quanta 250 FED was used.

Body colorations like bars, stripes, spots, ocelli, and so on appear more clearly after preservation in some fishes, namely, nemacheilines, barilines, minnows, and danionins. In such cases photography was done after a few days of preservation in formalin.

Morphological features

Morphological character of fishes, terms and methods of counting measures are followed after Hubbs and Lagler (1946) and Kottelat (2001). Morphological character of a fish is expressed basically in two ways. One is *morphometry* or body proportions, that is, how long is the head in relation to the standard length (SL), diameter of eye in relation to head length (HL), and so on. The proportion is expressed in percentages. Another one is the *meristic* characters, that is, counts, fin ray, scale counts, and so on.

Measurements

The following measurements are taken to express the morphometric character of a fish. All the lengths as mentioned below are the point-to-point straight distances, unless otherwise stated, and are not taken over the curve of the body (Figs. 2.2 and 2.3).

Total length (TL) is the distance between the most anteriorly projecting part of the head and the most posterior projecting tip of the caudal fin when the lobes are adpressed.

Standard length (SL) is the distance between the most anterior part head (tip of upper lip or snout) and the end of the vertebral column, that is, hypural plate (Fig. 2.4), usually marked by a crease when the caudal fin is flexed.

Body depth (BD) is the greatest distance between dorsal and ventral aspects of the body. It is usually taken at the origin of the pelvic fin in scaly fishes and at the origin of the anal fin in cat fishes. It should be clearly stated while reporting.

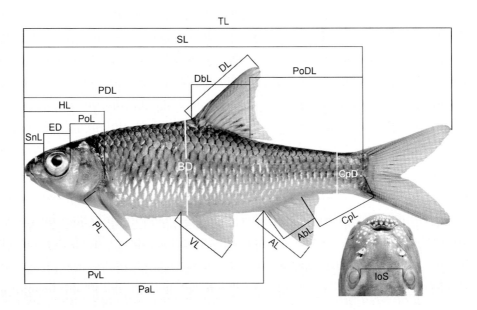

FIGURE 2.2

Measurements of a scaly fish: TL, total length; SL, standard length; HL, head length; BD, body depth; PDL, predorsal length; PoDL, postdorsal length, SnL, snout length; ED, eye diameter; PoL, postorbital length; IoS, interorbital space; DL, dorsal-fin length; PL, pectoral-fin length; VL, pelvic-fin length; AL, anal-fin length; DbL, dorsal-fin base length; AbL, anal-fin base length; CpL, caudal-peduncle length; CpD, caudal-peduncle depth; PvL, prepelvic length; PaL, preanal length; DsL, dorsal-spine length; PsL, pectoral-spine length; AdB, Adipose-fin base length; Nb, nasal barbel; MxB, maxillary barbel; MdB, mandibular barbel.

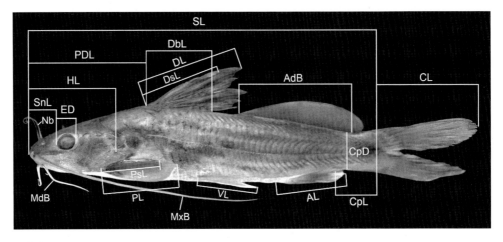

FIGURE 2.3

Measurements of a catfish: Additional legends as in Fig. 2.2: AdB, Adipose-fin base length; Nb, nasal barbel; MxB, maxillary barbel; MdB, mandibular barbel.

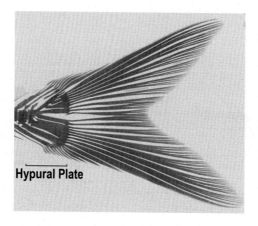

Hypural Plate

FIGURE 2.4

Caudal skeleton of teleost fish showing hypural plate.

Head length (HL) is the distance between the most anterior part of the head and the most posterior point on the operculum.

Caudal peduncle depth (CpD) is the least distance of the caudal peduncle measured between dorsal and ventral margins.

Caudal peduncle length (CpL) is the oblique distance between the posterior margin of the anal fin base and origin of median caudal fin rays

Predorsal length (PDL) is the distance between the most anterior part of the head and the structural base of the first dorsal fin ray.

Fins

The pectoral and pelvic fins are the paired fins. The dorsal, caudal, and anal fins are the unpaired or median fins. The adipose fin is a membranous fin devoid of rays usually in front of the caudal fin in some families. The origin of a fin is the point where the anterior-most ray is inserted into the body. A fin is supported by simple ray/rays (unbranched) anteriorly and branched rays posteriorly. The simple rays are unsegmented and may be osseous, which may be serrated or denticulated posteriorly or anteriorly or on both the sides (Fig. 2.5); the branched rays are segmented. There may be two dorsal fins, the anterior one with spines and the posterior, with a spine followed by soft rays (Fig. 2.6). In the caudal fin the principal ray on the dorsal margin and that on the ventral margin are simple. Principal rays are those which reach the posterior margins of the fin and procurrent rays are those which do not reach the posterior margins. Only the principal rays of caudal fin are counted. In dorsal and anal fins, the last branched ray is often split down to the base and may appear as two rays.

In the family Salmonidae and in many catfish families, an adipose dorsal fin, which is a membraneous fin without rays, is present behind the rayed dorsal fin and in front of the caudal fin (Fig. 2.7). It may be confluent with the caudal fin in some fishes.

Caudal fins are of various shapes. They are rounded, truncate, emarginate, or forked (Fig. 2.8).

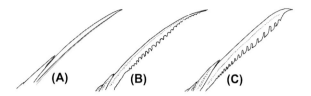

FIGURE 2.5

Dorsal spine: (A) finely serrated, (B) serrated, and (C) denticulated.

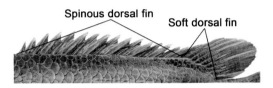

FIGURE 2.6

Dorsal fin of *Anabas* showing spinous and soft dorsal.

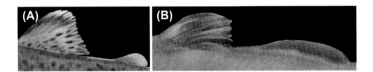

FIGURE 2.7

Rayed and adipose dorsal fins in (A) Salmonidae and (B) Catfish.

FIGURE 2.8

Shapes of caudal fin: (A) rounded, (B) truncate, (C) emarginate, (D) forked, (E) deeply forked.

Fins ray counts

Simple rays

Simple rays are unbranched and are median and not segmented. Simple rays may be hardened to be called spines (Fig. 2.9). Spines are sometimes serrate or denticulate posteriorly; anteriorly also in some genera. Simple rays are designated by Roman numerals.

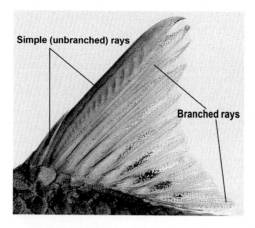

FIGURE 2.9

Dorsal fin showing simple and branched rays.

Branched rays

Branched rays are segmented and branched at the distal end in varying degrees. These are designated in Arabic numerals. In a fin having both simple and branched rays, normally simple rays preceding the latter, Roman numeral is followed by Arabic numeral after a coma, for example, II,9. The last two branched dorsal and anal rays are usually supported by a single pterygiophore and thus are counted as 1½.

Principal rays of caudal fin

Principal rays are the rays that reach the distal tip of the fin. It includes one simple ray followed by some branched rays in the upper lobe and some branched rays followed by a simple ray in the lower lobe (Fig. 2.10). Small rays, also called procurrent rays are not counted. Example: 10 + 9, 10 rays in the upper lobe which include one simple ray and nine branched rays and then nine rays in the lower lobe which include eight branched rays one simple ray.

Rays in paired fins

All rays are counted including the smallest one at the posterior end of the fin.

Last (most posterior) ray of dorsal and anal fins

Fin rays are supported by pterygiophores basally. A pterygiophore may support one or more rays. The last two branched rays, though appearing as two are in fact supported by a single pterygiophore (Fig. 2.11) and thus counted as 1½ (based on Kottelat, 2001).

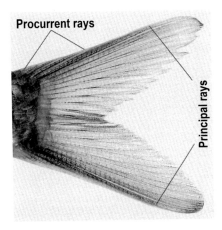

FIGURE 2.10

Caudal fin showing principal and procurrent rays.

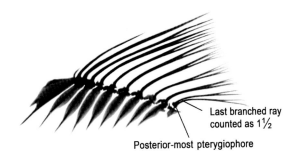

FIGURE 2.11

Dorsal fin ray showing posterior-most pterygiophore.

Scale counts

All scales including the smallest and interpolated scales are counted. Scales on the fin bases and on basal sheaths are not counted (Fig. 2.12).

Lateral line scales

The pored scales on the lateral line are counted. Lateral line is complete if the scales from the one in contact with the shoulder girdle to that on the posterior end of the structural base of the caudal fin are pored. The posterior margin of the vertebral column or the hypural plate is the end of the

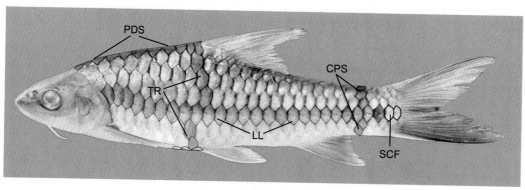

FIGURE 2.12

Counting of scales: PDS, predorsal scales; TR, transverse scales; LL, lateral longitudinal scales; CPS, circumpeduncular scales; SCF, scales on caudal fin.

structural base of the caudal fin, which is marked by a crease when the caudal fin is flexed. Pored scales from the scale behind shoulder shoulder girdle is counted when the lateral line is complete. Scales on the caudal fin beyond the point as indicated above are not counted even if they are pored or well developed. These may be indicated as + "..". Example: 24 + 2; means pored scales up to the end of the vertebral column and two on the caudal fin.

In fishes with incomplete lateral lines we may indicate the number of pored scales. We may designate the scales on the lateral body which would normally be occupied by a typical lateral line as "scales in lateral series."

In cases of fishes where pored scales are interrupted it may be stated clearly as "x" scales followed by "y" scales after a gap of "z" number of scales. This character is seen in some families, namely, Badidae and Nandidae.

Transverse scales

Transverse scales are counted from the origin of the dorsal fin to the lateral line following obliquely backward and from the origin of the pelvic fin to the lateral line following obliquely forward. Predorsal scale is counted as half and the preventral scale as half. The scale on the lateral line is not counted. Example: ½6/1/2½; means there are six scales above the lateral line, ½ scale is the predorsal scale; two scales below the lateral line and ½ is the preventral scale.

Scales

Scales may be cycloid, with smooth posterior margin and ctenoid, with comb-like projections "cteni" along the posterior margin.

Mouth position

Mouth position is an important taxonomic character. A fish mouth may be superior (upturned), terminal, subterminal, or inferior (Fig. 2.13).

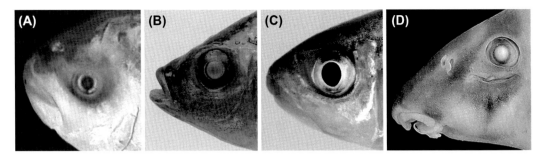

FIGURE 2.13

Different mouth positions: (A) superior, (B) terminal, (C) subterminal, and (D) inferior.

Fish coloration

Fish normally may be silvery-white on the sides, dark olivaceous dorsally, and pale white/yellowish ventrally. Coloration on the belly may vary depending on the food ingested. However, several fish species show marked variations in the coloration, some of which are cryptic, which are of interest.

Color marks

To avoid confusion between the bar (horizontal bar/vertical bar) and stripe, and so on, the bar is considered here as a vertical mark (Fig. 2.14) and stripe, the longitudinal mark (Fig. 2.15).

Colour mark smaller than the eye is called spot and those greater than the eye is blotch. Mark surrounded by pale area is called ocellus. Smaller marks are dots. Dots or spots when clustered may form bands (Fig. 2.16).

Gill raker

Gill arch consists of pharyngobranchial dorsally, epibranchial dorsolaterally, ceratobranchial ventrolaterally, and hypobranchial ventrally. Gill filaments are borne by hypobranchial (upper limb) and ceratobranchial (lower limb) muscles. Gill rakers are the bony projections from the gill arch on the opposite side of the gill filaments. The counts are given as $x + y$, that is, x, the numbers on the upper limb and y, on the lower limb. The rakers on the first gill arch are only counted (Fig. 2.17).

Osteology

Clearing and staining of bones for osteology followed Potthoff (1984) which involves trypsin digestion and alizarin red S staining. In case of specimens whose numbers are limited radiographs using digital X-ray model centricity CR SP100 were used. Osteological data were taken from either radiographs or from cleared and alizarin stained specimens and reproduced as photographs or line drawings (Fig. 2.18).

Preparation

Specimens used for osteology were cleared and stained with Alizarin Red S by using the modified method of Hollister (1934). The identified fishes were de-skinned and washed with water and then the fishes were put in freshly prepared 2% KOH. The specimen was rinsed with distilled water and

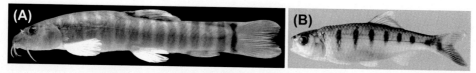

FIGURE 2.14

Colour bar in (A) Nemacheilines and in (B) Barilines.

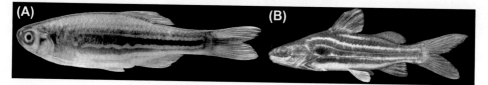

FIGURE 2.15

Colour stripe in (A) Danionin and in (B) Catfish.

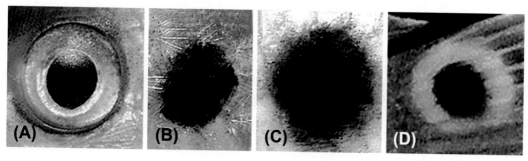

FIGURE 2.16

Colour marks: (A) eye to show its diameter in relation to B and C; (B), spot and (C) blotch; (D) ocellus.

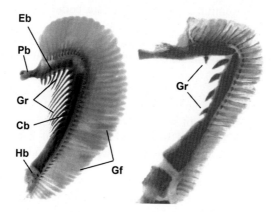

FIGURE 2.17

First gill arch showing gill rakers in the upper and lower limb; Pb: pharyngobranchial, Eb: epibranchial (upper limb), Cb: ceratobranchial and Hb: hyobranchial, in *Mystus dibrugarensis* (L) and *Batasio affinis* (R).

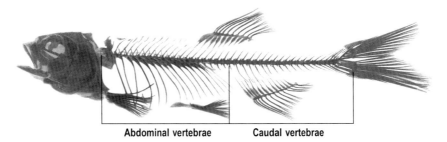

Abdominal vertebrae   Caudal vertebrae

FIGURE 2.18

Lateral view of fish skeleton showing abdominal and caudal vertebrae.

put in 1% KOH solution to which a few drops of 0.5% aqueous alizarin red S was added. The stained specimens were removed from the solution and finally rinsed with distilled water. The stained specimens are stored in glycerol. Some specimens were also dissected for detailed examination.

Recording

Alizarin stained bones were photographed from which drawings were prepared using tracing papers. Details of the individual bones were studied under a Zoom Stereoscopic microscope (Carl Zeiss, Germany). Fishes were also subjected to X-ray and radiographs were examined for accuracy of the counts, shapes, and drawings.

Terminology

The terminology of teleosts fish bones adopted here is mostly based on current usage and does not necessarily imply homology with similarly named tetrapods or even other groups of fishes. As it was pointed out by Romer (1947), the determination of precise homologies between the skull bones of different groups of fishes presents great difficulties. In counts of vertebrae the fused $PU_1 + U_1$ is considered as a single element. The vertebrae incorporated into the Weberian apparatus are counted as four elements.

The terminology of Weitzman (1962) was adopted with modifications here and there. In place of prevomer and opisthotic, vomer and intercalar have been used. In case of caudal skeleton, terminology of Fujita (1990) is adopted. In case of the genus *Puntius* the descriptions and terminology of orbital bones, Taki et al (1978) are followed and for other genera, those of Starks (1901) and Gregory (1933) are followed. The neurocranium length used in the description is measured from tip of mesethmoid to posterior-most tip of parietal. The width of parietal and frontal are taken at the broadest region. In case of Weberian apparatus, terminologies of Tilak (1963) and Howes (1980) are followed and for girdles, Saxena and Chandy (1965) is followed with some modifications. For the hypobranchial skeleton, Khanna (1961) is followed.

Figures

Line drawings which will give an idea of the character of a family, genus are inserted. Fresh color photographs of all available species, except for a few, are included. In cases where photographs of all the species are not available at least one representing a genus is included. In some cases line drawings redrawn from original drawings are reproduced.

Diagnosis

It consists of a combination of characters that are useful in identifying a genus or species. Thus the conventional use of the words "the genus/species is diagnosed by a unique combination of characters, and so on" is excluded. As far as possible, only those characters that are visible to the naked eye are included. A magnifying lens or a stereoscopic binocular microscope may be needed in case of small specimens.

Distribution

As far as possible, the occurrence of the taxa in drainage basin rather than political geographical boundaries is attempted.

Remarks

If there are variations in the character observed other than the known characters, additional observed characters, opinions based on the perusal of existing literature are included. Statements of previous workers may also be discussed.

Systematic index

1. **Family Notopteridae**
    Genus *Chitala* Fowler, 1934
        Chitala chitala (Hamilton, 1822)
    Genus *Notopterus* Lacepède, 1800
        Notopterus notopterus (Pallas, 1769)
2. **Family Anguillidae**
    Genus *Anguilla* Garsault, 1764
        Anguilla bengalensis (Gray, 1831)
3. **Family Engraulididae**
    Genus *Setipinna* Swainson, 1839
        Setipinna phasa (Hamilton, 1822)
4. **Family Clupeidae**
    Genus *Corica* Hamilton, 1822
        Corica soborna Hamilton, 1822
    Genus *Gonialosa* Regan, 1917
        Gonialosa manmina (Hamilton, 1822)
    Genus *Gudusia* Fowler, 1911
        Gudusia chapra (Hamilton, 1822)
    Genus *Tenualosa* Fowler, 1934
        Tenualosa ilisha (Hamilton, 1822)
5. **Family Cyprinidae**
    Genus *Amblypharyngodon* Bleeker, 1860
        Amblypharyngodon mola (Hamilton, 1822)
        Amblypharyngodon microlepis (Bleeker, 1853)
    Genus *Bangana* Hamilton, 1822
        Bangana dero (Hamilton, 1822)
        Bangana devdevi (Hora, 1936)
    Genus *Barilius* Hamilton, 1822
        Barilius barila (Hamilton, 1822)
        Barilius barna (Hamilton, 1822)
        Barilius barnoides (Vinciguerra, 1890)
        Barilius bendelisis (Hamilton, 1807)
        Barilius cosca (Hamilton, 1822)
        Barilius dogarsinghi Hora, 1921

Freshwater Fishes of the Eastern Himalayas. DOI: https://doi.org/10.1016/B978-0-12-823391-7.00008-4

Barilius lairokensis Arunkumar & Tombi Singh, 2000
Barilius profundus Dishma & Vishwanath, 2012
Barilius shacra (Hamilton, 1822)
Barilius tileo (Hamilton, 1822)
Barilius vagra (Hamilton, 1822)
Genus *Bengala* Gray, 1834
Bengala elanga (Hamilton, 1822)
Genus *Cabdio* Hamilton, 1822
Cabdio crassus Lalramliana, Lalronunga & Singh, 2019
Cabdio jaya (Hamilton, 1822)
Cabdio morar (Hamilton, 1822)
Cabdio ukhrulensis (Selim & Vishwanath, 2001)
Genus *Chagunius* Smith, 1938
Chagunius chagunio Hamilton, 1822
Chagunius nicholsi (Myers, 1924)
Genus *Chela* Hamilton, 1822
Chela cachius (Hamilton, 1822)
Genus *Cirrhinus* Oken, 1817
Cirrhinus cirrhosus (Bloch, 1795)
Cirrhinus reba (Hamilton, 1822)
Genus *Ctenopharyngodon* Steindachner, 1866
Ctenopharyngodon idella (Valenciennes, 1844)
Genus *Cyprinus* Linnaeus, 1758
Cyprinus carpio Linnaeus, 1758
Genus *Danio* Hamilton, 1822
Danio assamila Kullander, 2015
Danio dangila (Hamilton, 1822)
Danio jaintianensis (Sen, 2007)
Danio meghalayensis Sen & Dey, 1985
Danio quagga Kullander, Liao & Fang, 2009
Danio rerio (Hamilton, 1822)
Genus *Danionella* Roberts, 1986
Danionella priapus Britz, 2009
Genus *Devario* Heckel, 1843
Devario acuticephala (Hora, 1921)
Devario aequipinnatus (McClelland, 1839)
Devario coxi Kullander, Rahman, Norén & Mollah, 2017
Devario deruptotalea Ramananda & Vishwanath, 2014
Devario devario (Hamilton, 1822)
Devario horai (Barman, 1983)
Devario manipurensis (Barman, 1987)
Devario naganensis (Chaudhuri, 1912)
Devario yuensis (Arunkumar & Tombi Singh, 1998)

Genus *Esomus* Swainson, 1839
   Esomus danricus (Hamilton, 1822)
Genus *Garra* Hamilton, 1822
   Garra abhoyai Hora, 1921
   Garra annandalei Hora, 1921
   Garra arunachalensis Nebeshwar & Vishwanath, 2013
   Garra arupi Nebeshwar, Vishwanath & Das, 2009
   Garra biloborostris Roni & Vishwanath, 2017
   Garra bimaculacauda Thoni, Gurung & Mayden, 2016
   Garra birostris Nebeshwar & Vishwanath, 2013
   Garra chakpiensis Nebeshwar & Vishwanath, 2015
   Garra chathensis Ezung, Shangningam & Pankaj, 2020
   Garra chindwinensis Premananda, Kosygin & Saidullah, 2017
   Garra chivaensis Moyon & Arunkumar, 2020
   Garra clavirostris Roni, Thaosen & Vishwanath, 2017
   Garra compressa Kosygin & Vishwanath, 1998
   Garra cornigera Shangningam & Vishwanath, 2015
   Garra elongata Vishwanath & Kosygin, 2000
   Garra gotyla (Gray, 1830)
   Garra kalpangi Nebeshwar, Bagra & Das, 2012
   Garra khawbungi Arunachalam, Nandagopal & Mayden, 2014
   Garra kempi Hora, 1921
   Garra koladynensis Nebeshwar & Vishwanath, 2017
   Garra lamta (Hamilton, 1822)
   Garra lissorhynchus (McClelland, 1843)
   Garra litanensis Vishwanath, 1993
   Garra magnacavus Shangningam, Kosygin & Sinha, 2019
   Garra magnidiscus Tamang, 2013
   Garra manipurensis Vishwanath & Sarojnalini, 1988
   Garra matensis Nebeshwar & Vishwanath, 2017
   Garra mini Rahman, Mollah, Noren & Kullander, 2016
   Garra naganensis Hora, 1921
   Garra nambulica Vishwanath & Joyshree, 2005
   Garra namyaensis Shangningam & Vishwanath, 2012
   Garra nasuta (McClelland, 1838)
   Garra nepalensis Rayamajhi & Arunachalam, 2017
   Garra paralissorhynchus Vishwanath & Shanta Devi, 2005
   Garra parastenorhynchus Thoni, Gurung & Mayden, 2016
   Garra paratrilobata Roni, Chinglemba, Rameshori & Vishwanath, 2019
   Garra quadratirostris Nebeshwar & Vishwanath, 2013
   Garra rakhinica Kullander & Fang, 2004
   Garra rupicola (McClelland, 1839)
   Garra substrictorostris Roni & Vishwanath, 2018

Garra trilobata Shangningam & Vishwanath, 2015
Garra ukhrulensis Nebeshwar & Vishwanath, 2015
Genus *Gibelion* Heckel, 1843
  Gibelion catla (Hamilton, 1822)
Genus *Hypophthalmichthys* Bleeker, 1860
  Hypophthalmichthys molitrix (Valenciennes, 1844)
  Hypopthalmichthys nobilis (Richardson, 1845)
Genus *Hypsibarbus* Rainboth, 1996
  Hypsibarbus myitkyinae (Prashad & Mukerji, 1929)
Genus *Labeo* Cuvier, 1816
  Labeo angra (Hamilton, 1822)
  Labeo bata (Hamilton, 1822)
  Labeo boga (Hamilton, 1822)
  Labeo calbasu (Hamilton, 1822)
  Labeo dyocheilus (McClelland, 1877)
  Labeo gonius (Hamilton, 1822)
  Labeo nandina (Hamilton, 1822)
  Labeo pangusia (Hamilton, 1822)
  Labeo rohita (Hamilton, 1822)
Genus *Laubuka* Bleeker, 1859
  Laubuka khujairokensis (Arunkumar, 2000)
  Laubuka laubuca (Hamilton, 1822)
  Laubuka parafasciata Lalramliana, Vanlalhlimpuia & Singh, 2017
Genus *Neolissochilus* Rainboth, 1985
  Neolissochilus dukai (Day, 1878)
  Neolissochilus heterostomus Chen, Yang & Chen, 1999
  Neolissochilus hexagonolepis (McClelland, 1839)
  Neolissochilus kaladanensis Lalramliana, Lalronunga, Kumar & Singh, 2018
  Neolissochilus spinulosus (McClelland, 1845)
  Neolissochilus stevensonii (Day, 1870)
  Neolissochilus stracheyi (Day, 1871)
Genus *Oreichthys* Smith, 1933
  Oreichthys andrewi Knight, 2014
  Oreichthys cosuatis (Hamilton, 1822)
  Oreichthys crenuchoides Schäfer, 2009
Genus *Oreinus* McClelland, 1938
  Oreinus molesworthi (Chaudhuri, 1913)
Genus *Osteobrama* Heckel, 1842
  Osteobrama belangeri (Valenciennes, 1844)
  Osteobrama cotio (Hamilton, 1822)
  Osteobrama cunma (Day, 1888)
  Osteobrama feae Vinciguerra, 1890
Genus *Pethia* Pethiyagoda, Meegaskumbura & Maduwage, 2012
  Pethia ater (Linthoingambi & Vishwanath, 2007)

Pethia aurea Knight, 2013
Pethia canius (Hamilton, 1822)
Pethia conchonius (Hamilton, 1822)
Pethia expletiforis Dishma & Vishwanath, 2013
Pethia gelius (Hamilton, 1822)
Pethia guganio (Hamilton, 1822)
Pethia khugae (Linthoingambi & Vishwanath, 2007)
Pethia manipurensis (Menon, Rema Devi & Vishwanath, 2000)
Pethia meingangbii (Arunkumar & Tombi Singh, 2003)
Pethia phutunio (Hamilton, 1822)
Pethia rutila Lalramliana, Knight & Laltlanhlua, 2014
Pethia shalynius (Yazdani & Talukdar, 1975)
Pethia stoliczkana (Day, 1871)
Pethia ticto (Hamilton, 1822)
Pethia yuensis (Arunkumar & Tombi Singh, 2003)
Genus *Poropuntius* Smith, 1931
Poropuntius burtoni (Mukerji, 1934)
Poropuntius clavatus (McClelland, 1845)
Poropuntius margarianus (Anderson, 1879)
Poropuntius shanensis (Hora & Mukerji, 1934)
Genus *Puntius* Hamilton, 1822
Puntius chola (Hamilton, 1822)
Puntius sophore (Hamilton, 1822)
Puntius terio (Hamilton, 1822)
Genus *Raiamas* Jordan, 1919
Rasbora bola (Hamilton, 1822)
Rasbora guttatus (Day, 1870)
Genus *Rasbora* Bleeker, 1859
Rasbora daniconius (Hamilton, 1822)
Rasbora ornata Vishwanath & Laisram, 2005
Rasbora rasbora (Hamilton, 1822)
Genus *Salmostoma* Swainson, 1839
Salmostoma bacaila (Hamilton, 1822)
Salmostoma phulo (Hamilton, 1822)
Salmostoma sladoni (Day, 1870)
Genus *Schizothorax* Heckel, 1838
Schizothorax chivae Arunkumar & Moyon, 2016
Schizothorax progastus (McClelland, 1839)
Schizothorax richardsonii (Gray, 1832)
Genus *Securicula* Günther, 1868
Securicula gora (Hamilton, 1822)
Genus *Semiplotus* Bleeker, 1862
Semiplotus cirrhosus Chaudhuri, 1919

     Semiplotus modestus Day, 1870
     Semiplotus semiplotus (McClelland, 1839)
   Genus *Systomus* McClelland, 1838
     Systomus sarana (Hamilton, 1822)
   Genus *Tariqilabeo* Mirza & Saboohi, 1990
     Tariqilabeo burmanicus (Hora, 1936)
     Tariqilabeo latius (Hamilton, 1822)
   Genus *Tor* Gray, 1834
     Tor mosal (Hamilton, 1822)
     Tor putitora (Hamilton, 1822)
     Tor tor (Hamilton, 1822)
     Tor yingjiangensis Chen & Yang, 2004
6. Family Psilorhynchidae
   Genus *Psilorhynchus* McClelland, 1839
     Psilorhynchus amplicephalus Arunachalam, Muralidharan & Sivakumar, 2007
     Psilorhynchus arunachalensis (Nebeshwar, Bagra & Das, 2007)
     Psilorhynchus balitora (Hamilton, 1822)
     Psilorhynchus bichomensis Shangningam, Kosygin & Gopi, 2019
     Psilorhynchus chakpiensis Shangningam & Vishwanath, 2013
     Psilorhynchus hamiltoni Conway, Dittmer, Jezisek & Ng, 2013
     Psilorhynchus homaloptera (Hora & Mukerji, 1935)
     Psilorhynchus kaladanensis Lalramliana, Lalnuntluanga & Lalronunga, 2015
     Psilorhynchus khopai Lalramliana, Solo, Lalronunga & Lalnuntluanga, 2014
     Psilorhynchus konemi Shangningam & Vishwanath, 2016
     Psilorhynchus maculatus Shangningam & Vishwanath, 2013
     Psilorhynchus microphthalmus Vishwanath & Manojkumar, 1995
     Psilorhynchus nahlongthai Dey, Chaoudhuri, Mazumdar, Thaosen & Sarma, 2020
     Psilorhynchus nepalensis Conway & Mayden, 2008
     Psilorhynchus ngathanu Shangningam & Vishwanath, 2013
     Psilorhynchus nudithoracicus Tilak & Husain, 1980
     Psilorhynchus pseudecheneis Menon & Datta, 1964
     Psilorhynchus rowleyi (Hora & Misra, 1941)
     Psilorhynchus sucatio (Hamilton, 1822)
7. Family Botiidae
   Genus *Botia* Gray, 1831
     Botia dario (Hamilton, 1822)
     Botia histrionica Blyth, 1860
     Botia lohachata Chaudhuri, 1912
     Botia rostrata Günther, 1868
   Genus *Syncrossus* Blyth, 1860
     Syncrossus berdmorei (Blyth, 1860)
8. Family Cobitidae
   Genus *Acantopsis* van Hasselt, 1823
     Acantopsis spectabilis (Blyth, 1860)

Genus *Canthophrys* Swainson, 1938
  Canthophrys gongota (Hamilton, 1822)
Genus *Lepidocephalichthys* Bleeker, 1863
  Lepidocephalichthys alkaia Havird & Page, 2010
  Lepidocephalichthys annandalei Chaudhuri, 1912
  Lepidocephalichthys arunachalensis (Datta & Barman, 1984)
  Lepidocephalichthys berdmorei (Blyth, 1860)
  Lepidocephalichthys goalparensis Pillai & Yazdani, 1976
  Lepidocephalichthys guntea (Hamilton, 1822)
  Lepidocephalichthys irrorata Hora, 1921
  Lepidocephalichthys micropogon (Blyth, 1860)
Genus *Neoeucirrhichthys* Bănărescu & Nalbant, 1968
  Neoeucirrhichthys maydelli Bănărescu & Nalbant, 1968
Genus *Pangio* Blyth, 1860
  Pangio apoda Britz & Maclaine, 2007
  Pangio pangia (Hamilton, 1822)
9. **Family Balitoridae**
Genus *Balitora* Gray, 1830
  Balitora brucei Gray, 1830
  Balitora burmanica Hora, 1932
  Balitora eddsi Conway & Mayden, 2010
Genus *Hemimyzon* Regan, 1911
  Hemimyzon arunachalensis (Nath, Dam, Bhutia, Dey & Das, 2007)
  Hemimyzon indicus Lalramliana, Solo & Lalnuntluanga, 2018
Genus *Homalopteroides* Fowler, 1905
  Homalopteroides rupicola (Prashad & Mukerji, 1929)
10. **Family Nemacheilidae**
Genus *Aborichthys* Chaudhuri, 1913
  Aborichthys boutanensis (McClelland & Griffith, 1842)
  Aborichthys elongatus Hora, 1921
  Aborichthys garoensis Hora, 1925
  Aborichthys iphipaniensis Kosygin, Gurumayum, Singh & Chowdhury, 2019
  Aborichthys kailashi Shangningam, Kosygin, Sinha & Gurumayum, 2020
  Aborichthys kempi Chaudhuri, 1913
  Aborichthys pangensis Shangningam, Kosygin, Sinha & Gurumayum, 2020
  Aborichthys tikaderi Barman, 1985
  Aborichthys waikhomi Kosygin, 2012
Genus *Acanthocobitis* Peters, 1861
  Acanthocobitis pavonacea (McClelland, 1839)
Genus *Mustura* Kottelat, 2018
  Mustura chhimtuipuiensis Lalramliana, Lalhlimpuia, Solo & Vanramliana, 2016
  Mustura chindwinensis (Lokeshwor & Vishwanath, 2012)
  Mustura dikrongensis (Lokeshwor & Vishwanath, 2012)
  Mustura harkishorei (Das & Darshan, 2017)

Mustura prashadi (Hora, 1921)
Mustura tigrina (Lokeshwor & Vishwanath, 2012)
Mustura tuivaiensis (Lokeshwor & Vishwanath, 2012)
Mustura walongensis (Tamang & Sinha, 2016)
Genus *Nemacheilus* Bleeker, 1863
Nemacheilus corica (Hamilton, 1822)
Genus *Neonoemacheilus* Zhu & Guo, 1985
Neonoemacheilus assamensis (Menon, 1987)
Neonoemacheilus peguensis (Hora, 1929)
Genus *Paracanthocobitis* Grant, 2007
Paracanthocobitis abutwebi (Singer & Page, 2015)
Paracanthocobitis botia (Hamilton, 1822)
Paracanthocobitis linypha (Singer & Page, 2015)
Paracanthocobitis mackenziei (Chaudhuri, 1910)
Paracanthocobitis marmorata (Singer, Pfeiffer & Page, 2017)
Paracanthocobitis triangula (Singer, Pfeiffer & Page, 2017)
Genus *Physoschistura* Bănărescu & Nalbant, 1982
Physoschistura elongata Sen & Nalbant, 1982
Genus *Rhyacoschistura* Kottelat, 2019
Rhyacoschistura ferruginea (Lokeshwor & Vishwanath, 2011)
Rhyacoschistura maculosa (Lalronunga, Lalnuntluanga & Lalramliana, 2013)
Rhyacoschistura manipurensis (Chaudhuri, 1912)
Rhyacoschistura porocephala (Lokeshwor & Vishwanath, 2013)
Genus *Schistura* McClelland, 1838
Schistura aizawlensis Lalramliana, 2012
Schistura beavani (Günther, 1868)
Schistura chindwinica (Tilak & Husain, 1990)
Schistura devdevi (Hora, 1935)
Schistura fasciata Lokeshwor & Vishwanath, 2011
Schistura kangjupkhulensis (Hora, 1921)
Schistura khugae Vishwanath & Shanta, 2004
Schistura koladynensis Lokeshwor & Vishwanath, 2012
Schistura minuta Vishwanath & Shanta Kumar, 2006
Schistura mizoramensis Lalramliana, Lalronunga, Vanramliana & Lalthanzara, 2014
Schistura multifasciata (Day, 1878)
Schistura nagaensis (Menon, 1987)
Schistura papulifera Kottelat, Harries & Proudlove, 2007
Schistura paucireticulata Lokeshwor & Vishwanath, 2013
Schistura rebuw Choudhury, Dey, Bharali, Sarma & Vishwanath, 2019
Schistura reticulata Vishwanath & Nebeshwar, 2014
Schistura reticulofasciata (Singh & Banarescu, 1982)
Schistura rosammae (Sen, 2009)
Schistura savona (Hamilton, 1822)
Schistura scaturigina McClelland, 1839

Schistura sijuensis (Menon, 1987)
Schistura sikmaiensis (Hora, 1929)
Schistura singhi (Menon, 1987)
Schistura syngkhai Choudhury, Mukhim, Dey, Warbah & Sarma, 2019
Schistura tirapensis (Kottelat, 1990)
Schistura zonata McClelland, 1939

11. Family Amblycipitidae

Genus *Amblyceps* Blyth, 1858
Amblyceps apangi Nath & Dey, 1989
Amblyceps arunchalensis Nath & Dey, 1989
Amblyceps cerinum Ng & Wright, 2010
Amblyceps laticeps (McClelland, 1842)
Amblyceps mangois (Hamilton, 1822)
Amblyceps torrentis Linthoingambi & Vishwanath, 2008
Amblyceps tuberculatum Linthoingambi & Vishwanath, 2008
Amblyceps waikhomi Darshan, Kachari, Dutta, Ganguly & Das, 2016

12. Family Akysidae

Genus *Akysis* Bleeker, 1858
Akysis manipurensis (Arunkumar, 2000)
Akysis prashadi Hora, 1936

13. Family Sisoridae

Genus *Bagarius* Bleeker, 1853
Bagarius bagarius (Hamilton, 1822)
Bagarius yarrelli (Sykes, 1839)
Genus *Conta* Hora, 1950
Conta conta (Hamilton, 1822)
Conta pectinata Ng, 2005
Genus *Creteuchiloglanis* Zhou, Li & Thomson, 2011
Creteuchiloglanis arunachalensis Sinha & Tamang, 2014
Creteuchiloglanis bumdelingensis Thoni & Gurung, 2018
Creteuchiloglanis kamengensis (Jayaram, 1966)
Creteuchiloglanis payjab Darshan, Dutta, Kachari, Gogoi, Aran & Das, 2014
Creteuchiloglanis tawangensis Darshan, Abujam, Wangchu, Kumar, Das, Kumar & Imotomba, 2019
Genus *Erethistes* Müller & Troschel, 1849
Erethistes hara (Hamilton, 1822)
Erethistes horai Misra, 1976
Erethistes jerdoni Day, 1870
Erethistes koladynensis (Anganthoibi & Vishwanath, 2009)
Erethistes pusillus Müller & Troschel, 1849
Genus *Erethistoides* Hora, 1950
Erethistoides ascita Ng & Edds, 2005
Erethistoides cavatura Ng & Edds, 2005
Erethistoides infuscatus Ng, 2006

Erethistoides montana Hora, 1950
Erethistoides senkhiensis Tamang, Chaudhuri & Choudhury, 2008
Erethistoides sicula Ng, 2005
Genus *Exostoma* Blyth, 1860
  Exostoma barakensis Vishwanath & Joyshree, 2007
  Exostoma dujangensis Shangningam & Kosygin, 2020
  Exostoma kottelati Darshan, Vishwanath, Abujam & Das, 2019
  Exostoma labiatum (McClelland, 1842)
  Exostoma mangdechhuensis Thoni & Gurung, 2018
  Exostoma sawmteai Lalramliana, Lalronunga, Lalnuntluanga & Ng, 2015
  Exostoma tenuicaudatum Tamang, Sinha & Gurumayum, 2015
Genus *Gagata* Bleeker, 1858
  Gagata cenia (Hamilton, 1822)
  Gagata dolichonema He, 1996
  Gagata gagata (Hamilton, 1822)
  Gagata sexualis Tilak, 1970
Genus *Glyptothorax* Blyth, 1860
  Glyptothorax ater Anganthoibi & Vishwanath, 2011
  Glyptothorax botius (Hamilton, 1822)
  Glyptothorax burmanicus Prashad & Mukerji, 1929
  Glyptothorax caudimaculatus Anganthoibi & Vishwanath, 2011
  Glyptothorax cavia (Hamilton, 1822)
  Glyptothorax chimtuipuiensis Anganthoibi & Vishwanath, 2010
  Glyptothorax churamanii Rameshori & Vishwanath, 2012
  Glyptothorax dikrongensis Tamang & Chaudhry, 2011
  Glyptothorax giudikyensis Kosygin, Singh & Gurumayum, 2020
  Glyptothorax gopii Kosygin, Das, Singh & Roy Choudhuri, 2019
  Glyptothorax gracilis (Günther, 1864)
  Glyptothorax granulus Vishwanath & Linthoingambi, 2007
  Glyptothorax igniculus Ng & Kullander, 2013
  Glyptothorax indicus Talwar, 1991
  Glyptothorax jayarami Rameshori & Vishwanath, 2012
  Glyptothorax kailashi Kosygin, Singh & Mitra, 2020
  Glyptothorax maceriatus Ng & Lalramliana, 2012
  Glyptothorax manipurensis Menon, 1954
  Glyptothorax mibangi Darshan, Dutta, Kachari, Gogoi & Das, 2015
  Glyptothorax ngapang Vishwanath & Linthoingambi, 2007
  Glyptothorax pantherinus Anganthoibi & Vishwanath, 2007
  Glyptothorax radiolus Ng & Lalramliana, 2013
  Glyptothorax rugimentum Ng & Kottelat, 2008
  Glyptothorax scrobiculus Ng & Lalramliana, 2012
  Glyptothorax senapatiensis Premananda, Kosygin & Saidullah, 2015
  Glyptothorax striatus (McClelland, 1842)
  Glyptothorax telchitta (Hamilton, 1822)

Glyptothorax ventrolineatus Vishwanath & Linthoingambi, 2006
Glyptothorax verrucosus Rameshori & Vishwanath, 2012
Genus *Gogangra* Roberts, 2001
    Gogangra laevis Ng, 2005
    Gogangra viridescens (Hamilton, 1822)
Genus *Myersglanis* Hora & Silas, 1952
    Myersglanis blythii (Day, 1870)
    Myersglanis jayarami Vishwanath & Kosygin, 1999
Genus *Nangra* Day, 1877
    Nangra assamensis Sen & Biswas, 1994
    Nangra bucculenta Roberts & Ferraris, 1998
    Nangra nangra (Hamilton, 1822)
    Nangra ornata Roberts & Ferraris, 1998
Genus *Oreoglanis* Smith, 1933
    Oreoglanis majuscula Linthoingambi & Vishwanath, 2011
    Oreoglanis pangenensis Sinha & Tamang, 2015
Genus *Parachiloglanis* Wu, He & Chu, 1981
    Parachiloglanis benjii Thoni & Gurung, 2018
    Parachiloglanis bhutanensis Thoni & Gurung, 2014
    Parachiloglanis dangmechhuensis Thoni & Gurung, 2018
    Parachiloglanis drukyulensis Thoni & Gurung, 2018
    Parachiloglanis hodgarti (Hora, 1923)
Genus *Pseudecheneis* Blyth, 1860
    Pseudecheneis crassicauda Ng & Edds, 2005
    Pseudecheneis eddsi Ng, 2006
    Pseudecheneis koladynae Anganthoibi & Vishwanath, 2010
    Pseudecheneis serracula Ng & Edds, 2005
    Pseudecheneis sirenica Vishwanath & Darshan, 2007
    Pseudecheneis sulcata (McClelland, 1842)
    Pseudecheneis ukhrulensis Vishwanath & Darshan, 2007
Genus *Pseudolaguvia* Misra, 1976
    Pseudolaguvia assula Ng & Conway, 2013
    Pseudolaguvia ferruginea Ng, 2009
    Pseudolaguvia ferula Ng, 2006
    Pseudolaguvia flavida Ng, 2009
    Pseudolaguvia foveolata Ng, 2005
    Pseudolaguvia fucosa Ng, Lalramliana & Lalronunga, 2016
    Pseudolaguvia inornata Ng, 2005
    Pseudolaguvia jiyaensis Tamang & Sinha, 2014
    Pseudolaguvia magna Tamang & Sinha, 2014
    Pseudolaguvia muricata Ng, 2005
    Pseudolaguvia nubila Ng, Lalramliana, Lalronunga & Lalnuntluanga, 2019
    Pseudolaguvia ribeiroi (Hora, 1921)
    Pseudolaguvia shawi (Hora, 1921)

    Pseudolaguvia spicula Ng & Lalramliana, 2010
    Pseudolaguvia viriosa Tamang & Sinha, 2012
  Genus *Sisor* Hamilton, 1822
    Sisor barakensis Vishwanath & Darshan, 2005
    Sisor chennuah Ng & Lahkar, 2003
    Sisor rabdophorus Hamilton, 1822
    Sisor rheophilus Ng, 2003
    Sisor torosus Ng, 2003

14. Family Siluridae
  Genus *Ompok* Lacepède, 1803
    Ompok bimaculatus (Bloch, 1794)
    Ompok pabo (Hamilton, 1822)
    Ompok pabda (Hamilton, 1822)
  Genus *Pterocryptis* Peters, 1861
    Pterocryptis barakensis Vishwanath & Nebeshwar Sharma, 2006
    Pterocryptis berdmorei (Blyth, 1860)
    Pterocryptis gangelica Peters, 1861
    Pterocryptis indica (Datta, Barman & Jayaram, 1987)
    Pterocryptis subrisa Ng, Lalramliana & Lalronunga, 2018
  Genus *Wallago* Bleeker, 1851
    Wallago attu (Schneider, 1801)

15. Family Chacidae
  Genus *Chaca* Gray, 1831
    Chaca burmensis Brown & Ferraris, 1988
    Chaca chaca (Hamilton, 1822)

16. Family Clariidae
  Genus *Clarias* Scopoli, 1777
    Clarias gariepinus (Burchell, 1822)
    Clarias magur (Hamilton, 1822)
  Genus *Heteropneustes* Müller, 1840
    Heteropneustes fossilis (Bloch, 1794)

17. Family Ariidae
  Genus *Cochlefelis* Whitley, 1941
    Cochlefelis burmanicus (Day, 1870)

18. Family Ailiidae
  Genus *Ailia* Gray, 1830
    Ailia coila (Hamilton, 1822)
  Genus *Clupisoma* Swainson, 1838
    Clupisoma garua (Hamilton, 1822)
    Clupisoma montanum Hora, 1937
    Clupisoma pateri Hora, 1937
  Genus *Eutropiichthys* Bleeker, 1862
    Eutropiichthys burmanicus Day, 1877
    Eutropiichthys cetosus Ng, Lalramliana, Lalronunga & Lalnuntluana, 2014

Eutropiichthys murius (Hamilton, 1822)
Eutropiichthys vacha (Hamilton, 1822)

19. Family Horobagridae

Genus *Pachypterus* Swainson, 1838
Pachypterus atherinoides (Bloch, 1794)
Genus *Proeutropiichthys* Hora, 1937

20. Family Pangasiidae

Genus *Pangasianodon* Chevey, 1931
Pangasianodon hypophthalmus (Sauvage, 1878)

21. Family Bagridae

Genus *Batasio* Blyth, 1860
Batasio affinis Blyth, 1860
Batasio batasio (Hamilton, 1822)
Batasio convexirostrum Darshan, Anganthoibi & Vishwanath, 2011
Batasio fasciolatus Ng, 2006
Batasio macronotus Ng & Edds, 2004
Batasio merianiensis (Chaudhuri, 2013)
Batasio procerus Ng, 2008
Batasio spilurus Ng, 2006
Batasio tengana (Hamilton, 1822)
Genus *Hemibagrus* Bleeker, 1862
Hemibagrus menoda (Hamilton, 1822)
Hemibagrus microphthalmus (Day, 1877)
Hemibagrus peguensis (Boulenger, 1894)
Genus *Mystus* Scopoli, 1777
Mystus bleekeri (Day, 1877)
Mystus carcio (Hamilton, 1822)
Mystus cavasius (Hamilton, 1822)
Mystus cineraceus Ng & Kottelat, 2009
Mystus dibrugarensis (Chaudhuri, 1913)
Mystus falcarius Chakrabarty & Ng, 2005
Mystus ngasep Darshan, Vishwanath, Mahanta & Barat, 2011
Mystus prabini Darshan, Abujam, Kumar, Parhi, Singh, Vishwanath, Das & Pandey, 2019
Mystus pulcher (Chaudhuri, 1911)
Mystus rufescens (Vinciguerra, 1890)
Mystus tengara (Hamilton, 1822)
Genus *Olyra* McClelland, 1942
Olyra kempi Chaudhuri, 1912
Olyra longicaudata McClelland, 1842
Olyra parviocula Kosygin, Shangningam & Gopi, 2018
Olyra praestigiosa Ng & Ferraris, 2016
Olyra saginata Ng, Lalramliana & Lalthanzara, 2014

Genus *Rama* Bleeker, 1862
  Rama chandramara (Hamilton, 1822)
Genus *Rita* Bleeker, 1858
  Rita rita (Hamilton, 1822)
  Rita sacerdotum Anderson, 1879
Genus *Sperata* Holly, 1939
  Sperata acicularis Ferraris & Runge, 1999
  Sperata aor (Hamilton, 1822)
  Sperata aorella (Blyth, 1858)
  Sperata lamarrii (Valenciennes, 1840)
22. Family Salmonidae
Genus *Oncorhynchus* Suckley, 1861
  Oncorhynchus mykiss (Walbaum, 1792)
23. Family Mugilidae
Genus *Rhinomugil* Gill, 1863
  Rhinomugil corsula (Hamilton, 1822)
Genus *Sicamugil* Fowler, 1938
  Sicamugil cascasia (Hamilton, 1822)
24. Family Belonidae
Genus *Xenentodon* Regan, 1911
  Xenentodon cancila (Hamilton, 1822)
25. Family Aplocheilidae
Genus *Aplocheilus* McClelland, 1839
  Aplocheilus panchax (Hamilton, 1822)
26. Family Poeciliidae
Genus *Poecilia* Bloch & Schneider, 1801
  Poecilia reticulata Peters, 1859
27. Family Syngnathidae
Genus *Microphis* (Kaup, 1853)
  Microphis deocata (Hamilton, 1822)
28. Family Synbranchidae
Genus *Monopterus* Lacepède, 1800
  Monopterus cuchia (Hamilton, 1822)
  Monopterus ichthyophoides Britz, Lalremsanga, Lalrotluanga & Lalramliana, 2011
  Monopterus rongsaw Britz, Sykes, Gower & Kamei, 2018
  Monopterus javanensis Lacepède, 1800
29. Family Chaudhuriidae
Genus *Garo* Yazdani & Talwar, 1981
  Garo khajuriai (Talwar, Yazdani & Kundu, 1977)
Genus *Pillaia* Yazdani, 1972
  Pillaia indica Yazdani, 1972
30. Family Mastacembelidae
Genus *Macrognathus* Lacepède, 1800
  Macrognathus aral (Bloch & Schneider, 1801)

Macrognathus lineatomaculatus Britz, 2010
Macrognathus morehensis Arunkumar & Tombi Singh, 2000
Macrognathus pancalus Hamilton, 1822
Genus *Mastacembelus* Scopoli, 1777
Mastacembelus armatus (Lacepède, 1800)
Mastacembelus tinwini Britz, 2007

31. Family Ambassidae

Genus *Chanda* Hamilton, 1822
Chanda nama Hamilton, 1822
Genus *Parambassis* Bleeker, 1874
Parambassis baculis (Hamilton, 1822)
Parambassis bistigmata Geetakumari, 2012
Parambassis lala (Hamilton, 1822)
Parambassis ranga (Hamilton, 1822)
Parambassis serrata Dishma & Vishwanath, 2015
Parambassis waikhomi Geetakumari & Vishwanath, 2012

32. Family Sciaenidae

Genus *Johnius* Bloch, 1793
Johnius coitor (Hamilton, 1822)

33. Family Nandidae

Genus *Nandus* Valenciennes, 1831
Nandus andrewi Ng & Jaafar, 2008
Nandus nandus (Hamilton, 1822)

34. Family Badidae

Genus *Badis* Bleeker, 1853
Badis assamensis Ahl, 1937
Badis badis (Hamilton, 1822)
Badis blosyrus Kullander & Britz, 2002
Badis chittagongis Kullander & Britz, 2002
Badis dibruensis Geetakumari & Vishwanath, 2010
Badis ferrarisi Kullander & Britz, 2002
Badis kanabos Kullander & Britz, 2002
Badis rhabdotus Kullander, Norén, Rahman & Mollah, 2019
Badis singenensis Geetakumari & Kadu, 2011
Badis tuivaiei Vishwanath & Shanta, 2004
Genus *Dario* Kullander & Britz, 2002
Dario kajal Britz & Kullander, 2013
Dario dario (Hamilton, 1822)

35. Family Ciclidae

Genus *Oreochromis* Günther, 1889
Oreochromis mossambica (Peters, 1852)
Oreochromis niloticus (Linnaeus, 1758)

36. Family Gobiidae

 Genus *Awaous* Valenciennes, 1837
  Awaous grammepomus (Bleeker, 1849)
 Genus *Glossogobius* Gill, 1859
  Glossogobius giuris (Hamilton, 1822)

37. Family Anabantidae

 Genus *Anabas* Cuvier & Cloquet, 1816
  Anabas testudineus (Bloch, 1792)

38. Family Osphronemidae

 Genus *Trichogaster* Bloch & Schneider, 1801
  Trichogaster fasciata Bloch & Schneider, 1801
  Trichogaster labiosa Day, 1877
  Trichogaster lalius (Hamilton, 1822)
  Trichogaster sota (Hamilton, 1822)

39. Family Channidae

 Genus *Channa* Scopoli, 1977
  Channa amari Dey, Raychoudhury, Nur, Sarkar, Kosygin & Barat, 2019
  Channa amphibeus (McClelland, 1845)
  Channa andrao Britz, 2013
  Channa aurantimaculata Musikasinthorn, 2000
  Channa aurantipectoralis Lalhlimpuia, Lalrounga & Lalramliana, 2016
  Channa aurolineata (Day, 1870)
  Channa barca (Hamilton, 1822)
  Channa bleheri Vierke, 1991
  Channa gachua (Hamilton, 1822)
  Channa lipor Raveenraj, Uma, Moulitharan & Singh, 2019
  Channa marulius (Hamilton, 1822)
  Channa melanostigma Geetakumari & Vishwanath, 2011
  Channa pardalis Knight, 2016
  Channa punctata (Bloch, 1793)
  Channa stewartii (Playfair, 1867)
  Channa stiktos Lalramliana, Knight, Lalhlimpuia & Singh, 2018
  Channa striata (Bloch, 1793)
  Channa torsaensis Dey, Nur, Raychoudhury, Kosygin Singh & Barat, 2018

40. Family Tetraodontidae

 Genus *Leiodon* Swainson, 1839
  Leiodon cutcutia (Hamilton, 1822)

Systematic Account

Family Notopteridae
Featherfin knifefishes

Body compressed, back humped, cranio-dorsal profile concave or straight, abdomen edge in front of pelvic fin keeled; ventral scutes 25–52; dorsal fin small, its origin in the middle of back, pelvic fin small or absent; anal fin long, confluent with caudal fin, total rays more than 100; scales tiny, lateral line scales 120–180; swim bladder projected anteriorly to the ear lateral to the skull; maxilla, parasphenoid and tongue bones with teeth; caudal fin skeleton with large first ural centrum, one or more epurals fused with uroneurals, no urodermals (Fig. 4.1).

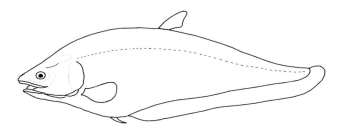

FIGURE 4.1

Outline diagram of Notopteridae.

Key to genera:

1a. Craniodorsal profile concave; preoperculat scale rows more than 10 .... *Chitala*
1b. Craniodorsal profile straight; preoperculat scale rows 6–8 .... *Notopterus*

Genus *Chitala* Fowler, 1934

Chitala Fowler, 1934a: 244 (subgenus of *Notopterus* Lacepède, 1800: 189; type species: *Mystus chitala* Hamilton, 1822). Gender: feminine.

Diagnosis. Body oblong, strongly compressed, tapers toward caudal fin; cranio-dorsal profile strongly concave; abdomen edge keeled and with prepelvic scutes; mouth wide; jaw lengths increase with age; preopercular scales present; body covered with minute scales; dorsal fin short, its origin midway between snout-tip and caudal-fin base; pelvic fin rudimentary; anal fin long and continuous with caudal fin.

Freshwater Fishes of the Eastern Himalayas. DOI: https://doi.org/10.1016/B978-0-12-823391-7.00006-0

Chitala chitala (Hamilton, 1822)
(Fig. 4.2)

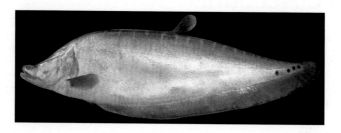

FIGURE 4.2

Chitala chitala, 540.0 mm SL, Jiri River, Manipur.

Mystus chitala Hamilton, 1822: 236, 382 (type locality: rivers of Bengal and Bihar, India).

Diagnosis. Maxilla extending considerably beyond posterior edge of eye, devoid of teeth; pelvic fin with four rays; preorbital smooth; body with a row of about 15 transverse silvery bars on back, caudal fins region with three or four rounded blue-black or black blotches. Maximum size ~1150 mm SL.

Distribution. India, Nepal, Bangladesh, Myanmar, Indonesia, Malaya, Pakistan, and Thailand.

Genus *Notopterus* Lacepède, 1800

Notopterus Lacepède, 1800: 189 (type species: *Gymnotus notopterus* Pallas, 1769). Gender: masculine.

Diagnosis. Body oblong, strongly compressed, tapers toward tail; cranio-dorsal profile straight or slightly concave; abdomen edge keeled, mouth wide, maxilla extends beyond posterior margin of eye; mandible with two rows of serrations; teeth present on tongue, palate and jaws; body covered with minute scales; dorsal fin short, its origin in the middle of back; pelvic fin rudimentary; anal fin long and continuous with caudal fin.

Notopterus notopterus (Pallas, 1769)
(Fig. 4.3)

FIGURE 4.3

Notopterus notopterus, 206.0 mm SL, Kharung pat, Chindwin basin, Manipur.

Gymnotus notopterus Pallas, 1769: 40, pl. 6 fig. 2 (type locality: Indonesia, most likely Java).

Diagnosis. Cranio-dorsal profile straight or slightly concave; maxilla extends to middle of eye; preorbital bone serrated; abdomen keeled with double serrations in front of pelvic-fin origin; back with no transverse bars; small wavy dark blue bars on the sides in small specimens. Maximum size ∼450 mm SL.

    Distribution. Widely distributed in the freshwaters of South and Southeast Asia.

Family Anguillidae

Freshwater eels

Body very elongated, eel-like, cylindrical anteriorly, becoming compressed posteriorly; pelvic fins and skeleton absent; dorsal, anal and caudal fins confluent; minute cycloid scales embedded in skin; gill opening lateral and crescentic; gill rakers absent; maxilla with teeth; premaxilla, vomer and ethmoid bones fused; united into a single bone; pectoral fin well developed; lateral line complete, present on body and head; vertebrae 100−119 (Fig. 4.4).

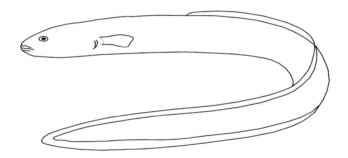

FIGURE 4.4

Outline diagram of Anguillidae.

Genus *Anguilla* Garsault, 1764

Anguilla Garsault, 1764: pl. 661 (type species: *Muraena anguilla* Linnaeus, 1758). Gender: feminine.

    Diagnosis. Body elongated, cylindrical anteriorly and compressed posteriorly; minute scales embedded under skin; lateral line complete; median fins confluent; dorsal-fin origin midway between pectoral-fin origin and vent; pectoral fin present.

Anguilla bengalensis (Gray, 1831)
(Fig. 4.5)

Muraena bengalensis Gray, 1831: pl. 95 fig. 5 (type locality: Ganges in Bihar, India).

    Diagnosis. Body robust, tail compressed; dorsal-fin origin about midway between the gill opening and anal-fin origin; vomerine teeth does not extend posteriorly so far as the maxillary one, mandibular teeth divided by a longitudinal groove; body with variegated yellow-olive color markings. Maximum size ∼1250 mm SL.

    Distribution. India: widely distributed in Northeast India, Pakistan, Sri Lanka, and Myanmar.

FIGURE 4.5

Anguilla bengalensis, 920.0 mm SL, Brahmaputra River, Chirang District, Assam.

Family Engraulididae
Anchovies

Body fusiform, compressed; scutes present along belly; snout blunt, projecting beyond tip of lower jaw; hind tip of maxilla extending behind eye; no fin spines; dorsal fin short, its origin equidistant between tip of snout and caudal-fin base; caudal fin low on body, sometimes with an outer filamentous ray; lateral line absent (Fig. 4.6).

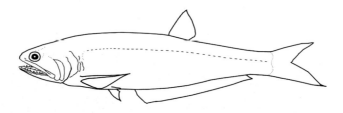

FIGURE 4.6

Outline diagram of Engraulididae.

Genus *Setipinna* Swainson, 1839

Setipinna Swainson, 1839: 186, 292 (type species: *Setipinna megalura* Swainson, 1839). Gender: feminine.

Diagnosis. Body compressed and elongated; abdomen edge keeled with 21—40 scutes; anal fin long, not confluent with caudal; caudal fin forked; jaws with minute teeth; first ray of pectoral fin produced into a filament.

Setipinna phasa (Hamilton, 1822)
(Fig. 4.7)
Clupea phasa Hamilton, 1822: 240, 382 (type locality: Brackish waters of Bengal).

FIGURE 4.7

Setipinna phasa, 125.0 mm SL, Brahmaputra River, Guwahati.

Diagnosis. Body compressed, oblong to elongate; abdomen edge keeled with 21−40 scutes; anal fin long, not contiguous with caudal fin; first ray of pectoral fin produced into a filament; body silvery, tip of caudal fin black. Maximum size ∼250 mm SL.

Distribution. Ganga, Brahmaputra, and Barak−Surma−Meghna drainages drainages.

Family Clupeidae
Herrings

Head devoid of scales; dorsal fin short, its origin halfway between tip of snout and caudal-fin base; caudal fin deeply forked; teeth small or absent; branchiostegal rays 5−10; anal fin rays less than 30; sharp ventral keel between the head and the anal-fin origin, keel with scales in the form of scutes; pectoral-fin origin close to the ventral profile; no fin spines; lateral line short or absent; recessus lateralis, in the form of otophysic connection, consisting of an intracranial space in the otic region into which may sensory canals open and separated from the internal ear by a thin elastic membrane present (Fig. 4.8).

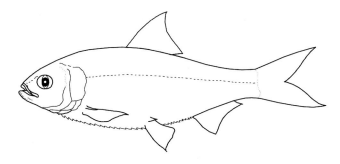

FIGURE 4.8

Outline diagram of Clupeidae.

Key to genera:

1a. Mouth inferior, dentary flared .... *Gonialosa*
1b. Mouth terminal, dentary normal, not flared 2

2a. Upper jaw without a notch at the center .... *Corica*
2b. Upper jaw with a notch at the center 3

3a. Scales on lateral series 37−47 .... *Hilsa*
3b. Scales on lateral series 77−91 .... *Gudusia*

Genus *Corica* Hamilton, 1822

Corica Hamilton, 1822: 253, 383 (type species: *Corica soborna* Hamilton, 1822: 253). Gender: feminine.

Diagnosis. Body oblong and compressed; abdomen keeled and serrated, 10−11 prepelvic scutes and 7−8 post pelvic scutes; lower jaw longer than the upper; teeth on jaws rudimentary; dorsal fin origin opposite pelvic fin origin; lateral line absent, 40−42 scales in lateral; caudal fin forked.

Corica soborna Hamilton, 1822

Corica soborna Hamilton, 1822: 253, 383 (type locality: Mahananda River, India).

Diagnosis. Small silvery fish of 1−2 in.; body strongly compressed, knife-like, abdomen keeled and serrated, 10−11 scute-like scales before and 7−8 behind the pelvic fin origin; dorsal fin with 12−13 rays, its origin behind the point between snout-tip and caudal-fin base; lower lobe of caudal fin longer; upper and lower edges and roots of fins dotted; lateral line straight; body translucent, bones, spine and gill covers green. Maximum size ~40 mm SL.

Distribution. West Bengal up to Mahananda River in northern Bengal and Orissa.

Genus *Gonialosa* Regan, 1917

Gonialosa Regan, 1917: 315 (type species: *Chatoessus modestus* Day, 1870a). Gender: feminine.

Diagnosis. Body compressed; profile above nape slightly concave, then a great rise to the dorsal fin origin; abdomen keeled with 17−18 scute-like scales, 11−12 behind ventral fin origin; mouth subterminal or inferior, lower jaw projecting over lower; dorsal-fin origin opposite pelvic-fin origin or slightly ahead; anal fin long with 22−28 rays; caudal fin forked; lateral line absent, 45−46 scales in lateral series; predorsal scales in pairs.

Gonialosa manmina (Hamilton, 1822)

Clupanodon manmina Hamilton, 1822: 247, 383 (type locality: freshwater branches of Ganga River, India).

Diagnosis. Body sword-like; scales on lateral line series 58−63, transverse 22−24; dorsal fin with 14 and anal fin with 24 rays; a black spot on the shoulder, dorsal side with bluish tint, dorsal and caudal fins with dark brown or blue outer edges; dirty green above, silvery below; fins clear and translucent. Maximum size ~110 mm SL.

Distribution. Ganga and Brahmaputra drainages in Assam and West Bengal.

Genus *Gudusia* Fowler, 1911

Gudusia Fowler, 1911: 207 (type species: *Clupanodon chapra* Hamilton, 1822). Gender: feminine.

Diagnosis. Body compressed, abdomen keeled with 18−19 prepelvic scutes and 8−10 post pelvic scutes; mouth oblique and slightly upturned; lower jaw projecting; dorsal-fin origin opposite pelvic-fin origin; anal fin long with 18−18 rays; caudal fin forked; lateral line absent, scales in lateral series 77−91.

Gudusia chapra **(Hamilton, 1822)**
(Fig. 4.9)

FIGURE 4.9

Gudusia chapra, 81.2 mm SL, Jiri River, Manipur.

Clupanodon chapra Hamilton, 1822: 248, 383 (type locality: upper parts of the rivers Ganga and Gandak).

Diagnosis. Abdominal profile more convex than of dorsal; gill rakers closely set, very numerous, shorter than eye; 18−19 scutes anterior to and 9−10 posterior to pelvic fin; body color silvery, shot with gold back darker, edge of caudal stained darkest; a dark spot may be present on shoulder. Maximum size ∼200 mm SL.

Distribution. Ganga and Brahmaputra, Barak drainages.

Genus *Tenualosa* Fowler, 1934

Tenualosa Fowler, 1934a, 1934b: 246 (type species: *Alosa reevesii* Richardson, 1846). Gender: feminine.

Diagnosis. Body deep and compressed; abdominal edge keeled with a row of scutes; upper jaw with a notch to fit lower jaw occlusion; scales small or of moderate size, regularly arranged.

Tenualosa ilisha **(Hamilton, 1822)**
(Fig. 4.10)

FIGURE 4.10

Tenualosa ilisha, 255.0 mm SL, Brahmaputra River, Goalpara, Assam.

Clupanodon ilisha Hamilton, 1822: 243, 382, pl. 19, fig. 73 (type locality: Bay of Bengal and the large saltwater estuaries of the Ganges, India).

Diagnosis. Body compressed; abdomen keeled and serrated with scales in the form of scutes directed backward, 17 before and 14−15 behind pelvic-fin origin; scutes directed backward;

posterior extremity of maxilla reaches below middle or hind edge of orbit; lower jaw not projecting beyond the upper; gill rakers numerous, as long as eye; body color silvery, shot with gold and purple; indistinct series of small spots along flanks; a dark blotch behind gill-opening in young ones, no spots in adult. Anadromous migrant. Maximum size ∼300 mm SL.

Distribution. Rivers in India, Sri Lanka, Bangladesh, Pakistan, and Myanmar.

Family Cyprinidae
Carps and minnows

Body cylindrical or slightly compressed, abdomen rounded or keeled in some; scales moderate to large; head without scales; barbels one to two pairs or absent; mouth and jaws toothless; lips thin and fleshy; pharyngeal teeth opposite enlarged posterior process of basioccipital bone, usually covered with tough horny pad; pharyngeal teeth in one to three rows; adipose fin absent; last simple dorsal fin osseous and posteriorly serrated in some; anal fin short; branchiostegal rays three; otophysic connection in the form of Weberian apparatus, a distinctive modification of anterior four to five vertebrae and associated ligaments for sound transmission from swim bladder to inner ear (Fig. 4.11).

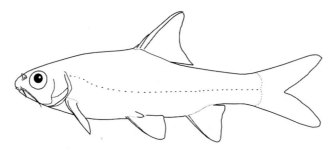

FIGURE 4.11

Outline diagram of Cyprinidae.

Key to genera:

1a. Vent and basal portion of anal fin encircled by a row of tile-like scales 2
1b. Vent and basal portion not encircled by a row of tile-like scales 3

2a. Lower jaw short and broad .... *Oreinus*
2b. Lower jaw long and narrower than the upper jaw .... *Schizothorax*

3a. Upper lip continuous with the skin of snout, mouth conspicuously inferior,
    lower lip with or without a suctorial disk 4
3b. Upper lip separated from the skin of snout by a deep groove,
    mouth superior or subinferior or inferior; lower lip without a suctorial disk 5

4a. Lower lip modified into a rounded or elliptical sucking disk,
    upper and lower lips continuous .... *Garra*

4b. Lower lip simple, not modified to form a suctorial disk,
upper and lower lip not continuous .... *Tariqilabeo*

5a. Abdomen keeled 6
5b. Abdomen rounded 13

6a. Abdomen keeled from below throat to vent, gill arches
fused into a spongious plate .... *Hypophthalmichthys*
6b. Abdomen keeled between belly or pelvic fin and vent,
gill arches not fused 7

7a. Lower jaw with no symphyseal knob 8
7b. Lower jaw with distinct symphyseal knob 10

8a. Mouth small, directed obliquely or almost vertically upward .... *Laubuka*
8b. Mouth inferior or subterminal 9

9a. Mouth subterminal and blunt; dorsal fin without spine, its origin
above or behind pelvic fins; anal fin short (7−12 branched rays) .... *Cabdio*
9b. Mouth directed forward and slightly upward; dorsal-fin origin
slightly behind pelvic-fin origin, its last simple ray osseous and
serrated posteriorly; anal fin long (17−33 branched rays) .... *Osteobrama*

10a. Dorsal surface of head with a muscular mass covered by skin and scutes;
abdominal keel anteriorly supported by an extension of pectoral girdle .... *Securicula*
10b. Dorsal surface of head with no muscular mass covered by skin and scutes;
abdominal keel not supported by pectoral girdle 11

11a. Mouth cleft extending beyond vertical through posterior margin of orbit .... *Raiamas*
11b. Mouth cleft never reaching to posterior rim of orbit 12

12a. Pelvic-fin origin close to pectoral-fin base than to anal-fin base;
outer pelvic-fin rays produced .... *Chela*
12b. Pelvic-fin origin close to anal-fin base than to pectoral-fin base;
outer pelvic-fin rays not produced .... *Salmostoma*

13a. Lower jaw with a well-developed symphyseal knob 14
13b. Lower jaw with no symphyseal knob 19

14a. Body almost transparent .... *Danionella*
14b. Body not transparent 15

15a. Barbels absent 16
15b. Barbels present 17

16a. Lateral line complete .... *Rasbora*
16b. Lateral line incomplete .... *Amblypharyngodon*

17a. Anal fin short with seven rays .... *Bengala*
17b. Anal fin long with 10−19 rays 18

18a. Body and median fins with stripes, no cleithral spot .... *Danio*
18b. Body with bars or stripes, cleithral spot present .... *Devario*

19a. Dorsal-fin origin behind pelvic-fin origin, often extending over anal-fin or
dorsal-fin origin, slightly ahead of anal-fin origin 20
19b. Dorsal fin with 7−30 branched rays, with or without an osseous simple ray,
its origin before or opposite pelvic-fin origin 21

20a. Maxillary barbel long, extending up to anal fin; lateral line present
or absent, curved downward if present, 27−34 scales in lateral series;
body without color bars .... *Esomus*
20b. Barbels present or absent, if present always minute and short,
sublaterally placed lateral line; body often with color bars .... *Barilius*

21a. Interorbital region, cheeks, and opercle with rows of minute pores;
lateral line scales larger than others .... *Oreichthys*
21b. Interorbital region, cheeks, and opercle with rows of minute pores;
lateral line scales larger than others 22

22a. Last simple anal-fin ray osseous and serrated posteriorly .... *Cyprinus*
22b. Last simple anal-fin ray not osseous 23

23a. Sector mouth (wide inferior mouth with exposed cornified mandibular
cutting edge) present .... *Semiplotus*
23b. Sector mouth absent 24

24a. Lower jaw with a small post symphyseal knob or tubercle,
last simple dorsal fin ray smooth, not spiny,
upper lip not continuous with lower lip .... *Cirrhinus*
24b. Lower jaw without post symphyseal knob or tubercle; upper and lower lip
continuous at the angle of mouth; upper and lower lip continuous 25

25a. Lower jaw without post symphyseal knob or tubercle; upper and lower lip
continuous at the angle of mouth;
25b. Upper and lower lip continuous 26

26a. Lower lip with continuous posterior groove around corners of mouth .... *Ctenopharyngodon*
26b. Lower lip with interrupted posterior groove in the middle 27

27a. Snout without any lobe, last simple dorsal-fin ray not serrated posteriorly 28
27b. Snout with tubercles or open pores, last simple dorsal-fin ray spiny
and strongly serrated posteriorly 34

28a. Lower lip with fleshy lobe below mandibular symphysis; cheek with no tubercles .... *Tor*
28b. Lower lip without any lobe, cheeks with or without tubercles 29

29a. Scales large, cheek with rows of tubercles or open pores copper colored longitudinal band laterally .... *Neolissochilus*
29b. Scales medium, cheek without tubercles or pores, colored longitudinal band absent laterally 30

30a. Lips with no horny covering 30
30b. Lips with horny covering on the inner side of one or both 31

31a. Last unbranched dorsal-fin ray serrated posteriorly 32
31b. Last unbranched dorsal-fin ray smooth posteriorly .... *Puntius*

32a. Adult size small, usually less than 50 mm SL; rostral barbels absent; maxillary barbels minute if present; a black blotch on caudal peduncle and frequently also other black blotches, spots or bars on side of body .... *Pethia*
32b. Adult size medium to large; barbels two pairs; no black blotch on caudal peduncle .... *Hypsibarbus*

33a. Snout with deep groove transverse groove, lower lip thin, postlabial groove broadly interrupted .... *Bangana*
33b. Snout plain, lips continuous at the angle of mouth .... *Labeo*

34a. Snout with median and two lateral lobes, no horny covering on lower jaw, one pores on snout not in grooves .... *Chagunius*
34b. Snout without groove and lobes, horny covering present on lower jaw 35

35a. Accessory pore present on the canals of the lateral-line system .... *Poropuntius*
35b. Accessory pore absent on the canals of the lateral-line system .... *Systomus*

Genus *Amblypharyngodon* Bleeker, 1860

Amblypharyngodon Bleeker, 1860: 433 (type species: *Cyprinus mola* Hamilton, 1822). Gender: masculine.

Diagnosis. Body elongated, compressed, abdomen rounded; eyes large, centrally placed; no discrete upper lip, no rostral cap and no horny jaw sheaths; a mandibular symphyseal knob which fits in notch of upper jaw (danionin notch); barbels absent; dorsal fin origin opposite behind pelvic fins; dorsal-fin with 7½ branched rays; anal fin with seven rays; scales small; lateral line incomplete, pored scales 6−7; lateral row of scales 55−75 (Fig. 4.12).

Amblypharyngodon mola (Hamilton, 1822)
(Fig. 4.13)

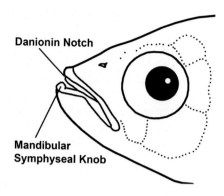

FIGURE 4.12

Mandibular symphyseal knob and Danionin notch.

FIGURE 4.13

Amblypharyngodon mola, 45.0 mm SL, Loktak Lake, Manipur.

Cyprinus mola Hamilton, 1822: 334, 392 (type locality: Gangetic provinces).

Diagnosis. Lateral line incomplete, extends up to posterior part of pectoral-fin; lateral rows of scales 65−75, rows between lateral line and pelvic-fin origin 9−10; dorsal-fin origin midway between anterior margin of eye and caudal-fin base; mandibular symphyseal knob present; a broad silvery lateral band on body. Maximum size ~55 mm SL.

Distribution. India, Pakistan, Sri Lanka, Nepal, and Bangladesh.

Amblypharyngodon microlepis (Bleeker, 1853)

Leuciscus microlepis Bleeker, 1853: 141 (type locality: Hooghly River, Calcutta, India).

Diagnosis. Lateral line incomplete, extends only up to three to four scales; lateral rows of scales 55−60; five rows between lateral line and origin of pelvic fin; dorsal-fin origin midway between posterior edge of eye and caudal-fin base. Maximum size ~9 cm. SL

Distribution. North Bengal, Nepal, Bangladesh, and Pakistan.

Genus *Bangana* Hamilton, 1822

Bangana Hamilton, 1822: 277, 385 (type species: *Cyprinus dero* Hamilton, 1822). Gender: feminine.

Diagnosis. Mouth inferior, semicircular; snout smooth, blunt and slightly projecting; rostral fold smooth and is separated from the upper lip by a deep groove and is disconnected from the lower lip around the corner of mouth; lateral lobe of upper lip smooth or slightly papillose and is laterally connected with lower lip, lower lip with a free anterior margin containing numerous papillae on dorsal surface, anteriorly separated from the lower jaw by a transverse groove extending along length of entire lower jaw; lower jaw heavily cornified with sharp cutting edge, postlabial groove interrupted and present on the side of lower jaw; dorsal fin falcate; 40–43 lateral line scales; 5–9½ scale rows between lateral line and origin of dorsal fin; presence of a supracleithral blotch, that is, a diamond-shaped blotch above pectoral fin, at level of lateral line (Fig. 4.14).

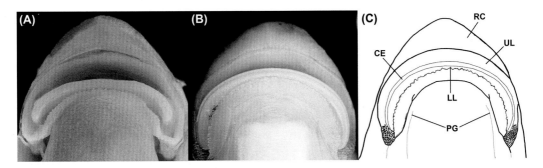

FIGURE 4.14

Oromandibular structure of: (A) *Bangana dero*, (B) *B. devdevi*; (C) drawing showing different parts of oromandibular structures of *Bangana*: CE, cornified edge of lower lip; LL, lower lip; PG, postlabial groove; RC, rostral cap; UL, upper lip.

Bangana dero (Hamilton, 1822)
(Fig. 4.15)

FIGURE 4.15

Bangana dero, 332.0 mm SL, Noa Dehing River, Arunachal Pradesh.

Photo courtesy: A. Darshan.

Cyprinus dero Hamilton, 1822: 277, 385, (type locality: Brahmaputra, India).

Diagnosis. Lateral line scales 38–40; circumpeduncular scale rows 16; branched dorsal-fin rays nine; snout inferior; mouth opening C-shaped; post labial groove interrupted medially; lower lip thin with slight papillation; 11 branched pectoral-fin rays; branched pelvic fin rays eight; transverse scale rows above lateral line 7½ and below lateral line 4½; a supracleithral blotch only above the pectoral fin at fifth scales of lateral line; opercular region with black blotch, ventral and anal fin region light brown; dorsal fin darker than pelvic and anal fin. Maximum size ~400 mm SL.

Distribution. Brahmaputra and Barak-Surma-Meghna drainage, India; Pakistan and China.

Bangana devdevi **(Hora, 1936)**
(Fig. 4.16)

FIGURE 4.16

Bangana devdevi, (A) 106.5 mm SL, (B) 210.0 mm SL, Thoubal River at Nongpok Keithelmanbi, Manipur.

Labeo devdevi Hora, 1936a: 324 (type locality: Naga Hills, India).

Diagnosis. Lateral line scales 40–41, circumpeduncular scale rows 20–21, branched dorsal-fin ray 11–12; dorsal surface of the snout with conspicuous transverse notch with large sized uni- to tetracuspid tubercles, presence of ethmoidal furrow; snout inferior, C-shaped mouth opening; post-labial groove interrupted medially; supracleithral blotch above pectoral fin at fifth scales of lateral line, at level of lateral line and base of the caudal peduncle. Max. size ~35 cm SL.

Distribution. Chindwin drainage, India; Upper Irrawaddy drainages, China, Myanmar and Thailand.

Genus *Barilius* Hamilton, 1822
Barilius (subgenus of *Cyprinus*) Hamilton, 1822: 266, 384 (type species: *Cyprinus barila* Hamilton, 1822). Gender: masculine.

Diagnosis. Compressed body with rounded belly. Body compressed, moderately elongated, shallow to deep; head pointed, tubercles often present; mouth terminal, obliquely directed upward; mandibular symphyseal process poorly developed; gape of mouth not extending up to middle of orbit; eyes large, superior; lips thin, simple; dorsal-fin inserted behind pelvic fin origin, often extending over anal fin; muscular pads often present in front of bases of pectoral-fin; barbels one to two pairs; caudal-fin forked; body silvery, with dark blue spots or bars; lateral line sublaterally placed.

***Barilius barila* (Hamilton, 1822)**
(Fig. 4.17)

FIGURE 4.17

Barilius barila, 86.1 mm SL, Dikrong River, Arunachal Pradesh.

Cyprinus barila Hamilton, 1822: 267, 384 (type locality: Northern Bengal).
Diagnosis. Body elongated, compressed, depth body greater than head length; mouth pointed; rostral and maxillary barbels rudimentary; head length almost equal to pectoral-fin length; pectoral-fin end well before pelvic-fin origin; scales large and easily removed; body with 24−26 indistinct bars, bars not reaching the lateral line; predorsal scales 19−20, lateral line scales 41−43; lower lobe of caudal-fin longer than upper lobe. Maximum size ~120 mm SL.
Distribution. Nepal, India, China, Myanmar, Pakistan, and Sri Lanka.

***Barilius barna* (Hamilton, 1822)**
(Fig. 4.18)

FIGURE 4.18

Barilius barna, 98.2 mm SL, Brahmaputra River near Dhubri, Assam.

Cyprinus barna Hamilton, 1822: 268, 384 (type locality: Yamuna and Brahmaputra rivers, India).
Diagnosis. Body elongated and compressed; lateral head length longer than body depth; pectoral fins almost reach pelvic-fin origin when adpressed; presence of one pair of rudimentary maxillary

barbels, pectoral lobe and pelvic axillary scale; pelvic-fin axillary scales longer than and thinner than pectoral-fin axillary lobe; lateral line complete with 36−38 pored scales; predorsal scales 15−16; sides of body with 10−11 bars, first two to three bars descend beyond lateral line, others shorter; body color in preservative: ground color creamish, bars and other markings brownish to blackish, a faint stripe on the subterminal margin of dorsal fin; other fins hyaline or creamish. Maximum size ∼85 mm SL.

Distribution. Ganga and Brahmaputra drainages in India, Bangladesh, Nepal, and Bhutan.

Barilius barnoides (Vinciguerra, 1890)
(Fig. 4.19)

FIGURE 4.19

Barilius barnoides, 59.7 mm SL, Chatrikong River, Manipur.

Barilius barnoides Vinciguerra, 1890: 307, Pl. 9, fig. 9 (type locality: Irrawaddy drainage, Kachin State, upper Myanmar).

Barilius chatricensis Selim and Vishwanath, 2002: 267 (type locality: Chatrikong River, Ukhrul District, Manipur, India).

Diagnosis. Body moderately deep and laterally compressed, its depth almost equals lateral head length; lateral line complete with 34−36 pored scales; predorsal scales 14−16; tubercles around snout; barbels absent; sides of the body with seven to eight bars, not extending to lateral line, bars dark brown to black, about two scales width, becomes fainter and tapering ventrally, interspaces wider than bars width; dorsum of the head above orbit darker, becoming creamish below; a blackish stripe extends posteriorly along the subterminal region of dorsal fin branched rays, other fins hyaline. Maximum size ∼100 mm SL.

Distribution. Chatrikong and Challou rivers, Ukhrul District, Manipur, India; Myanmar.

Barilius bendelisis (Hamilton, 1807)
(Fig. 4.20)

Cyprinus bendelisis Hamilton, 1807: 345 (type locality: Vedawati Stream, headwaters of Krishna River near Heriuru, Mysore, India).

Diagnosis. Body elongated and compressed; its depth almost equal to lateral head length; mouth blunt, its gap narrow, not reaching the anterior margin of orbit; rostral and maxillary barbels rudimentary; pectoral-fin much shorter than lateral head length; matured males with tubercles on tip of snout; pectoral-fin axillary lobes fleshier than pelvic-fin axillary scales; body with 8−12 bars on the sides; bars thin, interspaces about twice the width of bars; scales on body spotted with black markings; lateral line scales 40−45; sexual dimorphism seen: males with expanded paired fins, cushion-like pectoral-fin base and chest, tubercles on snout and cheek. Maximum size ∼135 mm SL.

Distribution. Nepal, Bhutan, India, Pakistan, and Sri Lanka.

FIGURE 4.20

(A) *Barilius bendelisis*, 92.4 mm SL, male, Likhailok, Barak drainage, Manipur; (B) *Barilius bendelisis*, 86.1 mm SL, female, Likhailok, Barak drainage, Manipur.

Barilius cosca **(Hamilton, 1822)**
(Fig. 4.21)

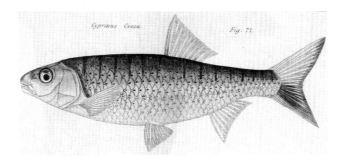

FIGURE 4.21

Barilius cosca, Hamilton's (1822), pl. 3, fig. 77 reproduced.

Cyprinus cocsa Hamilton, 1822: 272, 385, pl. 3, fig. 77 (type locality: northern rivers of Bengal and Bihar, especially the Mahananda).

Diagnosis. Two pairs of barbels, rostral barbel longer than maxillary barbel; scales large having several dots at the posterior margin; a row of blackish spots on the side of the bod; tips of both lobes of caudal fin pointed; presence of both pectoral axillary lobe and pelvic axillary scales; dorsal fin with nine rays, anal fin with 10. Maximum size ~90 mm SL.

Distribution. Ganga River drainage in Bihar and North Bengal, India.

Barilius dogarsinghi **Hora, 1921**
(Fig. 4.22)

FIGURE 4.22

Barilius dogarsinghi, 70.4 mm SL, Khuga River, Manipur.

Barilius dogarsinghi Hora, 1921a: 191, fig. 3 (type locality: Ethok stream near Chandrakhong, Manipur, India).

Diagnosis. Body elongated and moderately compressed; depth of the body longer than lateral head length; barbels in two pairs, rostral barbel longer than maxillary barbels; paired fins provided with muscular lobes at base; pectoral fin shorter than lateral head length; dorsal fin origin far posterior to middle of the body; pectoral fin does not reach pelvic-fin origin, pelvic fin nearly extends to anal-fin origin; caudal fin deeply forked with 10 + 9 rays, lower lobe slightly longer; scales of moderate size; lateral line complete with 38−39 scales; body silvery with the dorsum darker, pale yellow-white ventrally; sides of the body with eight to nine bars; bars prominent which is about two scales width and touch lateral line; eight to nine blackish bars on the sides of the body, the last bar at caudal-fin base, the deepest; dorsal fin with a blackish bar in the middle. Maximum size ~ 100 mm SL.

Distribution. Chindwin River drainage, Manipur, India and Myanmar.

***Barilius lairokensis* Arunkumar & Tombi Singh, 2000**
(Fig. 4.23)

FIGURE 4.23

Barilius laorokensis, 98.3 mm SL, Maklang River, Manipur.

Barilius lairokensis Arunkumar
Barilius ngawa Vishwanath & Manojkumar, 2002: 86 (description from Chakpi River, a tributary of Manipur River, Chandel District, Manipur, India).

Diagnosis. Body elongated and compressed; body depth almost equal to lateral head length; pectoral fin shorter than lateral head length; both maxillary and rostral barbels rudimentary; tubercles developed on snout and pectoral-fin base of matured male during breeding season; caudal fin forked, lower lobe longer than the upper; scales moderate; lateral line downwardly curved with 41 + 3 pored scales; predorsal scales 21; dorsal-fin rays ii + 8½; presence of 12−13 bars on the sides of the body, of which two to three anterior bars longer and touching lateral line, others slender and never touches the lateral line, bars about 1½ scale width. Maximum size ~140 mm SL.

Distribution. Chindwin drainage, Manipur, India.

***Barilius profundus* Dishma & Vishwanath, 2012**
(Fig. 4.24)

FIGURE 4.24

Barilius profundus, 71.1 mm SL, paratype, Kaladan River at Kolchaw, Mizoram.

Barilius profundus Dishma & Vishwanath, 2012: 2364 (type locality: Koladyne River at Kolchaw, Lawntlai District, Mizoram, India).

Diagnosis. Body short and deep; rostral and maxillary barbels rudimentary; presence of pectoral axillary lobes and pelvic axillary scales; mouth gap narrow, slightly longer than eye diameter; pelvic-fin reaching the anal aperture; 17−18 predorsal scales; lateral line complete with 30−32 pored scales; 7−10 brownish or blackish bars on the side of body. Maximum size ~70 mm SL.

Distribution. Kaladan River, Mizoram, India.

***Barilius shacra* (Hamilton, 1822)**
(Fig. 4.25)

FIGURE 4.25

Barilius shacra, 65.2 mm SL, Aie River, Chirang District, Brahmaputra drainage, Assam.

Cyprinus shacra Hamilton, 1822: 271 (type locality: Kosi River, Uttar Pradesh, India).

Diagnosis. Body elongated stouter than other known species of the genus; body depth shorter than lateral head length; scales small; barbels in two pairs, rostral longer than maxillary; brownish or blackish bars on the side of body 15−16, first two to three bars extend below lateral line, interspace twice the width of the bars; pectoral-fin lobe and pelvic-fin axillary scale absent; dorsal-fin rays. ii + 7; predorsal scales 31−32; lateral line scales 55−56. Maximum size ~100 mm SL.

Distribution. Ganga and Brahmaputra River drainages, India, Nepal, and Bangladesh.

Barilius tileo (Hamilton, 1822)
(Fig. 4.26)

FIGURE 4.26

Barilius tileo, 93.2 mm SL, Dikrong River, Arunachal Pradesh.

Cyprinus tileo Hamilton, 1822: 276, 385 (type locality: Kosi River, Uttar Pradesh, India).

Diagnosis. Body elongated and laterally compressed; body depth less than head length; presence of 10−11 bars on the sides of the body; first two to three bars almost touch the lateral line, others not touching the lateral line; mouth cleft extends up to the vertical axis passing through the middle of the orbit; presence of two pairs of barbels, rostral barbels much longer than maxillary barbels; lateral-head length greater than pectoral-fin length; interspaces width about twice that of bars. Maximum size ~150 mm SL.

Distribution. India: Arunachal Pradesh: Papum Pare District: Dikrong River; Bangladesh, Nepal and Pakistan.

Barilius vagra (Hamilton, 1822)
(Fig. 4.27)

FIGURE 4.27

Barilius vagra, 67.6 mm SL, Dikrong River, Arunachal Pradesh.

Cyprinus vagra Hamilton, 1822: 269, 385 [type locality: Ganges River about Patua (Patna), India].

Diagnosis. Body elongated, body depth smaller than lateral head length; 13−14 prominent brownish or blackish bars of about one scale width on the side of the body; interspace about twice the width of the bars; rostral barbel much longer than maxillary barbel; predorsal scales 22−23; lateral line complete with 43−45 pored scales; anal fin length longer than dorsal fin length; presence of pectoral fin axillary lobe and pelvic fin axillary scale. Maximum size ~120 mm SL.

Distribution. Ganga and Brahmaputra drainages: India, Bhutan, Nepal, Bangladesh.

Genus *Bengala* Gray, 1834

Bengala Gray, 1834: pl. 96, fig. 3 (type species: *Cyprinus elanga* Hamilton, 1822). Gender: feminine.

Diagnosis. Body elongated and abdomen rounded; snout pointed, mouth cleft oblique; lower jaw slightly produced with a symphyseal knob; one pair of rostral barbels; anal-fin short with seven rays; caudal-fin forked; lateral line complete with 40−44 scales; scales with striae and ridges.

Bengala elanga **(Hamilton, 1822)**
Cyprinus elanga Hamilton, 1822: 281, 386 (type locality: Ganges River, Elang, India).
Diagnosis. Body elongated, slender; mouth small, lower jaw longer, symphyseal knob, barbels short rostral pair, pectoral fin as long as head length, caudal fin formed, lateral line scales complete with 40−44 scales. Maximum size ∼20 mm SL.
Distribution. Ganga and Brahmaputra drainages in India and Bangladesh and Rakhine State, Myanmar.

Genus *Cabdio* **Hamilton, 1822**
Cabdio (subgenus of *Cyprinus*) Hamilton, 1822: 333, 392 (type species: *Cyprinus (Cabdio) jaya* Hamilton, 1822). Gender: masculine.
Diagnosis. Small-sized fish with considerably compressed body; abdomen keeled; barbels absent; mouth subterminal and blunt; suborbital process well developed; lips absent, lower jaw with crescentic edge; no barbels; dorsal-fin without spine, its origin above or behind pelvic-fins; lateral line decurved, running in lower half of caudal peduncle; scales of moderate size, 36−60 along the lateral line.

Cabdio crassus **Lalramliana, Lalronunga & Singh, 2019**
(Fig. 4.28)

FIGURE 4.28

Cabdio crassus, 60.4 mm SL, Ka-ao River, Kaladan drainage, Saiha District, Mizoram.

Cabdio crassus Lalramliana, Lalronunga & Singh, 2019: 162, figs. 2, 4a (type locality: Kaladan River in the vicinity of Kawlchaw village, Mizoram, India).
Diagnosis. Mid ventral surface rounded from isthmus to fifth to sixth scales and then keeled up to anal fin origin; body depth 25.8%−31.3% SL; presence of prominent pectoral and pelvic axillary scales; 45−52 lateral line pored scales; 20−25 predorsal scales; 16 circumpeduncular scales; 11 anal-fin base scales; ii + 7½ dorsal-fin rays; ii + 11½ anal-fin rays. Maximum size ∼100 mm SL.
Distribution. Ka-ao River and Kaladan River, Saiha District, Mizoram, India.

Cabdio jaya **(Hamilton, 1822)**
(Fig. 4.29)

FIGURE 4.29

Cabdio jaya, 61.6 mm SL, Dikrong River, Arunachal Pradesh.

Cyprinus jaya Hamilton, 1822: 333, 392 (type locality: Northern Bihar, India).

Diagnosis. Smallest of all the known *Cabdio* species; body elongated and compressed; abdomen keeled; barbels absent; body depth at dorsal-fin origin 28.6% SL; mouth moderately inferior, upper lip thicker than lower lip; dorsal-fin origin slightly posterior to vertical line through pelvic-fin origin; pectoral-fin longer than lateral head length; lateral line complete with 52−63 pored scales; predorsal scales 23; circumpeduncular scales 18; anal-fin base scales 15; transverse scale rows 15. Maximum size ∼110 mm SL.

Distribution. Ganga and Brahmaputra drainages, Nepal and Bangladesh.

Cabdio morar (Hamilton, 1822)
(Fig. 4.30)

FIGURE 4.30

Cabdio morar, 57.8 mm SL, Barak River at Silchar, Assam.

Cyprinus morar Hamilton, 1822: 264, 384 (type locality: Yamuna and Tista rivers, India).

Diagnosis. Body elongated and compressed; body depth at dorsal fin origin 24.2%−25.3% SL; barbels absent; mouth inferior, upper lip thicker than lower lip; scales moderately large and easily removed; dorsal-fin with ii + 7 rays, its origin slightly posterior to vertical line through pelvic-fin origin; axillary pectoral-fin lobe more fleshy and longer than axillary pelvic-fin scale; pectoral-fin longer than lateral head length; 18 predorsal scales; lateral line complete with 36 pored scales; nine scale rows between dorsal-fin origin and pelvic-fin origin; circumpeduncular scales 12, anal-fin base scales eight. Maximum size ∼100 mm SL.

Distribution. Ganga and Brahmaputra drainages: India, Nepal, Bhutan, Bangladesh, and China.

Cabdio ukhrulensis (Selim & Vishwanath, 2001)
(Fig. 4.31)
Aspidoparia ukhrulensis Selim & Vishwanath, 2001: 254 (type locality: Chatrickong River, Ukhrul District, Manipur, India).

FIGURE 4.31

Cabdio ukhrulensis, 102 mm SL, Maklang River, Kamjong District, Manipur.

Diagnosis. Body elongated, compressed and moderately deep; body depth at dorsal-fin origin 26.8%−28.5% SL; upper lip present, lower lip absent; barbels absent; two rows of pharyngeal teeth; dorsal-fin origin slightly posterior to vertical line through pelvic-fin origin; lateral line complete with 35−37 pored scales; predorsal scales 14; dorsal fin with ii + 7 rays, its first simple rays strong and osseous; pectoral-fin not reaching pelvic-fin and pelvic fin not reaching anal fin; caudal fin forked. Maximum size ∼ 100 mm SL.

Distribution. Chindwin drainage, Manipur, India.

Genus *Chagunius* Smith, 1938

Chagunius Smith, 1938 (type species: *Chagunius chagunio* Hamilton, 1822). Gender: masculine.

Chagunius: Rainboth, 1986: 5 (revision).

Diagnosis. Body medium to large size, abdomen rounded; head deep, cheek compressed; head deep and compressed; eyes large and superior, mouth subterminal, lips fleshy with loose rough skin due to dense covering of minute papillae; lower jaw blunt, not keratinized; postlabial groove incomplete, lower lip and lower jaw appear continuous; barbels four, long; snout and cheek tuberculated, more densely and strongly usually in males; dorsal fin origin ahead of pelvic fin origin, last simple dorsal fin ray osseous, stout and denticulated posteriorly; caudal fin forked; lateral line complete with 42−45 + 2−3 scales; first row of scales behind operculum black from mid-dorsal line to origin of pectoral fin; tips of dorsal and caudal fin red in color.

Chagunius chagunio Hamilton, 1822

(Fig. 4.32)

FIGURE 4.32

Chagunius chagunio, 190.0 mm SL, Lohit River, Lohit District, Arunachal Pradesh.

Photo courtesy: A. Darshan.

Chagunius chagunio Hamilton, 1822: 295, 387 (type locality: Yamuna River and northern rivers of Bihar and Bengal).

Chagunius chagunio: Rainboth, 1986: 6 (revision).

Diagnosis. Circumferential scales 40−44, circumpeduncular scales 23−25; nuptial tubercles densely developed in males covering snout and cheek, thinning in preopercle, small tubercles on head spreading up to scales in dorsum, last two anal fin rays elongated in males, extending to base of caudal fin; anal fin in females uniformly curved, last rays not elongated. Maximum size ∼250 mm SL.

Distribution. Ganga and Brahmaputra drainages.

Chagunius nicholsi (Myers, 1924)
(Fig. 4.33)

FIGURE 4.33

Chagunius nicholsi, 194.0 mm SL, Manipur River near Myanmar border, Manipur.

Barbus nicholsi Myers, 1924: 3 (type locality: Monywa, Sagaing, Myanmar).

Diagnosis. Body elongated; mouth narrow and subterminal; barbels two pairs, maxillary and rostral; scales medium, diamond-shaped, lateral line 45−46, lateral transverse rows 8/1/5, circumferential scales 34, circumpeduncular 19−20, predorsal 15−16; last simple dorsal-fin ray strong and weakly denticulated posteriorly and recurved; muscular lobes present on pelvic-fin base, rarely on anal-fin base; anal fin short and caudal fin forked; dorsum greyish, silvery below with a pinkish tinge, edges of scales black, a black band on the margin of opercle; caudal fin lobe red; sexual dimorphism unknown. Maximum size ∼250 mm SL.

Distribution. Chindwin drainage, Manipur, Myanmar.

Genus *Chela* Hamilton, 1822

Chela Hamilton, 1822: 258, 383 [type species: *Cyprinus* (*Chela*) *cachius* Hamilton, 1822]. Gender: feminine.

Diagnosis. Body compressed, moderate to deep, ventral profile more convex than that of dorsal; mouth large, oblique, upper jaw protractile; no symphyseal process on lower jaw; dorsal-fin origin slightly behind anal-fin origin; pectoral-fin elongated, its origin nearer to pectoral-fin origin; anal-fin origin long with 13−26 rays; caudal-fin forked; lateral line complete with 34 or more scales, curved downward.

Chela cachius Hamilton, 1822
(Fig. 4.34)

FIGURE 4.34

Chela cachius, 42.0 mm SL, Poma River at Ramgath, Arunachal Pradesh.

Cyprinus cachius Hamilton, 1822: 258, 384 (type locality: Ganges River about commencement of the Delta).

Diagnosis. Body moderately deep and greatly compressed, its depth 21%−31% SL; abdomen keeled behind pelvic-fin origin, outer ray of pelvic-fin produced into a filament reaching up to two-thirds of anal-fin; lateral line complete with 51−56 scales; color in fresh: body translucent, shining brilliant silver and a greenish stripe from level of dorsal-fin; in preservative: a black narrow strip from behind the opercle to caudal-fin base in the upper half of body. Maximum size ∼50 mm SL.

Distribution. Ganga and Brahmaputra drainage, India.

Genus *Cirrhinus* Oken, 1817

Cirrhinus Oken, 1817: 1182 (type species: *Cyprinus cirrhosus* Bloch, 1795). Gender: masculine.

Diagnosis. The genus is characterized in having a subinferior mouth; thick and smooth rostral cap separated from the upper lip by a deep groove; thin upper lip, not covered by rostral fold; rostral fold without lateral lobe; a very thin lower lip, closely adnate to lower jaw; and absence of barbell or a rostral pair, if present.

Cirrhinus cirrhosus (Bloch, 1795)
(Fig. 4.35)

FIGURE 4.35

Cirrhinus cirrhosus, 277.0 mm SL, Loktak Lake, Manipur.

Cyprinus cirrhosus Bloch, 1795: 52 (type locality: Malabar Coast, India)

Diagnosis. Characterized in having 42−44 lateral line scales; 12½ branched dorsal fin rays; 5½−6½ branched anal fin rays; rostral barbel present, maxillary barbel absent if present very minute; subterminal mouth; rostral cap without fimbriae or papillae covering the upper jaw, separated from the upper jaw by a deep groove; absence of upper lip; upper jaw connected to lower jaw in the corner of the mouth; lower lip weakly developed in the form of transverse series of small papillae; transverse scale rows above lateral line 7½, between lateral line and pelvic fin origin 6½; 14−15 predorsal scales. In preservative, head, dorsum dark gray. Lateral side above lateral line dark gray and lateral surface below lateral line creamy white. Ventral surface creamy white. Dorsal and caudal fin darker than pectoral, pelvic, and anal fin. Black humeral spot at the base of the caudal-fin base. Maximum size ~800 mm SL.

Distribution. Native to the major rivers of India; widely introduced and cultured in Asia.

Cirrhinus reba (Hamilton, 1822)
(Fig. 4.36)

FIGURE 4.36

Cirrhinus reba, 294.0 mm SL, Barak River, Silchar, Assam.

Cyprinus reba Hamilton, 1822: 280, 386 (type locality: India: Bengal and Behar).

Diagnosis. Body elongated, its depth much longer than head length; barbels one pair of short rostral; mouth broad, lower jaw with a thin cartilaginous cover; pectoral fin as long as head; scales hexagonal and moderate, lateral line with 34−38 scales, transverse scale rows 7/5−6. Dorsal fin ii−iii, eight rays. Anal fin with iii, five rays; pectoral fin with i, 15 rays; pelvic fin i, eight rays; caudal fin deeply forked. In live, dorsum of body dark gray, flanks and belly silver; edges of scales with melanophore, forming bluish longitudinal bands above lateral line. Maximum size ~250 mm SL.

Distribution. India, Bangladesh, Nepal, Pakistan.

Genus *Ctenopharyngodon* Steindachner, 1866

Ctenopharyngodon Steindachner, 1866: 782 (type species: *Ctenopharyngodon laticeps* Steindachner, 1866). Gender: masculine.

Diagnosis. Body elongated, subcylindrical; head depressed; mouth terminal, cleft not extending to anterior margin of orbit; lips moderate, lower lip distinct at corners of mouth; barbels absent; pharyngeal teeth in two rows; gill rakers biserial; dorsal fin short, its origin slightly ahead of pelvic fins; dorsal spine smooth and weak; lateral line complete with 40−42 scales.

Ctenopharyngodon idella (Valenciennes, 1844)
(Fig. 4.37)

FIGURE 4.37

Ctenopharyngodon idella, 458.0 mm SL, Loktak Lake, Manipur.

Leuciscus idella Valenciennes, in Cuvier and Valenciennes, 1844: 362 (type locality: China).

Diagnosis. Snout short, its length less than or equals eye diameter; barbels absent; postorbital length more than half the length of head; dorsal fin with three simple and seven to eight branched rays; and anal fin with three simple and 7−11 branched rays. Maximum size ∼1300 mm SL.

Distribution. Flatland rivers of China and the middle and lower reaches of river Amur in Russia; introduced throughout the world.

Genus *Cyprinus* Linnaeus, 1758

Cyprinus Linnaeus, 1758: 320 (type species *Cyprinus carpio* Linnaeus, 1758). Gender: masculine.

Diagnosis. Body compressed, robust; head small, snout rounded; eyes placed relatively high on head; mouth small, directed forward and protrusible, lips thick and fleshy; barbels two pairs, rostral and maxillary; pharyngeal teeth in three rows; dorsal fin long with 16−22½ simple rays, its origin opposite pelvic fin origin, its last simple ray serrated posteriorly; anal fin short, the last simple ray osseus and serrated posteriorly.

Cyprinus carpio Linnaeus, 1758
(Fig. 4.38)

FIGURE 4.38

Cyprinus carpio: (A) common carp, 253 mm SL; (B) mirror carp, 196 mm SL; (C) golden carp, 215 mm SL; wetlands of Imphal valley, Manipur.

Cyprinus carpio Linnaeus, 1758: 320 (type locality: Europe).

Diagnosis. Last simple ray osseous and serrated posteriorly; barbels 4; branched dorsal fin rays 17−20½; lateral line scales 33−37. Maximum size ∼1200 mm SL.

Distribution. Western Europe, native of basins around Black Sea, Caspian Sea, and Aral Sea; widely introduced worldwide.

Genus *Danio* Hamilton, 1822

Danio Hamilton, 1822: 321, 390 [type species: *Cyprinus (Danio) dangila* Hamilton, 1822]. Gender: masculine.

Celestichthys Roberts, 2007: 132 (type species: *Celestichthys margaritatus* Roberts, 2007). Gender: masculine (Fig. 4.39).

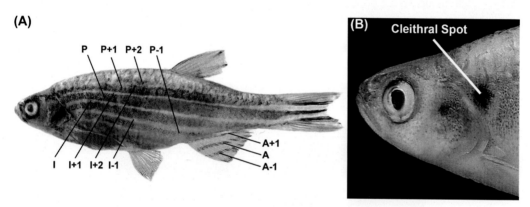

FIGURE 4.39

(A) Lateral view of *Danio quagga* showing stripes: P, stripes on body; I, interstripes, A, stripes on anal fin; (B) cleithral spot.

Diagnosis. Body compressed and elongated, belly rounded; snout obtuse, mouth anterior and cleft oblique, lower jaw with a symphyseal knob and a notch to fit it in the upper jaw; barbels small, 1−2 pairs; dorsal fin origin nearer to caudal-fin base than to tip of snout, opposite the middle between pelvic fin and anal fin origins; lateral line curved downward, with 36−40 scales; specific color patterns in the form of stripes and interstripes, stripes often coalescing anteriorly, more rarely with light or dark spots or bars, body stripes often continuing to caudal fin, anal fin with stripes.

Danio assamila Kullander, 2015
(Fig. 4.40)

Danio assamila Kullander, 2015: 362 (type locality: tributary of Dibru River at Amguri, Brahmaputra drainage, Assam, India).

Diagnosis. First ray in pectoral and pelvic fins produced; pattern of dark rings enclosing light interspaces on the side, rings in series elongated, rings not clear in smaller specimens; large round or slightly oval cleithral spot; anterior interstripe Ia usually present; ring pattern not extending onto caudal peduncle; lateral scale 32−34. Maximum size ∼70 mm SL.

Distribution. Brahmaputra drainage in Assam and Arunachal Pradesh, India.

FIGURE 4.40

Danio assamila, 60.0 mm SL, Namdapha River, Arunachal Pradesh.

Danio dangila (Hamilton, 1822)

Cyprinus dangila Hamilton, 1822: 321, 390 (type locality: Mountain streams of Mongher, Bihar, India).

Diagnosis. First rays of pectoral and pelvic fins produced; large cleithral spot vertically extended; dark rings in series enclosing light interspaces on the side; rings rounded; absence of complete anterior inter-stripe Ia; ring pattern extending onto caudal peduncle; lateral line 32−34. Maximum size ∼90 mm SL.

Distribution. Ganga and Brahmaputra basins in Bangladesh, India and Nepal; probably also in Meghna and Godavari basins.

Danio jaintianensis (Sen, 2007)

Brachydanio jaintianensis Sen, 2007: 28, pl. I, fig. A; pl. 2, fig. A. (type locality: Rangriang Jowai, Jaintia Hills District, Meghalaya, India).

Danio jaintianensis: Kullander and Fang, 2009: 47 (placement in *Danio* and treated as valid).

Diagnosis. P-stripe extending from behind opercle to caudal-fin, a thinner P + 1 stripe; dorsal-fin with seven branched rays and anal-fin with 9−10 branched rays; lateral line absent, lateral rows of scales 31−32; two pairs of barbels, rostral short extending to middle or to posterior edge of eye, maxillary up to preopercle, occasionally up to opercle; predorsal scales 16−17. Maximum size ∼80 mm. SL.

Distribution. Streams of Jaintia Hills, Meghalaya, India.

Danio meghalayensis Sen & Dey, 1985

(Fig. 4.41)

FIGURE 4.41

Danio meghalayensis, 76.0 mm SL, Amtapoh, Barak drainage, West Jaintia Hills, Meghalaya.

Danio meghalayensis Sen & Dey, 1985: 62 (type locality: a hill stream at Barapani, Meghalaya).

Diagnosis. Unbranched ray in pectoral and pelvic fins not elongated; presence of cleithral spot, usually difficult to make out at anterior end of P stripe; complete lateral line; rostral barbel not reaching beyond gill cover margin, maxillary barbel not reaching beyond pectoral-fin base; presence or absence of multiple rows of round or slightly elongated rings on side, P, P + 1, and P-1 stripes distinct, occasionally P and P + 1 stripes anastomosing. Maximum size ~ 100 mm SL.

Distribution. Umroi stream near Barapani, Khasi Hills, Meghalaya, India.

Danio quagga Kullander, Liao & Fang, 2009
(Fig. 4.42)

FIGURE 4.42

Danio quagga, 45.3 mm SL, Yu River, Tamu, Myanmar.

Danio quagga Kullander, Liao & Fang, 2009: 193−199 (type locality: a small river in Saw Bwa Ye Shan Village, Sagaing Division, Kamphat River drainage, Myanmar).

Diagnosis. Body with four distinct dark stripes (P + 1, P, P-1, P-2) along middle of the side; snout short, rounded, shorter than eye diameter; mouth terminal, obliquely upward directed; small bony knob at dentary symphysis; maxillary barbel much longer than rostral; lateral line incomplete, pored scales up to five, 28−30 scales in the row; predorsal scales 15−16; circumpeduncular scales 10; a row of scales along anal-fin base; dorsal-fin with ii + 7 rays; anal fin with iii + 12−13 rays; pectoral fin with i + 11−13 rays; pelvic fin with i + 7 rays; caudal fin with 10 + 9 rays; six dark brown stripes; short narrow band of black pigment on chest just ventral to pectoral-fin base; caudal fin hyaline, with narrow dark stripes continuing body stripes P + 1, P, and P-1. Maximum size ~ 40 mm SL.

Distribution. Yu drainage, Manipur, India and Myanmar.

Danio rerio (Hamilton, 1822)
(Fig. 4.43)

FIGURE 4.43

Danio rerio, 21.3 mm SL Dikrong River, Arunachal Pradesh.

Cyprinus rerio Hamilton, 1822: 323, 390 (type locality: Kosi River, Uttar Pradesh, India).

Diagnosis. Small-sized *Danio* of a maximum of 50.8 mm length, lateral line absent, if present short, extending up to pelvic-fin base, scales on lateral series 28−30; sides of body with four well defined colored stripes extending up to caundal fin; oblique colored bars on anal fin, paired fins hyaline; barbels two pairs, rostral usually longer than eye diameter, maxillary extending beyond half of pectoral fin. Maximum size ∼40 mm SL

Distribution. Ganga and Brahmaputra drainages, northern and northeastern India, Bhutan, and Bangladesh.

Genus *Danionella* Roberts, 1986

Danionella Roberts, 1986a: 232 (type species: *Danionella translucida* Roberts, 1986a). Gender: feminine.

Diagnosis. Most tiny vertebrate organism of 10−20 mm SL, commonly called micro glass fish; almost completely transparent except for its eye and some distinctive black spots on the lower half of its body; scales absent; lateral lines absent.

Danionella priapus Britz, 2009
(Fig. 4.44)

FIGURE 4.44

Danionella priapus, 13.8 mm SL, Jorai River, Alipurduar District, northern Bengal.

Photo courtesy: H. Choudhury.

Danionella priapus Britz, 2009: 53 (type locality: outskirts of Barobisha town, Jorai River, a tributary of the Sankosh at Laskarpara, Brahmaputra River drainage, Jalpaiguri District, West Bengal, India).

Diagnosis. A conical projection of the genital papilla situated between the pelvic fins which form a funnel-like structure in adult males; eight pectoral-fin rays; 20−21 anal-fin rays; last anal-fin ptery-giophore inserted in front of hemal spine of vertebra 27 or 28; two paramedian rows of pigment cells on the dorsal side of the body; seven to eight dorsal and six to eight ventral caudal procurrent rays; two upper jaw bones; edges of jaw bones entire, without processes; a well-developed lateral stripe extending from ear capsule to caudal peduncle. Maximum size ∼20 mm SL.

Distribution. Jorai River in West Bengal, India.

Genus *Devario* Heckel, 1843

Devario Heckel, 1843: 1015 (type species: *Cyprinus devario* Hamilton, 1822). Gender: masculine.

Diagnosis. Body compressed; abdomen rounded; mouth anterior; cleft of mouth shallow, not protractile, directed obliquely upwards; lower jaw with symphyseal knob; dorsal fin with 10−19

rays, its origin opposite between pelvic and anal fins, nearer to caudal-fin base than to tip of snout; a distinct cleithral spot; body with colored stripes or bars or both; fins without stripes; P stripe extending to the median caudal-fin rays, infraorbital five not or slightly reduced; maxillary barbel shorter than eye diameter; short and wide ascending process of premaxillary with a minute apophysis contacting the kinethmoid.

Devario acuticephala (Hora, 1921)
(Fig. 4.45)

FIGURE 4.45

Devario acuticephala, 37.1 mm SL, Sekmai River, Pallel, Manipur.

Danio (Brachydanio) acuticephala Hora, 1921: 193 (type locality: Yairibok, Chindwin drainage, Manipur, India).

Diagnosis. Body small; wide P-stripe tapering posteriorly and ending at caudal-fin base, its anterior portion less pigmented and margins diffuse up to vertical level of first anal-fin origin; lateral line absent; dorsal fin with seven rays and anal with 9−20 rays; distinct cleithral spot. Maximum size ∼40 mm SL.

Distribution. Chindwin drainage in Manipur, India and Myanmar.

Devario aequipinnatus (McClelland, 1839)
(Fig. 4.46)

Perilampus aequipinnatus McClelland, 1839: 393 (type locality: India).

Devario aequipinnatus: Kullander et al., 2017: 11, 37 (redescription based on specimens from Bangladesh).

Devario assamensis Barman, 1984: 163 (Kullander et al., 2017: 31 treats it as a possible synonym of *D. aequipinnatus*).

Diagnosis. Unique color pattern: presence of distinct and straight P + 1, P and P-1, stripes, P stripe wider than others, equal in width from anterior to posterior end, or slightly tapering posteriorly; anterior part of interstripe I below P may be interrupted at one or two points anteriorly; stripes not confluent anteriorly; dorsal-fin rays 9−10, circumpeduncular scales 12, body depth 26.7%− 30.8% SL; presence of infraorbital process. Maximum size ∼ 100 mm SL.

Distribution. Teesta, Brahmaputra and Barak drainages in India; Karnafuli, Feni and Meghna drainages in Bangladesh.

Devario coxi Kullander, Rahman, Norén & Mollah, 2017
(Fig. 4.47)

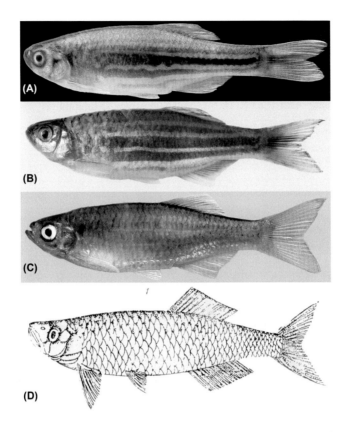

FIGURE 4.46

Devario aequipinnatus; (A) 78.2 mm SL, Barak River, Khunphung, Manipur; (B) 72.3 mm SL. Namdapha River, Arunachal Pradesh; (C) 56.4 mm SL, Tuivai River at Likhailok, Manipur; (D) *Perilampus aequipinnatus*, McClelland's (1839) Pl. 51, fig. 1, reproduced.

FIGURE 4.47

Devario coxi, 68.0 mm SL, Karnafuli River, Mizoram.

Devario coxi Kullander, Rahman, Norén & Mollah,2017: 15, fig. 4d and e (type locality: Majerchora stream, 10 km south of Cox's Bazar, Chittagong Division, Bangladesh).

Diagnosis. Distinct P + 1, P and P stripes along the side, P and P-1 stripes confluent anteriorly and containing light spots; 11−12 dorsal-fin rays; 14 circumpeduncular scales; body depth 29.9%−33.2% SL. Maximum size ∼70 mm SL.

Distribution. Stream near Cox's Bazar, and the lower Matamuhuri River in Bangladesh, Karnafuli River, Mizoram.

Devario deruptotalea **Ramananda & Vishwanath, 2014**
(Fig. 4.48)

FIGURE 4.48

Devario deruptotalea, 63.4 mm SL; Dutah Stream, Chindwin drainage, Chandel District, Manipur.

Devario deruptotalea Ramananda & Vishwanath, 2014: 79, fig. 1 (type locality: Dutah Stream, a tributary of Yu River, Chandel District, Manipur).

Diagnosis. Lateral line complete with 32−34 pored scales; color pattern consisting of bars and reticulations anteriorly, that is, four to six dark brown irregularly shaped and arranged bars, each of which is partly confluent with adjacent bar at different levels on anterior one-third of the side of the body, followed by three distinct dark brown stripes posteriorly; P-stripe darker, about twice as broad as other stripes, extending onto median caudal-fin rays; blackish stripe on dorsal fin; series of three to six yellowish patches formed along path of I−1 by joining P and P−1 stripes. Maximum size ∼65 mm SL.

Distribution. Tributaries of the Yu River, Chindwin drainage in Manipur.

Devario devario **(Hamilton, 1822)**
(Fig. 4.49)
Cyprinus devario Hamilton, 1822: 341, 393 (type locality: Bengal).
Devario devario: Kullander et al., 2017: 6, fig. 1 (redecription).

Diagnosis. Body deep, body depth at dorsal-fin origin 33%−44% SL; dorsal fin with iii, 14−17 rays, its base long; infraorbital and suborbital processes absent; barbels absent or rudimentary; P-stripe short, starting from level of middle of dorsal-fin base becoming broader behind and extending to caudal-fin base, and further to lower part of upper lobe of caudal fin; narrow irregular light bars anteriorly and above and below P stripe, often coalescing and enclosing lighter spots; scales small, lateral line complete with 38−50 scales, 16 circumpeduncular scales. Maximum size ∼70 mm SL.

Distribution. Large rivers and beels of northern and north-eastern India; Pakistan, Bangladesh and Nepal.

FIGURE 4.49

Devario devario, (A) 57.5 mm SL, Barak River, Silchar, Assam; (B) 49.0 mm SL, Surma basin, Bangladesh.

Devario horai (Barman, 1983)

Danio (Brachydanio) horai Barman, 1983: 177 (type locality: Namdapha River, Tirap District, Arunachal Pradesh, India).

Diagnosis. Lateral line absent, scales on lateral line row 28−30; lateral transverse scales seven; predorsal scales 14; circumpeduncular scales eight; barbels absent; body depth 26.0%−31.2% SL; body and fins with no bars and stripes. Maximum size ∼30 mm SL.

Distribution. Arunachal Pradesh, Brahmaputra basin, Arunachal Pradesh.

Devario manipurensis (Barman, 1987)

(Fig. 4.50)

FIGURE 4.50

Devario manipurensis, 46.8 mm SL, Challou River, tributary of Tizu River, Ukhrul, Manipur.

Danio manipurensis Barman, 1987: 173 (type locality: Manipur, India).

Devario manipurensis Kottelat, 2013: 99 (valid).

Diagnosis. Body elongated and compressed; a prominent stripe having a width greater than one scale width extending from cleithral spot to subterminal region of median caudal-fin rays; stripe becoming darker posteriorly; a blackish streak along the mid predorsal region, scales on dorsum spotted with brownish spots; caudal-fin lobes with orange tinge excluding its free margin which is hyaline; dorsal and anal fins brownish basally and extends posteriorly up to tips of two to three branched rays; lateral line complete with 38—40 pored scales; predorsal scales 14—16; pelvic-fin axillary scale longer and more prominent than pectoral-fin axillary scale. Maximum size ~40 mm SL.

Distribution. Challou River, Chindwin drainage, Ukhrul District, Manipur, India.

Devario naganensis **(Chaudhuri, 1912)**
(Fig. 4.51)

FIGURE 4.51

Devario naganensis, 54.8 mm SL, Momo stream, tributary of Tizu River, Ukhrul, Manipur.

Danio naganensis Chaudhuri, 1912: 441 (type locality: Langting River, Manipur, Chindwin River drainage, India).

Devario naganensis: Kottelat, 2013: 99 (valid).

Diagnosis. Body elongated and compressed; mouth cleft shallow, extends up to vertically through anterior margin of orbit; indistinct P stripe on the body extending onto the median caudal-fin rays; barbels in two pairs, maxillary and rostral; dorsal-fin origin behind the middle of the body, extends over three-fourth of anal fin; presence of cleithral spot; lateral head length greater than pectoral fin; caudal fin forked with 10 + 9 principal rays; scales moderate, predorsal scale 18—20, lateral line complete with 40—42 pored scales, circumpeduncular scales rows 10—12; total vertebrae 37—38. Maximum size ~60 mm SL.

Distribution. Chindwin River drainage, Manipur.

Devario yuensis **(Arunkumar & Tombi Singh, 1998)**
(Fig. 4.52)

FIGURE 4.52

Devario yuensis, 57.0 mm SL, Lokchao River, Tamu, Myanmar.

Danio yuensis Arunkumar and Tombi Singh, 1998: 3, fig. 1 (type locality: Moreh Bazar, near the border areas of Manipur, Yu River System, India and Myanmar).

Diagnosis. Body deep, its depth 31.0%−35.7% SL; osseous spine on supraorbital bone and an osseous process on the first infraorbital bone; lateral line scales 37−40; circumpeduncular scales 14−16; three to five distinct stripes from the level of one-third of the body on the side of the body: P, P + 1, P + 2, P-1and P-2, stripes expanded anteriorly at irregular intervals and are joined together at some points interrupting I + 1 and I-1 interspaces in varying degrees. Maximum size ∼60 mm SL.

Distribution. India: Yu River System in Manipur and Myanmar.

Genus *Esomus* Swainson, 1839

Esomus (subgenus of *Leuciscus*) Swainson, 1839: 185, 285 (type species: *Esomus vittatus* Swainson, 1839). Gender: masculine.

Diagnosis. Body compressed and elongated with rounded abdomen; mouth oblique, symphyseal knob absent, lower jaw prominent; pharyngeal teeth in single row; maxillary barbel long, extending upto anal fin; pectoral fin often long, dorsal fin origin slightly ahead of anal fin origin, caudal fin forked; lateral line present or absent, curved downwards in present, 27−34 scales in lateral series.

Esomus danrica (Hamilton, 1822)
(Fig. 4.53)

FIGURE 4.53

Esomus danricus, 38.0 mm SL, Lilakhong, Imphal, Chindwin basin, Manipur.

Cyprinus danrica Hamilton, 1822: 325, 390, pl. 16, fig. 88 (type locality: Bengal).

Diagnosis. A broad silvery blue green stripe on the side; lateral line incomplete with four to six pored scales, 27−30 scales in lateral series; maxillary barbel reaching anal fin; predorsal scales 16−17, circumpeduncular scales 14. Maximum size ∼60 mm SL.

Distribution. Water bodies of India, Pakistan, Nepal, Sri Lanka and Myanmar.

Genus *Garra* Hamilton, 1822

Garra Hamilton, 1822: 343, 393 (subgenus of *Cyprinus* Linnaeus, 1758; type species: *Cyprinus lamta* Hamilton, 1822).

Diagnosis. Lower lip modified forming a mental adhesive disk whose posterior margin is discontinuous with the mental region; crescentic anteromedian fold of the mental adhesive disk is similar to or wider than the width of the central callous pad and the lateral end of the anteromedian fold on each side usually reaches the anterolateral lobe of the mental adhesive disk; snout with or without a proboscis, often with tubercles.

Garra abhoyai **Hora, 1921**
(Fig. 4.54)

FIGURE 4.54

Garra abhoyai, 45.0 mm SL, Chakpi River, Manipur.

Garra abhoyai Hora, 1921a, 1921b, 1921c: 664 (type locality: Ukhrul District, Manipur, India).
Garra abhoyai: Vishwanath and Linthoingambi, 2008: 101 (redescription).
Diagnosis. Smoothly rounded snout tip; rostral lobe absent; proboscis absent; predorsal scales present but those toward head very reduced, irregularly arranged and covered by mucus almost making it appear to be absent; chest and abdominal region naked, however, area just in front of pelvic fin scales covered by mucus; papilliferous tissue absent along the upper jaw; papillations present at an angle of upper and lower lip; lateral line scales 30−33 + 1−3; dorsal fin with a submarginal black band, band present only on the rays; caudal fin with a distinct W-shaped black band. Maximum size ∼70 mm SL.
Distribution. Chindwin drainage, Manipur, India.

Garra annandalei **Hora, 1921**
Garra annandalei Hora, 1921b: 657 (type locality: Assam and streams at the base of the Darjeeling Himalayas, India).
Diagnosis. Nearly cylindrical body shape; smooth snout with no proboscis, transverse grove, rostral lobe; small mouth with well-developed, yet small mental disk; presence of a pleated papilliferous fold on the lower lip. Maximum size ∼120 mm SL.
Distribution. Brahmaputra River drainage, Bhutan, India, Nepal.

Garra arunachalensis **Nebeshwar & Vishwanath, 2013**
(Fig. 4.55)

FIGURE 4.55

Garra arunachalensis, 121.0 mm SL, holotype, Deopani River, Arunachal Pradesh.

Garra arunachalensis Nebeshwar & Vishwanath, 2013: 101, fig. 3 (type locality: lower Divang valley District, Deopani River at Roing, Brahmaputra basin Arunachal Pradesh, India) (Fig. 4.56).

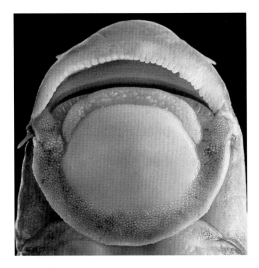

FIGURE 4.56

Mental disk of *Garra arunachalensis.*

Diagnosis. A transverse lobe with 8−24 small to medium-sized tubercles; a prominent quadrate proboscis, slightly tapering anteriorly, moderately elevated upward, the anterior margin of the proboscis truncate, sharply delineated from the depressed rostral surface by a narrow transverse groove; each anterolateral marginal corner of the proboscis with one large unicuspid, acanthoid tubercle, one small tubercle in between; the lateral margin of the proboscis with three to six small tubercles in one row. Lateral line complete with 35 scales. Circumpeduncular scales rows with 12; predorsal scales with 10−12; three or four narrow faint black stripes on caudal peduncle; base of last six branched dorsal-fin rays spotted with black. Maximum size ~130 mm SL.

Distribution. East Kameng, West Siang and Lower Divang Districts of the upper Brahmaputra drainage in Arunachal Pradesh, India.

Garra arupi Nebeshwar, Vishwanath & Das, 2009
(Fig. 4.57)

FIGURE 4.57

Garra arupi, 60.0 mm SL, holotype, Deopani River at Roing, Arunachal Pradesh.

Garra arupi Nebeshwar, Vishwanath & Das, 2009: 198 (Deopani River at Roing, lower Divang valley, Arunachal Pradesh).

Diagnosis. Presence of two pairs of barbels; anterior position of vent, vent to anal distance 52.6%– 60.0% pelvic to anal distance; a band of six to nine prominent horny tubercles on the tip of the snout, a submarginal black band on the dorsal fin, and 16 circumpeduncular scales; a distinct submarginal band on the dorsal fin; absence of a w-shaped color band on the caudal fin; presence of a transverse band of tubercles on the snout tip. Maximum size ∼70 mm SL.

Distribution. Upper Brahmaputra basin in the Lower Divang Valley and Lohit District of Arunachal Pradesh; Serichhu River and Aringkhola River, Bhutan.

Garra biloborostris **Roni & Vishwanath, 2017**
(Fig. 4.58)

FIGURE 4.58

Garra biloborostris, 92.35 mm SL, holotype: Kanamakra River, Assam.

Garra biloborostris Roni & Vishwanath, 2017: 134, figs. 1–4 (type locality: Kanamakra River, Chirang District, Assam, India) (Fig. 4.59).

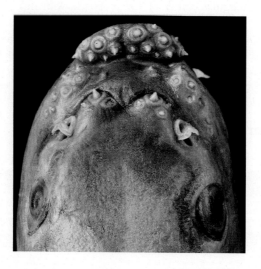

FIGURE 4.59

Snout of *Garra biloborostris* showing bilobed proboscis.

Diagnosis. A proboscis represented by two separate, slightly elevated arch-shaped lobes; each lobe demarcated from depressed rostral surface by a shallow groove, with three acanthoid tubercles on its anterodorsal marginal aspect, medial tubercle large-sized and lateral ones medium-sized; lateral surface of the snout bulgy giving lobe-like appearance; transverse lobe appearing as a prominent knob as the demarcating transverse groove extending laterally on each side, its dorsal surface covered with 10−16 small to medium-sized acanthoid tubercles; 30 + 3 lateral line scales; 8ó branched dorsal-fin rays; 5½ branched anal-fin rays. Maximum size ∼100 mm SL.

Distribution. Kanamakra River, Chirang District, Assam, India.

Garra bimaculacauda **Thoni, Gurung & Mayden, 2016**

Garra bimaculacauda Thoni, Gurung & Mayden, 2016: 118 (type locality: Dangmechhu River, Bhutan: Zhemgang Dzongkhag: Marangdud, Bhutan).

Diagnosis. Two distinct black spots on the caudal fin, one on the tip of the dorsal lobe and the other on the tip of the ventral lobe, weak proboscis forming a tuberculate ridge, 32−34 lateral-line scale rows, 12 circumpeduncular-scale rows, 11−12 predorsal-scale rows, six scales from dorsal origin to lateral line, three anal scales, and the absence of a dark lateral band along the length of the body. Maximum size ∼100 mm SL.

Distribution. Dangmechhu drainage in central and eastern Bhutan from near Panbang Village to the confluence of the Sherichhu River.

Garra birostris **Nebeshwar & Vishwanath, 2013**
(Fig. 4.60)

FIGURE 4.60

Garra birostris, 102.0 mm SL; holotype. Dikrong River, Arunachal Pradesh.

Garra birostris Nebeshwar & Vishwanath, 2013: 104, fig. 5 (type locality: Dikrong River at Doimukh, Papum Pare District, Brahmaputra drainage, Arunachal Pradesh, India).

Garra ranganensis Tamang et al., 2019: 61 (description from Ranga River, Yazli (Yachuli), lower Subansiri District, Brahmaputra drainage, Arunachal Pradesh) (Fig. 4.61).

Diagnosis. Transverse lobe on snout with 11−19 small to large-sized uni- to tctracuspid acanthoid tubercles; a prominent bilobed proboscis, moderately elevated upward, each lobe forwardly protruding and tapering; the tip of each lobe with a large, anteriorly directed tri- or tetracuspid acanthoid tubercle; the anterior margin of the proboscis sharply delineated by a deep groove from the depressed rostral surface; one to two small tubercles on the lateral margin of the proboscis. Maximum size ∼120 mm SL.

Distribution. West Kameng and Papum Pare Districts of the Upper Brahmaputra drainage, India and Bhutan.

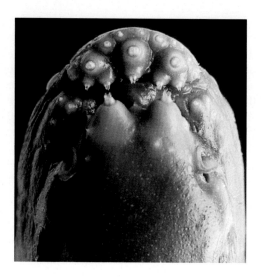

FIGURE 4.61

Snout of *Garra birostris* showing bilobed rostrum.

Garra chakpiensis Nebeshwar & Vishwanath, 2015
(Fig. 4.62)

FIGURE 4.62

Garra chakpiensis, 83.0 mm SL, holotype, Chakpi River at Tangpal, Manipur.

Garra chakpiensis Nebeshwar & Vishwanath, 2015: 306, fig. 1 (type locality: Chakpi River at Tongpal, Chandel District, Chindwin basin, Manipur, India).

Diagnosis. Smooth snout; 38–40 lateral-line scales; one faint blackish midlateral stripe on body; 16 circumpeduncular scale rows; irregularly arranged 11–14 predorsal scales; presence of pleated papilliferous fold and distinct upper lip as one transverse flap-like nonpapilliferous fleshy tissue; chest and abdomen with deeply embedded scales; central callous pad of adhesive disk small, its width 15%–18% HL; length 22%–28% HL, its anterior and posterior halves equally rounded, and relatively anteriorly placed (a transverse line drawn through the base of maxillary barbels crossing the median region of the central callous pad); and anteromedian fold of adhesive disk moderately arched and extending anteriorly to mid-length of lower jaw. Maximum size ~134 mm SL.

Distribution. Chakpi River at Chakpikarong in Chandel District, Chindwin River basin, Manipur.

? *Garra chathensis* Ezung, Shangningam & Pankaj, 2020

Garra chathensis Ezung, Shangningam & Pankaj, 2020: 1333 (type locality: Chathe River, Brahmaputra basin, Nagaland)

Diagnosis. A member of the snout with a proboscis species group, characterized by a combination of characters: a bilobed proboscis, a black spot at the upper angle of the gill opening; 32−33 lateral line scales; head length 26.8%−29.0% SL; mental disk width and transverse scale rows 3½/1/3.

Distribution. Chathe River, tributary of Dhansiri River near Dimapur, Nagaland, Brahmaputra drainage.

Garra chindwinensis Premananda, Kosygin & Saidullah, 2017

Garra chindwinensis Premananda, Kosygin & Saidullah, 2017: 191, fig. 1 (type locality: Laniye River near Laii, Chindwin basin, Senapati District, Manipur).

Diagnosis. Distinct from congeners in having the anterior margin of callous pad with a papillated narrow transverse lobe which is demarcated posteriorly by a transverse groove; a bilobed proboscis, protruding beyond the level of transverse groove; acanthoid tubercles on snout unicuspid; predorsal scales 10; lateral line scales 34; circumpeduncular scales 12; presence of anterolateral lobe and five faint longitudinal color stripes on the body. Maximum size ∼120 mm SL.

Distribution. Rivers draining into the Tizu River, Chindwin drainage in Senapati and Ukhrul Districts of Manipur.

? *Garra chivaensis* Moyon & Arunkumar, 2020

Garra chivaensis Moyon & Arunkumar, 2020: 33 (type locality: Manipur: Chandel District, Khongjon Village, Chiva River, Chindwin drainage, Manipur).

Diagnosis. Body covered with scales, the margins of which are greyish black, minute tubercles on snout and cheek; branched dorsal-fin rays 6, lateral line scales 34−36, scales between dorsal fin origin and lateral line 5½ and between lateral line and pelvic-fin origin 4½; predorsal scales 16; chest scaled, belly scales poorly developed, dorsal-fin origin close to caudal-fin base than to the tip of snout, predorsal length 52.5%−53.8% SL; vent close to anal-fin origin than to the pelvic-fin origin; snout length 45.2%−47.0% HL, disk width 45.4%−46.5% HL, disk length 27.4%−28.0% HL, callous pad width 34.1%−34.2% HL, a rounded black patch or blotch at the upper base of caudal fin. Maximum size ∼90 mm SL.

Distribution. Chindwin drainage in Chandel District, Manipur.

Garra clavirostris Roni, Sarbojit & Vishwanath, 2017
(Fig. 4.63)

FIGURE 4.63

Garra clavirostris, 117.5 mm SL, holotype, Dilaima River at Boro Chenam Village, Assam.

Garra clavirostris Roni, Sarbojit & Vishwanath, 2017: 368, fig. 1 (type locality: Dilaima River at Boro Chenam village below the confluence of Dilaima and Dihandi, Brahmaputra drainage, Dima Hasao District, Assam, India) (Fig. 4.64).

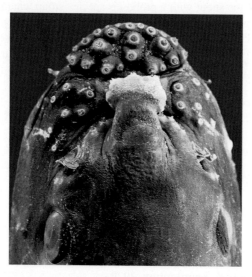

FIGURE 4.64

Dorsal view of S\snout of *Garra clavirostris* showing club-shaped unilobed proboscis.

Diagnosis. A transverse lobe bearing 17−25 small to large-sized multicuspid tubercles; a prominent club-shaped unilobed proboscis prominently protruding forward and overhanging the depressed rostral surface (Fig. 4.74), with a row of five to nine medium- to large-sized uni- to multi-cuspid tubercles on its anterior margin; the antero-ventral marginal aspect of the proboscis with 5−6 medium-sized uni- to multicuspid tubercles, distinctly differentiated from the depressed rostral surface by a distinct transverse groove; 8½ or 9½ branched dorsal-fin rays; 5½ branched anal-fin rays; 33−34 lateral-line scales; a black spot at the upper angle of gill opening; and four to five narrow black stripes laterally, more distinct toward the caudal peduncle. Maximum size ∼130 mm SL.

Distribution. Brahmaputra drainage, Dima Hasao District, Assam, India.

Garra compressa Kosygin & Vishwanath, 1998
(Fig. 4.65)

FIGURE 4.65

Garra compressa, 85.0 mm SL, Wanze stream at Khamsom, tributary of Tizu River, Manipur.

Garra compressus Kosygin & Vishwanath, 1998: 45 (type locality: Wanze stream at Khamsom, Ukhrul District, Manipur, India).

Diagnosis. A species of *Garra* with no proboscis and transverse groove on the snout, tip if snout studded with large tubercles; laterally compressed body; seven branched dorsal fin rays; 39–40 lateral line scales; 13 predorsal scales; two pair of barbels; pentagonal shaped suctorial disk; position of vent slightly nearer to pelvic fin origin than to anal fin origin; a distinct transverse black bar on dorsal fin and a horizontal black bar on median rays of caudal fin. Maximum size ∼90 mm SL.

Distribution. Wanze stream, Khamsom, Chindwin basin, Ukhrul District, Manipur, India.

Garra cornigera **Shangningam & Vishwanath, 2015**
(Fig. 4.66)

FIGURE 4.66

Garra cornigera, 76.0 mm SL, holotype, Sanalok River, Chindwin drainage, Manipur.

Garra cornigera Shangningam & Vishwanath, 2015: 264, fig. 1 (type locality: Sanalok River, Chindwin basin, Ukhrul District, Manipur, India) (Fig. 4.67).

FIGURE 4.67

Dorsal view of snout of *Garra cornigera* showing bilobed proboscis.

Diagnosis. Slightly pointed snout with a depressed rostral surface with three to four transverse ridges, a prominent bilobed proboscis (Fig. 4.77), each with large unicuspid acanthoid tubercles, 33 lateral line scales, 14 circumpeduncular scales and 3½/1/2½ lateral transverse scale rows. Maximum size ~80 mm SL.

Distribution. Known from the Sanalok River, Ukhrul District, Chindwin drainage, Manipur, India.

Garra elongata **Vishwanath & Kosygin, 2000**
(Fig. 4.68)

FIGURE 4.68

Garra elongata, 85.5 mm SL, stream near Tolloi, Ukhrul District, Chindwin drainage, Manipur.

Garra elongata Vishwanath & Kosygin, 2000a: 408 (type locality: hill stream near Tolloi, Ukhrul District, Chindwin basin, Manipur).

Diagnosis. Lateral line scales 39−40, predorsal scales 13, scale rows between lateral line and pelvic fin origin 2½, transverse groove on snout, weakly developed proboscis, no scales on chest, seven branched dorsal fin rays, 11−12 branched pectoral fin rays and position of vent situated midway between pelvic and anal fin origins. It is also distinct in having a dorsal fin with a transverse black bar, and caudal fin with a longitudinal black band in the middle. Maximum size ~98 mm SL.

Distribution. Hill stream at Tolloi, Ukhrul District, Chindwin basin, Manipur, India.

Garra gotyla **(Gray, 1830)**
(Fig. 4.69)

FIGURE 4.69

Garra gotyla, 104.3 mm SL, neotype, Teesta River at Rangpo, Sikkim.

Cyprus gotyla Gray, 1830: pl. 88, fig. 3 (type locality: mountain streams of India, probably in the Himalayan foothills, but unknown.

Garra gotyla: Nebeshwar and Vishwanath, 2013: 110 (redescription, neotype designated, neotype locality: Tista River at Rangpo, Sikkim, India) (Fig. 4.70).

Diagnosis. A transverse lobe with 9−13 small or medium sized tubercles; a prominent quadrate proboscis (Fig. 4.80) with or without a shallow depression in the middle to present a bilobed

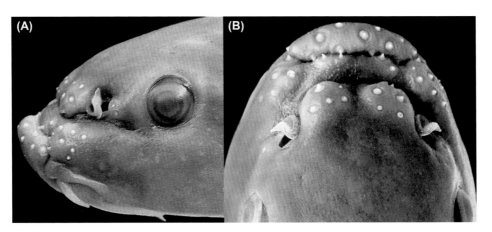

FIGURE 4.70

Snout of *Garra gotyla*, (A) lateral view, (B) dorsal view.

appearance, moderately elevated upward; the anterior region of the proboscis depressed with 4−13 small- to medium-sized tubercles; and the anterior margin of the proboscis delineated from the depressed rostral surface by a continuous or medially interrupted narrow groove; mental disk width 51%−57% HL and length 35%−40% HL; lateral line complete with 33−34 scales; circumpeduncular scale rows with 16; predorsal scales 10−12; a black spot at upper angle of gill cover. Maximum size ∼130 mm SL.

Distribution. Teesta River, Sikkim, India and Bhutan.

Garra kalpangi Nebeshwar, Bagra & Das, 2012
Garra kalpangi Nebeshwar, Bagra & Das, 2012: 2355 (type locality: Kalpangi River at Yachuli, Brahmaputra River system, Lower Subansiri District, Arunachal Pradesh, India).

Diagnosis. Two pairs of barbels; a poorly developed proboscis represented by a squarish area in front of the nostrils and 16 circumpeduncular scales; branched dorsal-fin rays 8; absence (vs presence) of indistinct black spots at the bases of the branched dorsal-fin rays, absence (vs presence) of lateral stripes on side of body. branched pectoral-fin rays 10−12 (vs 13−15), lateral line scales 32−33. Maximum size ∼72 mm SL.

Distribution. Kalpangi River, Brahmaputra drainage, Lower Subansiri District, Arunachal Pradesh.

? *Garra khawbungi* Arunachalam, Nandagopal & Mayden, 2014
Garra khawbungi Arunachalam, Nandagopal & Mayden, 2014: 59, figs. 1, 3 and 4 (type locality: Tuipui River, Khawbung Village, Champhai District, Mizoram).

Diagnosis. *Garra* species with a rounded snout in dorsal profile, no proboscis, a weakly developed transverse groove, groove with pointed tubercles arranged irregularly but approximately in three rows across; barbels in pairs, rostral and maxillary; lateral line scales 36−37;

circumpeduncular scales 16; scales between dorsal fin origin and lateral line 4½ and between lateral line and pelvic-fin origin 3½. Maximum size ∼98 mm SL.

Distribution. Tuipui River, Barak River drainage, Mizoram.

Garra kempi **Hora, 1921**
(Fig. 4.71)

FIGURE 4.71

Garra kempi, 62.2 mm SL, Dikrong River, Brahmaputra drainage, Arunachal Pradesh.

Garra kempi Hora, 1921b: 665 (type locality: Siyom R., below Damda, among the Abor Hills, Brahmaputra drainage, Assam, India) (Fig. 4.72).

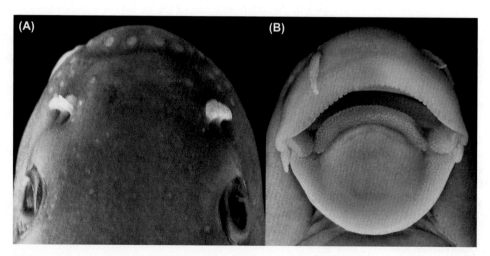

FIGURE 4.72

Garra kempi: (A) dorsal view of snout, (B) mental disk.

Diagnosis. Snout with transverse lobe with eight to nine minute tubercles (Fig. 4.82); 38−39 lateral line scales; 8½ branched dorsal fin rays; 5½ branched anal fin rays; 12 circumpedunclar scale rows; regularly arranged; 14 predorsal scale rows; presence of pleated papilliferous fold; mental adhesive disk posteriorly positioned; central callous pad of mental adhesive disk large; distal margin of the rostral cap slightly fimbriae; anteromedian fold of mental adhesive disk rounded with regularly arranged papilla (Fig. 4.82); chest and belly scaled; vent position midway between pelvic fin and anal fin origins. Maximum size ∼115 mm SL.

Distribution. Brahmaputra drainage, India.

***Garra koladynensis* Nebeshwar & Vishwanath, 2017**
(Fig. 4.73)

FIGURE 4.73

Garra koladynensis, 130.6 mm SL, holotype, Kaladan River at Kawlchaw, Mizoram.

Garra koladynensis Nebeshwar & Vishwanath, 2017: 18 (type locality: Kaladan River at Kawlchaw, Lawngtlai District, Mizoram) (Fig. 4.74).

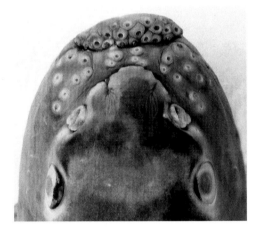

FIGURE 4.74

Dorsal view of snout of *Garra koladynensis* showing trilobed proboscis.

Diagnosis. Transverse lobe on snout with 11−23 small-sized unicuspid or medium- to large-sized bi- to hexacuspid acanthoid tubercles; a prominent trilobed proboscis (Fig. 4.84), its median lobe squarish, slightly projecting forward with three large-sized bi- to hexacuspid acanthoid tubercles on anterior marginal aspect and two medium to large-sized bi- to pentacuspid tubercles on anteroventral marginal aspect; multicuspid tubercles variously protruding; 30−31 + 3 lateral-line scales; mental adhesive disk medially positioned, extending anteriorly to three-fourths of length of lower jaw; anteromedian fold of mental adhesive disk slightly arched medially, highly arched laterally on each side, its lateral extension on each side reaching the imaginary vertical line through the lateral margin of the central callous pad; and central callous pad of mental adhesive disk relatively large (width 34%−42% HL; length 25%−29% HL), its anterior and posterior halves equally rounded, width 1.2−1.5 times its length. Maximum size ∼130 mm SL.

Distribution. Kaladan River basin, Mizoram, India.

Garra lamta **(Hamilton, 1822)**
(Fig. 4.75)

FIGURE 4.75

Garra lamta, 88.0 mm SL, Rangit River, tributary of Teesta River, Sikkim.

Cyprinus lamta Hamilton, 1822: 343 (type locality: Bihar province and Rapti River, Gorakhpur District, Uttar Pradesh, India).

Diagnosis. Head oval, blunt and of moderate size; dorsal fin with 10 rays, seven in the anal fin, 13 in pectoral fins, ventral fin with nine rays and caudal fin with 19 distinct rays, beside short compacted ones. Maximum size ~120 mm SL.

Distribution. Ganga and Brahmaputra River drainages, India, Nepal, and Bangladesh.

Garra lissorhynchus **(McClelland & Griffith, 1842)**
(Fig. 4.76)

FIGURE 4.76

Garra lissorhynchus, 63.7 mm SL, stream at Kharam Palen, Barak drainage, Manipur.

Platycara lissorhynchus McClleland & Griffith, in McClelland, 1842: 587 (type locality: Khasi Hills, Meghalaya, India).

Diagnosis. *Garra* with a smooth snout; a small rostral flap with 8−18 small minute conical tubercles; 34−35 lateral line scales; 6½ branched dorsal fin rays; 4½ branched anal fin rays; mental adhesive disk medially positioned; anteromedian fold of mental adhesive disk slightly arched; no fimbriation on rostral cap; chest and belly naked; papilla on anteromedian fold, lateroposterior flap and lateroposterior fold regularly arranged; a distinct W-shaped black band on caudal fin; distinct black submarginal band on dorsal fin; a black spot at the upper angle of the gill opening. Maximum size ~90 mm SL.

Distribution. Barak and Brahmaputra drainages, northeast India.

Garra litanensis **Vishwanath, 1993**
(Fig. 4.77)

FIGURE 4.77

Garra litanensis, 73.9 mm SL, Litan stream, Chindwin drainage, Manipur.

Garra litanensis Vishwanath, 1993: 62 (type locality: Litan stream at Litan, Manipur, India).

Diagnosis. An elongated form of *Garra* with a snout which has a tuberculated proboscis marked off by a transverse groove, black spots along the base of dorsal fin, a long suctorial disk and distance between vent and anal fin origin 28.6 (25.6−30.3) percent of the distance between pelvic and anal fin origins; lateral line complete with 32 scales. Predorsal scales 10. Body dark green on sides, black dorsally, yellowish ventrally; black spots at the base of dorsal fin; fins pale white with no markings. Maximum size ∼98 mm SL.

Distribution. Litan stream, tributary of Thoubal River, Chindwin drainage, Manipur, India.

Garra magnacavus Shangningam, Kosygin & Sinha, 2019

Garra magnacavus. Shangningam, Kosygin & Sinha, 2019: 149 (type locality: Ranga River, Brahmaputra drainage, Lower Subansiri District, Arunachal Pradesh).

Diagnosis. *Garra* with a transverse lobe and a weakly developed proboscis; crescentic anteromedian fold of the lower lip slightly wider than the width of the callous pad, the lateral end of the anteromedian fold of the lower lip reaching the anterolateral lobe of the mental adhesive disk; three rows of pharyngeal teeth 2, 3, 4/4, 3, 2, first row widely separated; body elongated with 42 lateral-line scales, 14−16 predorsal scales, 15−19 rounded large pits on the snout and a weakly developed nonfleshy central callous pad. Maximum size ∼82 mm SL.

Distribution. Ranga River, Lower Subansiri District, Brahmaputra drainage, Arunachal Pradesh.

Garra magnidiscus Tamang, 2013
(Fig. 4.78)

FIGURE 4.78

Garra magnidiscus, 80.5 mm SL, Siang River, Upper Siang District, Arunachal Pradesh.

Photo courtesy: A. Darshan.

Garra magnidiscus Tamang, 2013: 32 (type locality: tributary of Siang River, Bomdo village, Upper Siang District, Arunachal Pradesh, India) (Fig. 4.79).

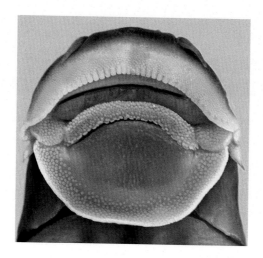

FIGURE 4.79

Mental disk of *Garra magnidiscus*.

Photo courtesy: A. Darshan.

Diagnosis. Distinct in having a combination of characters: a large mental adhesive disk (60%−68% HL), its posterior margin almost reaches the level of pectoral-fin origin; visible ventral gill opening groove short and deeply inclined from the pectoral-fin origin toward ventral mid-line; about half of the gill opening groove extends to the base of the callous pad below the adhesive mental disk (Fig. 4.89); a series of tubercles on transverse lobe on snout; 40−42 perforated lateral line scales; faint black to dusky blotch at caudal-fin base on margin of hypural complex, and head U-shaped when viewed dorsally. Maximum size ∼85 mm SL.

Distribution. Siang River, tributary of the Brahmaputra River, Upper Siang District, Arunachal Pradesh, India.

Garra manipurensis **Vishwanath & Sarojnalini, 1988**
(Fig. 4.80)

FIGURE 4.80

Garra manipurensis, 68.3 mm SL, Iyei River at Noney, Barak drainage, Manipur.

Garra manipurensis Vishwanath & Sarojnalini, 1988: 124 (type locality: Sherou, Manipur, India).
Garra manipurensis: Nebeshwar & Vishwanath, 2017: 35 (redescription and type locality correction) (Fig. 4.81).

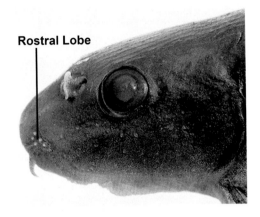

FIGURE 4.81

Lateral view of snout of *Garra manipurensis* showing rostral lobe.

Diagnosis. A member of the rostral lobe species group, distinguished from other members of this group in possessing the following combination of characters: a rostral lobe with three to six small conical tubercles (Fig. 4.91); 31−32 + 3 lateral-line scales; 7½ branched dorsal-fin rays; 4½ branched anal-fin rays; mental adhesive disk posteriorly positioned, extending anteriorly to base of lower jaw; anteromedian fold of mental adhesive disk slightly arched, its lateral extension reaching beyond imaginary vertical line through lateral margin of central callous pad; central callous pad of mental adhesive disk small (width 33%−36% HL; length 20%−22% HL), its posterior half more rounded than anterior half, width 1.8−2.0 times its length (Fig. 4.66); caudal fin slightly emarginated; a pleated papilliferous fold at corner of mouth; variable length of plicae on posterior surfaces of anterolateral lobe and anteromedian fold; and distinct upper lip as a distinct transverse flap-like nonpapilliferous fleshy tissue. Maximum size ∼80 mm SL.

Distribution. Barak-Meghna drainage in northeast India and the Koladyne and its tributaries in Mizoram to the rivers flowing in the western slope of the Rakhine Yoma in Myanmar.

Garra matensis **Nebeshwar & Vishwanath, 2017**
(Fig. 4.82)

FIGURE 4.82

Garra matensis, 73.8 mm SL, paratype, Mat River at Thualthu, tributary of Kaladan River, Mizoram.

Garra matensis Nebeshwar & Vishwanath, 2017: 18 (type locality: Mat River, a tributary of Koladyne River at Thualthu, Mizoram).

Diagnosis. Small rostral flap with four to seven small conical tubercles; dorsolateral and ventro-lateral free margins of the rostral flap equally extended; 27−28 + 3 lateral-line scales; 6½ branched dorsal-fin rays; 4½ branched anal-fin rays; mental adhesive disk anteriorly positioned, extending anteriorly half the length of lower jaw; anteromedian fold of mental adhesive disk highly arched, its lateral extension reaching beyond imaginary vertical line through lateral margin of central callous pad; central callous pad of mental adhesive disk relatively small (width 34%−38% HL; length 22%−24% HL), its anterior half more rounded than posterior half, width 1.4−1.7 times its length; caudal fin forked, with a distinct W-shaped black band; and a distinct black submarginal band on dorsal fin. Maximum size ∼76 mm SL.

Distribution. Kaladan River drainage, Mizoram, India.

Garra mini Rahman, Mollah, Noren & Kullander, 2016

Garra mini Rahman, Mollah, Noren & Kullander, 2016: 174, figs. 1−2 (type locality: pool at bottom of Shuvolong waterfall, Borokal District, Karnafuli River drainage, Bangladesh).

Diagnosis. Small species of *Garra* with no proboscis on snout, maximum of 50 mm SL, diagnosed by a combination of characters: smoothly rounded snout tip; very slight transverse groove; presence of lateral lobe; absence of scales on chest; 30−31 + 1−2 lateral line scales; 15−16 very small, deeply embedded predorsal scales; 16 circumpeduncular scales; no black spots on dorsal-fin base; caudal fin with no transverse dark bars; contrasting dark band from head to caudal-fin base. Maximum size ∼45 mm SL.

Distribution. Karnafuli drainage, western Lunglei District, Mizoram, Karnafuli and Sangu River drainages, Bangladesh.

Garra naganensis Hora, 1921
(Fig. 4.83)

FIGURE 4.83

Garra naganensis, 97.6 mm SL, Likhailok, Barak drainage, Manipur.

Garra naganensis Hora, 1921a, 1921b, 1921c: 667 (type locality: Senapati stream near Karong, Manipur, India).

Diagnosis. Smooth snout; 38 lateral line scales; 8½ branched dorsal fin rays; 5½ branched anal fin rays; 19 circumpeduncular scale rows; 13−14 predorsal scales; mental adhesive disk oval, posteriorly positioned; presence of pleated papilliferous fold; upper lip present with thin transverse band of papillae; central callous pad of adhesive disk small; scales absent near bases of pectoral fins and are greatly reduced on the chest and in the middle of abdomen; large scales on the postpelvic region; black spot at the upper angle of the gill opening. Maximum size ∼110 mm SL.

Distribution. *Garra naganensis* is distributed in the Barak-Meghna drainage, India.

Garra nambulica Vishwanath & Joyshree, 2005
(Fig. 4.84)

FIGURE 4.84

Garra nambulica, 62.7 mm SL, Nambul River, Singda Village, Chindwin drainage, Manipur.

Garra nambulica Vishwanath & Joyshree, 2005: 1832 (type locality: Irenglok stream; Singda Village, Manipur, India).

Diagnosis. *Garra* with smooth snout; 34−35 lateral line scales; 16−29 predorsal scales; chest and belly naked; posteriorly positioned mental adhesive disk; dorsal fin with six branched rays, pectoral fin with 12, pelvic with six, and anal fin with four; circumpeduncular scale rows 19; caudal fin with distinct W-shaped black band; a dark spot at the upper angle of the gill opening; dorsal fin with broad transverse black bar near the free margin. Maximum size ∼76 mm SL.

Distribution. Nambul River, Chindwin drainage, Manipur.

Garra namyaensis Shangningam & Vishwanath, 2012
(Fig. 4.85)

FIGURE 4.85

Garra namyaensis, 72.5 mm SL, holotype, Namya River, Indo-Myanmar border, Manipur.

Garra namyaensis Shangningam & Vishwanath, 2012: 10 (type locality: Namya River, Kamjong District, Manipur India, close to Indo-Myanmar border).

Diagnosis. A species of *Garra* without proboscis and transverse groove on the snout and with a smoothly rounded snout tip, prominent triangular-shaped rostral flap with 10−11 strong, conical tubercles; scales in the predorsal region, chest and abdomen; two pairs of barbels, 31 lateral line scales, 14 circumpeduncular scales; 3½/1/3 ½ transverse-scale rows; nine gill rakers on first branchial arch; 32 vertebrae; a broad black band near the posterior margin of the dorsal fin and a distinct W-shaped broad black band in the caudal fin; dark brown spot is immediately posterior to dorsal gill opening; dorsal fin is basally yellowish white, dark black band near the posterior margin. Pectoral fin is basally yellowish with greyish submarginal band on anterior rays. Pelvic and anal-fins are greyish white; caudal fin is with a distinct W-shaped broad black band with lines of black spots anterior to its base. Maximum size ~73 mm SL.

Distribution. Rivers in the Indo-Myanmar border, Kamjong District, Manipur, draining into the Chindwin River.

Garra nasuta **(McClelland, 1838)**
(Fig. 4.86)

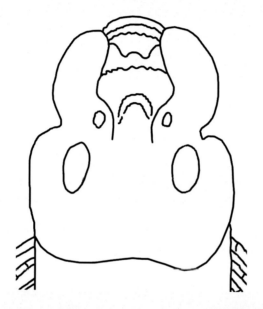

FIGURE 4.86

Dorsal view of head of *Garra nasuta* showing a pit between the nares (McClelland's, 1838, pl. 55, fig. 2a reproduced).

Platycara nasuta McClelland, 1838: 947, pl. 55, fig. 2 (type locality: Khasi Hills, Meghalaya, India).

Diagnosis. Head and body greatly depressed, short and indistinct proboscis with anteriorly truncated margin and a transverse lobe; tip of snout marked off into a rounded lobe; characteristic in having a pit between the nares; lateral line scales 34; caudal fin deeply emarginated, lower lobe longer. Max. size ~150−200 mm? SL.

Distribution. Khasi hills, Meghalaya, India.

Garra nepalensis **Rayamajhi & Arunachalam, 2017**
Garra nepalensis Rayamajhi & Arunachalam, 2017: 403, figs. 1−4 (type locality: Mardi River, upstream from Kaligandaki River, Kaski District, Nepal).

Diagnosis. Garra with a smoothly rounded snout without proboscis and transverse groove; lateral line scales 35−36; scales between dorsal fin origin and lateral line 4½ and that between lateral line and pelvic-fin origin 3½; absence of a W-shaped band on the caudal fin; mental disk wider than long, disk length 26.1%−29.1% HL, width 38.5%−45.7% HL, callous pad elliptical. Maximum size ∼94 mm SL.

Distribution. Kaligandaki River drainage, Nepal.

Garra paralissorhynchus **Vishwanath & Shanta Devi, 2005**
(Fig. 4.87)

FIGURE 4.87

Garra paralissorhynchus, 63.5 mm SL, Khuga River, Chindwin drainage, Manipur.

Garra paralissorhynchus Vishwanath & Shanta Devi, 2005: 86 (type locality: Khuga River, Churachandpur District, Manipur, India).

Diagnosis. *Garra* with a smooth snout; a small rostral flap with eight to nine small conical tubercles; dorsolateral and ventrolateral free margins of the rostral flap equally extended; 30−31 lateral line scales; 11−12 predorsal scales; medially positioned mental adhesive disk; scales absent on chest and belly, a dark streak near the free margin of the dorsal fin, a thin and light black W-shaped band on the posterior half of the caudal fin and one or two dark vertical lines anterior to the W-shaped band. Maximum size ∼70 mm SL.

Distribution. Khuga River, Churachandpur District, Chindwin drainage, Manipur.

Garra parastenorhynchus **Thoni, Gurung & Mayden, 2016**
(Fig. 4.88)

FIGURE 4.88

Garra parastenorhynchus, 74 mm SL, Kanamakra River, Chirang District, Assam.

Garra parastenorhynchus Thoni, Gurung & Mayden, 2016: 125, figs. 9−10 (type locality: Dungsamchhu River at Samdrup Jongkhar town, Bhutan).

Diagnosis. A transverse lobe bearing 11−15 small- to medium- sized uni- to bicuspid tubercles; a prominent club-shaped unilobed proboscis, with a row of six to seven small- to medium-sized uni- to bicuspid tubercles on its anterior margin; the antero-ventral margin of the proboscis with five to size small sized tubercles, distinctly differentiated from the depressed rostral surface by a distinct transverse groove; 8½ branched dorsal-fin rays; 5½ branched anal-fin rays; 31−33 lateral-line scales; a black spot at the upper angle of gill opening; and four to five narrow black stripe laterally, more distinct toward caudal peduncle. Maximum size ∼57 mm SL.

Distribution. Kanamakra stream, Chirang District, Assam, India and Dungsamchhu and Martangchhu rivers in southeastern Bhutan.

Garra paratrilobata Roni, Chinglemba, Rameshori & Vishwanath, 2019
(Fig. 4.89)

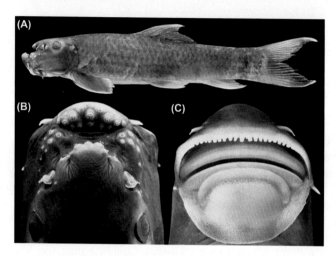

FIGURE 4.89

Garra paratrilobata, 137 mm SL, Leimatak River, Awangkhul, Tamenglong District, Manipur; (A) lateral view; (B) dorsal view of head showing trilobed proboscis, and (C) mental disk.

Garra paratrilobata Roni, Chinglemba, Rameshori & Vishwanath, 2019: 546 (type locality: Leimatak River, a tributary of the Irang River, Barak drainage, Awangkhul Village, Tamenglong District, Manipur).

Diagnosis. *Garra* with a transverse lobe on the snout, the lobe with 13−17 small- to medium sized, bi- to tetracuspid tubercles; proboscis prominent and trilobed, the median lobe squarish, slightly projecting forward, 5−7 uni- to tricuspid tubercles on its anterior margin and 4−6 minute tubercles on its antero-ventral margin; lateral lobes of the proboscis smaller, each with three to four minute tubercles; lateral surfaces of the snout slightly elevated, lobular and with six to nine minute tubercles; lateral line scales 33−34; mental disk elliptical, posteriorly positioned; anterolateral lobe of the lower lip absent. Maximum size ∼137 mm SL.

Distribution. Barak drainage, Tamenglong District, Manipur.

Garra quadratirostris Nebeshwar & Vishwanath, 2013
(Fig. 4.90)

FIGURE 4.90

Garra quadratirostris, holotype, 108.0 mm SL, Teesta River, Sikkim, India.

 Garra quadratirostris Nebeshwar & Vishwanath, 2013; 107, fig. 6 (type locality: Teesta River at Rangpo, Teesta drainage) (Fig. 4.91).

FIGURE 4.91

Dorsal view of snout of *Garra quadratirostris* showing quadrate shaped proboscis.

 Diagnosis. *Garra* with a transverse lobe and proboscis; transverse lobe with 13−20 small- to large-sized tubercles; a prominent quadrate proboscis (Fig. 4.99), moderately elevated upward; the anterior margin of the proboscis truncate, and sharply delineated from the depressed rostral surface by a narrow transverse groove; and small- to medium-sized tubercles on the margins of the proboscis in a single row (three to four on anterior and two to four on lateral margins); lateral line with 37 scales. Maximum size ∼ 132 mm SL.
 Distribution. Rivers of East Siang and Lower Divang Districts, Arunachal Pradesh.

Garra rakhinica **Kullander & Fang, 2004**
(Fig. 4.92)
Garra rakhinica Kullander & Fang, 2004, fig. 9 (type locality: Yan Khaw Chaung, Thade River drainage, Taunggok, Rakhine state, Myanmar).
 Garra tyao Arunachalam et al., 2014: 60 (description from Tyao River, Tyao Village, Kaladan River drainage, Champhai District, Mizoram).
 Diagnosis. *Garra* with no transverse lobe and proboscis on snout, snout with a shallow rostral furrow; rostral lobe well developed, studded with minute tubercles; predorsal region, chest and abdomen scaled, barbels in two pairs; lateral line with 28−29 scales; body color uniform greyish; caudal fin grayish, tips of the upper and lower lobe back. Maximum size ∼80 mm SL.

FIGURE 4.92

Garra rakhinica, 68.0 mm SL, Kaladan River near confluence with Mat River, Mizoram.

Distribution. River in the Rakhine Yoma, Myanmar and Kaladan drainage, Mizoram, India.

Garra rupicola (McClelland, 1839)

Gonorhynchus rupicolus McClelland, 1839: 281, 373, figs. 4, 5 (type locality: Lareeh River, few miles beyond Bramacuna, in Mishmi Hills, Arunachal Pradesh, India).

Diagnosis. Snout thick and smooth; 35−36 lateral line scales; dorsal fin with eight branched rays, pectoral fin with 10 rays, ventral fin with nine rays and caudal fin with 20 rays. Maximum size ∼70 mm SL.

Distribution. Brahmaputra drainage, Arunachal Pradesh, India.

Garra substrictorostris Roni & Vishwanath, 2018
(Fig. 4.93)

FIGURE 4.93

Garra substrictorostris, 173.0 mm SL, holotype, Leimatak River, Barak drainage, Manipur.

Garra substrictorostris Roni & Vishwanath, 2018: 264 (type locality: Leimatak River, Barak River drainage, Churachandpur District, Manipur, India) (Fig. 4.94).

Diagnosis. A robust, large sized *Garra*, (maximum 173 mm SL), distinguished in having a prominent transverse lobe on the snout with 14−20 multicuspid tubercles; a narrow antrorse uni-lobed proboscis on the snout (Fig. 4.94), with three to five medium to large multicuspid tubercles on its anterior margin and four to six multicuspid tubercles on its anteroventral margin; lateral surface of the snout slightly elevated, lobular with 7−11 small tubercles; 5½ transverse scales rows above the lateral line, 3½ or 4½ rows between the lateral line and anal-fin origin. Distance between anus and anal-fin origin 15%−27% of that between pelvic-fin origin and anal-fin origin. Maximum size ∼173 mm SL. Maximum size ∼173 mm SL.

Distribution. Barak River drainage, Churachandpur District, Manipur, India.

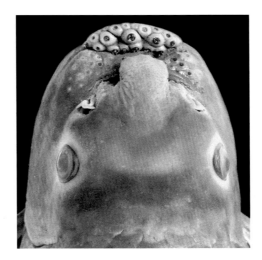

FIGURE 4.94

Dorsal view of snout of *Garra substrictorostris*.

Garra trilobata Shangningam & Vishwanath, 2015
(Fig. 4.95)

FIGURE 4.95

Garra trilobata, 118.5 mm SL, holotype, Sanalok River, tributary of Chindwin River, Manipur.

Garra trilobata Shangningam & Vishwanath, 2015: 267, fig. 3 (type locality: Sanalok River, Chindwin basin, Ukhrul District, Manipur, India).

Diagnosis. Snout broadly rounded with transverse lobe covered with 12–28 small to large uni- to tetracuspid tubercles on each lateral surface of snout, demarcated posteriorly by a shallow groove; a distinct trilobed proboscis (Fig. 4.96); two to three small sized tubercles on posteroventral region to nostril on each side. Proboscis highly elevated upward, width equals internarial space, anterior median lobe quadrate, its inferior surface not in contact with depressed rostral surface, larger than lateral lobes with 8–15 anteriorly directed tri- or tetracuspid tubercles; lateral lobes more depressed than median lobe with three to seven small to large uni- to tetracuspid tubercles. Lateral line complete with 31–32 scales. Circumpeduncular scale rows with 14 scales. Predorsal scales with 9–10. Bases of third branched anal-fin and third branched dorsal fin rays spotted with black. A black spot at upper angle of gill opening. Maximum size ∼135 mm SL.

Distribution. Streams draining to Chindwin River, Ukhrul District, Manipur, India.

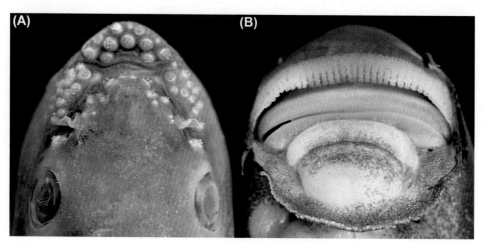

FIGURE 4.96

Dorsal view of snout (A) showing proboscis and (B) mental disk of *Garra trilobata*.

Garra ukhrulensis Nebeshwar & Vishwanath, 2015
(Fig. 4.97)

FIGURE 4.97

Garra ukhrulensis, 110.5 mm SL, Challou River, Chindwin drainage, Manipur.

Garra ukhrulensis Nebeshwar & Vishwanath, 2015: 310, fig. 3 (type locality: Challou River at Khamsom, Ukhrul District, Chindwin basin, Manipur, India).

Diagnosis. A *Garra* with smooth snout; 40−41 lateral-line scales; one faint blackish midlateral stripe on body; 16 circumpeduncular scale rows; predorsal scales very small and irregularly arranged; presence of pleated papilliferous fold and distinct upper lip as a thick transverse band of nonpapilliferous and papilliferous fleshy tissue arranged in two to four transverse ridges; chest and median region of abdomen naked; central callous pad of the adhesive disk small (width 12%−16% HL; length 26%−30% HL), its posterior half more rounded than anterior half (Fig. 4.98), and relatively posteriorly placed (a transverse line drawn through the base of the maxillary barbels crossing the anterior margin of the central callous pad); and anteromedian fold of the adhesive disk slightly arched, and extending anteriorly to base of lower jaw; lateral line completes with 40−41; one thin faint black midlateral stripe over scale row of lateral line, stripe absent in larger specimens. Maximum size ∼140 mm SL.

Distribution. Rivers draining to the Tizu River, Chindwin drainage, Ukhrul District, Manipur.

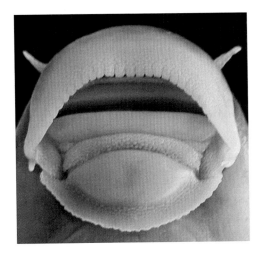

FIGURE 4.98

Mental disk of *Garra ukhrulensis* showing posteriorly placed callous pad.

Genus *Gibelion* Heckel, 1843

Gibelion Heckel, 1843: 1014 (type species: *Cyprinus catla* Hamilton, 1822). Gender: neuter.

Diagnosis. Body deep, deeper than head length; compressed and short; mouth superior or upturned, lower jaw protruding; upper lip absent, lower lip thick; barbels absent; dorsal-fin origin in advance of pelvic-fin origin; anal fin short; lateral line complete.

Gibelion catla (Hamilton, 1822)
(Fig. 4.99)

FIGURE 4.99

Gibelion catla, 320.00 mm SL, Loktak Lake, Manipur.

Cyprinus catla Hamilton, 1822: 287, 387 (type locality: rivers and tanks of Bengal).

Diagnosis. Body deep; snout bluntly rounded with thin skin covering; mouth wide, upper lip absent, lower lip very thick; barbels absent; dorsal fin with 14−16 branched rays, its origin opposite pelvic fin origin; anal fin short with five branched rays; caudal fin forked. Maximum size ∼1200 mm SL.

Distribution. Widely distributed in India, Nepal, Bangladesh, Pakistan, Thailand.

Genus *Hypophthalmichthys* Bleeker, 1860

Hypopthalmichthys Bleeker, 1860: 433 (type species: *Leuciscus molitrix* Valenciennes, 1844). Gender: masculine.

Diagnosis. Body compressed, elongated; head large, 28%−33% SL; abdomen edge keeled; mouth terminal, operculum large, deeper than long; its outer surface strongly ridged; dorsal fin small, origin slightly behind pelvic fin origin; caudal feel forked; mid-dorsal scales extending forward to eyes; lateral line complete with 75−120 + 4−5 scales.

Hypophthalmichthys molitrix (Valenciennes, 1844)
(Fig. 4.100)

FIGURE 4.100

Hypophthalmichthys molitrix, 345.00 mm SL, Pumlen Lake, Manipur.

Leuciscus molitrix Valenciennes, in Cuvier and Valenciennes, 1844a, 1844b: 360 (type locality: not stated, probably China).

Diagnosis. Body compressed, abdomen keeled from throat to vent; dorsal fin short, its origin slightly behind pelvic fin origin; lateral complete with 110−115 scales; body silvery with red spots scattered all over. Maximum size ∼1200 mm SL.

Distribution. Originally China and eastern Siberia, introduced throughout.

Hypophthalmichthys nobilis (Richardson, 1845)
(Fig. 4.101)

FIGURE 4.101

Hypophthalmichthys nobilis, 326.0 mm SL, wetlands near Mayang Imphal, Manipur.

Leuciscus nobilis Richardson, 1845: 140, pl. 63, fig. 3 (type locality: Canton, China).

Diagnosis. Body compressed; mouth terminal, oblique; a knob at the symphysis of lower jaw and a notch in the upper jaw to receive it; abdomen keeled from pelvic fin origin to vent; outer surface of operculum strongly ridged; dorsal fin small with no osseous ray, its origin slightly behind pelvic-fin origin; caudal fin forked, lateral line complete with 80−100 scales, curved downward. Maximum size ∼1400 mm SL.

Distribution. Originally China, introduced throughout the world.

Genus *Hypsibarbus* **Rainboth, 1996**

Hypsibarbus Rainboth, 1996: 20 (type species: *Acrossocheilus malcolmi* Smith, 1945). Gender: masculine.

Diagnosis. Medium to large barbins inhabiting large upland rivers of Southeast Asia; deep bodied, strongly compressed; dorsal-fin origin midway between tip of snout and caudal-fin base, opposite pelvic fin origin; pelvic fin with well-developed axillary scale; suborbital bones fairly large, mouth subterminal with well-developed groove separating lips from both upper and lower jaws; barbels two pairs; scales large, 23−32 scales on lateral series, 14−16 circumpeduncular scales, dorsal fin with four simple and 8½ branched rays; last simple dorsal fin ray osseous with heavy retrose serrations; anal fin with iii, 5; pelvic fin with i, 8 and pectoral fin with I, 13−16 rays.

Hypsibarbus myitkyinae **(Prashad & Mukerji, 1929)**
(Fig. 4.102)

FIGURE 4.102

Hypsibarbus myitkyinae, 114.0 mm SL, Chakpi River, Manipur.

Puntius myitkyinae Prashad & Mukerji, 1929: 198 (type locality: Indawgyi Lake, Myanmar).
Hypsibarbus myitkyinae: Kottelat, 2013: 112.
Puntius jayarami Vishwanath & Tombi Singh, 1986a: 129 (Chakpi stream, Chindwin drainage, Manipur).

Diagnosis. Body compressed; two pairs of barbels, rostral and maxillary, lengths equals eye diameter; last unblanched dorsal fin ray osseous and denticulated posteriorly, dorsal spine length 64.1%−75.2% depth of body; head length 17.0%−19.6% SL; lateral line complete with 28−30 pored scales. Maximum size ∼140 mm SL.

Distribution. Chindwin and Irrawaddy drainage, India and Myanmar.

Genus *Labeo* **Cuvier, 1816**

Labeo Cuvier, 1816: 194 (type species: *Cyprinus niloticus* Forskål in Niebuhr, 1775).

Diagnosis. The genus *Labeo* is characterized in having inferior mouth; thick and smooth rostral cap; upper lip broad, partly or entirely covered by rostral cap, entirely separated from upper jaw; lower lip separated from the lower jaw by a transverse groove with highly papillose on anterior and posterior surface; postlabial continuous forming a deep transverse groove or narrowly interrupted.

Labeo angra (Hamilton, 1822)

Cyprinus angra Hamilton, 1822: 331, 391 (type locality: India: Brahmaputra River).
Diagnosis. Small species of *Labeo* of about 6–7 in. in length; yellowish in color with a broad dark brown to black stripe on lateral sides; two minute barbels; dorsal fin rays 12, the first simple minute ray absent; pectoral fin with 12, pelvic with nine and anal with eight rays. Maximum size ~170 mm SL.
Distribution. Brahmaputra drainage, Assam.

Labeo bata (Hamilton, 1822)
(Fig. 4.103)

FIGURE 4.103

Labeo bata, 175.0 mm SL, Imphal River, Manipur.

Cyprinus bata Hamilton, 1822: 283, 386 (type locality: Rivers and ponds of Bengal: India) (Fig. 4.104).

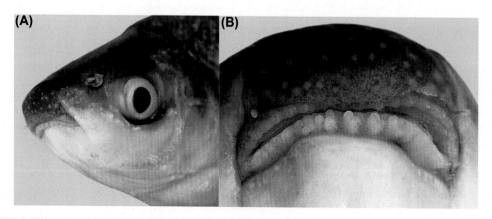

FIGURE 4.104

Labeo bata: (A) side view of snout; (B) mouth and lip structure.

Diagnosis. Body elongated. Mouth inferior, lips thin, lower lips slightly fringed and folded back and joined to isthmus by narrow bridge; snout slightly projecting beyond mouth, often studded with pores; maxillary barbel minute; dorsal fin origin nearer snout-tip than to caudal-fin base, fin with ii−iv, 9−11 rays; pectoral fin with i, 13−15; pelvic fin with i, 8; and anal fin with ii−iii, 5 rays; scales moderate, lateral line with 40−41 scales; Tubercles inside lower jaw above symphysis, horny covering inside the lower jaw, lips very thin, black blotch on fifth and sixth scales on the shoulder; in live, body color golden yellow above and on dorsal half of flanks, silvery on lower half of abdomen, an irregular black blotch present on fourth to sixth scales of lateral line, pelvic- and anal-fin tips orange red. Maximum size ∼450 mm SL.

Distribution. Ganga and Brahmaputra drainages in India, Nepal, Bangladesh, and Myanmar.

***Labeo boga* (Hamilton, 1822)**
(Fig. 4.105)

FIGURE 4.105

Labeo boga, 193.0 mm SL, Brahmaputra River, Guwahati, Assam.

Cyprinus boga Hamilton, 1822: 286, 386 (type locality: Brahmaputra: India).

Diagnosis. Snout moderately projecting beyond jaw, devoid of lateral lobe, occasionally covered with large pores; mouth fairly narrow; a thin layer of cartilaginous covering to inner surface of lower jaw, lower lip plain; a pair of minute maxillary pair only; dorsal fin with ii−iii, 9−10 rays, its origin nearer to tip of snout than to base of caudal fin, placed above or slightly anterior to tip of pectoral fin; pectoral fin with i, 15 rays, do not extend to pelvic fins; pelvic fin with i, 8 rays; anal fin with ii, 5 rays; caudal fin forked with 9 + 8 rays; scales moderate, scales along lateral line 37−39 + 2, transverse scale rows 6−7½/1/5½, predorsal scales 10−11, circumpeduncular scales 20; body silvery, fins orange with a reddish tinge. Maximum size ∼240 mm SL.

Distribution. Ganga and Brahmaputra drainages, India, Bangladesh, and Nepal.

***Labeo calbasu* (Hamilton, 1822)**
(Fig. 4.106)

FIGURE 4.106

Labeo calbasu, 220.0 mm SL, Jiri River, Manipur.

Cyprinus calbasu Hamilton, 1822: 297, 387 (type locality: Rivers and ponds of Bengal: India).

Diagnosis. Body stout and deep; snout depressed and fairly pointed, devoid of lateral lobe, studded with pores; eyes moderate, mouth inferior, lips thick and conspicuously fringed, both lips with a distinct inner fold; barbels two pairs, rostral and maxillary; dorsal fin with iii–iv, 14–16 rays, its origin midway between snout-tip and caudal-fin base; pectoral fin with i, 16–18 rays, pelvic fin with i, 8; anal fin with ii–iii, 5 and. Caudal fin with 19–20, deeply forked. Scales moderate, lateral line scales 43–44, transverse scale row 9/1/6, predorsal scales 16–18; in live, body color blackish-green, lighter ventrally; flank with scarlet spots at dark edges; fins black, upper lobe of caudal fin usually tipped with white. Maximum size ~700 mm SL.

Distribution. Widely distributed in India, Bangladesh, Myanmar, Nepal, Pakistan, and Thailand.

Labeo dyocheilus (McClelland, 1839)
(Fig. 4.107)

FIGURE 4.107

Labeo dyocheilus, 248.0 mm SL, Namdapha River, Arunachal Pradesh.

Cyprinus dyocheilus McClelland, 1839: 268, 330 (type locality: Brahmaputra, India).

Diagnosis. Snout conical projecting beyond mouth, with a distinct lateral lobe. Mouth wide and inferior. A pair of maxillary barbel. Dorsal fin with ii–iii, 11–13 rays, inserted equidistant between snout tip and base of caudal fin. Pectoral fin with i, 16 rays, extends to pelvic fins. Pelvic fin i, 8 rays. Anal fin with ii, 5 rays. Caudal fin deeply forked. Scales moderate, lateral line scales 40–43, transverse scale rows 5–6. Snout blunt, rostral fold thick, often with a depression arising from below eyes and joining at the middle of the snout, snout often coarse with fine tubercles in both male and female. Maximum size ~700 mm SL.

Distribution. Rivers of the Himalayan foothills of Nepal, India Bangladesh.

Labeo gonius (Hamilton, 1822)
(Fig. 4.108)

FIGURE 4.108

Labeo gonius, 243.0 mm SL, Loktak Lake, Manipur.

Cyprinus gonius Hamilton, 1822: 293, 387 (type locality: Ganges drainage: India).

Diagnosis. Snout slightly projecting beyond mouth, devoid of lateral lobe, studded with numerous pores; mouth narrow and subinferior, lips thick and fringed, with a distinct inner fold in circumference; two short pairs of rostral and maxillary barbels; dorsal fin with ii–iii, 13–16 rays, its origin nearer to tip of snout than to base of caudal fin; pectoral fins with i, 16 rays, equal to head length; pelvic fin with i, 8 and anal fin with ii, five to six rays; caudal fin deeply forked; scales small, lateral line scales 71–84, transverse scale rows 9/1/13, preanal scales 44–57. In live, greenish black on dorsum, becoming dull white on flanks and belly. Maximum size ~1100 mm SL.

Distribution. India, Pakistan, Bangladesh, and Myanmar.

Labeo nandina (Hamilton, 1822)

Cyprinus nandina Hamilton, 1822: 300, 388, pl. 8 fig. 84 (type locality: India: "Mahananda River, and in the large adjacent marshes or lakes which surround the ruins of ancient Gaur"/ Gorakhpur District).

Diagnosis. A large sized species of *Labeo* which grows to 2–3 ft in length; distinguished from other species of *Labeo* in having a long dorsal fin with 23–26; two pairs of barbels; lips fringed; snout plain; body covered with large scales; 37–38 lateral line scales. Maximum size ~800 mm SL.

Distribution. Ganga River drainage, Eastern Uttar Pradesh to North Bengal.

Labeo pangusia (Hamilton, 1822)
(Fig. 4.109)

FIGURE 4.109

Labeo pangusia, 196.0 mm SL, Barak River, Silchar, Assam.

Cyprinus pangusia Hamilton, 1822: 285, 386 (type locality: Kosi River, India)

Diagnosis. Body elongated, its dorsal profile more convex than the ventral. Snout subterminal, with lateral lobes; lips thick, slightly fimbriated; postlabial groove interrupted. Dorsal fin with 11–13 branched fin rays; lateral line with 40–42 scales, 5½–6 transverse scale rows between lateral line and pelvic fin. Caudal fin deeply forked. In live, dorsum and body dark green. Edges of scales with black mark. Dorsal and caudal fins gray. Pelvic and anal fins tinged with red. Maximum size ~700 mm SL.

Distribution. India, Pakistan, Nepal, Bangladesh, and Myanmar.

Labeo rohita (Hamilton, 1822)
(Fig. 4.110)

FIGURE 4.110

Labeo rohita, 346.0 mm SL, Loktak Lake, Manipur.

Cyprinus rohita Hamilton, 1822: 301, 388 (type locality: Gangetic provinces and Ava, India).
Diagnosis. Relatively small snout with no lateral lobe; isthmus rounded; a depression toward postlabial grove; barbels two pairs; lips fimbriated. Maximum size ~1500 mm SL.
Distribution. Native of Ganga drainage; introduced widely in Asia.

Genus *Laubuka* Bleeker, 1859

Laubuka Bleeker, 1859a: 261 [type species: *Cyprinus (Chela) laubuca* Hamilton, 1822]. Gender: feminine.
Laubuka: Pethiyagoda et al., 2008: 21 (discussion on the status of the genus).
Diagnosis. Premaxillae broadly in contact at symphysis; maxilla without a palatine process, its distal portion long; fifth ceratobranchial with $5 + 4 + 2$ teeth; 14 precaudal vertebrae; much larger and less numerous scales: $31-37 + 1-2$ lateral-line scales; $7-11$ scales in transverse line on body; $17-21$ predorsal scales and $12-14$ circumpeduncular scales; $14-20$ branched anal-fin rays and five branched pelvic-fin rays.

Laubuka khujairokensis Arunkumar, 2000
(Fig. 4.111)

FIGURE 4.111

Laubuka khujairokensis, 56.1 mm SL, Lokchao River near Moreh, Manipur, India.

Chela khujairokensis Arunkumar, 2000a: 122, fig. 1 (type locality: Khujairok stream, a tributary of the Yu River, at Moreh, near the adjoining borderland areas of Manipur, India and Myanmar).
Diagnosis. Lateral line incomplete, $32-34$ scales in the row, pored scales $14-15$; predorsal scales $16-18$; branched dorsal-fin rays nine; branched anal-fin rays $17-18$; ventral profile more arched than that of dorsal; dorsal profile of cranium straight up to occiput; snout pointed; mouth upturned, upper jaw slightly longer than upper jaw; abdominal keeled from the region of three to

four scale rows up to vent; a cleithral spot; pectoral fin extends beyond origin of pelvic-fin origin; scales are spotted with brownish or blackish dots. Maximum size ~50 mm SL.

Distribution. Khujairok stream, tributary of Yu River, India-Myanmar border.

Laubuka laubuca **(Hamilton, 1822)**
(Fig. 4.112)

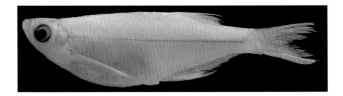

FIGURE 4.112

Laubuka laubuca, 54.4 mm SL, Brahmaputra river near Dhubri, Assam.

Cyprinus laubuca Hamilton, 1822: 260, 342 (type locality: Northern Bengal, India).

Diagnosis. Body deep and greatly compressed; abdomen keeled between and behind pelvic fins; mouth slightly oblique; lateral line incomplete with 35−36 scales; predorsal scales 18−20; dorsal fin with ii + 9 rays, inserted opposite to the anal fin; mouth upturned; lower lip longer than the upper; lateral end of operculum acute; axillary pectoral lobes present; caudal fin with 17−20 rays, deeply forked, lobes equal in length; silver to greenish gray with a violet luster on caudal peduncle, a green to clack stripe from the vertical level of dorsal-fin origin along the flank to caudal-fin base, a deep black blotch around operculum and caudal fin each; fins orange. Maximum size ~80 mm SL.

Distribution. Brahmaputra River near Dhubri, Assam and Ganga basin, India; Bangladesh and Nepal.

Laubuka parafasciata **Lalramliana, Vanlalhlimpuia & Singh, 2017**
(Fig. 4.113)

FIGURE 4.113

Laubuka parafasciata, 37.8 mm SL holotype, Sala River, Kaladan drainage, Mizoram.

Photo courtesy: Lalramliana.

Laubuka parafasciata Lalramliana, Vanlalhlimpuia & Singh, 2017: 270, figs. 1, 2A−B, 3 (type locality: Sala River, a tributary of Kaladan River, in the vicinity of Lungpuk, Siaha District, Mizoram, India).

Diagnosis. Precaudal vertebrae 16; premaxilla not in contact; broad, dark-brown midlateral stripe from the posterior rim of the orbit to the middle of the caudal-fin base; pelvic fin, not reaching the anal-fin origin; branched dorsal-fin rays 8½; branched anal-fin rays 16½−19½; minute tubercles scattered on the lower jaw; 28−33 lateral-line scales; 16−18 predorsal scales, 8½ branched dorsal-fin rays, 2½ scales between lateral line and pelvic-fin origin. Maximum size ~60 mm SL.

Distribution. Sala River, Kaladan drainage, Mizoram.

Genus *Neolissochilus* Rainboth, 1985

Neolissochilus Rainboth, 1985: 25 (type species: *Barbus stracheyi* Day, 1871); Gender: masculine.

Diagnosis. Body deep anteriorly, smoothly tapers posteriorly, caudal peduncle strongly compressed; head broad, snout blunt; mouth subterminal or inferior; lower lip with no fleshy lobe and without notches; postlabial groove interrupted medially; lips thick; barbels in two pairs, one rostral and one maxillary; eyes in the upper half of head; infraorbital bone broad; cheeks with numerous tubercles; dorsal fin origin midway between tip of snout and base of caudal fin, its last simple ray osseous and entire; scales large.

Neolissochilus dukai (Day, 1878)
(Fig. 4.114)

FIGURE 4.114

Neolissochilus dukai, 268.0 mm SL, confluence of Teesta and Rangit rivers, Sikkim. This photograph was wrongly inserted as *Tor tor* in Vishwanath et al. (2007).

Barbus dukai Day, 1878: 564, pl. 143, fig. 3 (type locality: Teesta River, near Darjeeling, India).

Diagnosis. Sides of snout below eyes with large open tubercles; lower labial fold interrupted; rostral barbell slightly longer than orbit, maxillary almost reach angle of preopercle; lateral line complete with 25−27 scales; transverse scales 4½/1/2½; predorsal scales 9. Body color coppery violet, with golden tinge in the dorsal half, pale white below, scale margins dark brown, fins dark brown with yellowish tinge on the margins. Maximum size ~90 mm SL.

Distribution. Teesta and Brahmaputra drainage, northeastern Bengal, Arunachal Pradesh and northern Assam.

Neolissochilus heterostomus Chen & Yang, 1999
(Fig. 4.115)

Neolissochilus heterostomus Chen & Yang in Chen et al., 1999: 84 (type locality: Nabang, Yinjiang, Yunnan, China).

Diagnosis. Mouth terminal, gape horizontal in male; last unbranched dorsal ray smooth with its upper 1/3 articulated; dorsal fin origin anterior to pelvic fin origin; no dark band on sides of body;

FIGURE 4.115

Neolissochilus heterostomus, 166.0 mm SL, Tizu River, Chindwin drainage, Manipur.

gill rakers 13−19; longest caudal fin ray length about two times that of the shortest; lateral line scales 28−29. Maximum size ∼230 mm SL.

Distribution. Irrawaddy drainage, Yunnan, China; Chindwin and Irrawaddy drainage, Manipur, India and Myanmar.

Neolissochilus hexagonolepis (McClelland, 1839)
(Fig. 4.116)

FIGURE 4.116

Neolissochilus hexagonolepis, 288.0 mm SL, Barak River, Tamenglong, Manipur.

Barbus hexagonolepis McClelland, 1839: 270, 336 (type locality: all large rivers on the eastern frontier, Assam).

Diagnosis. Body elongated, tubercles below the eye, rostral barbel reaching anterior margin of orbit or beyond, maxillary barbel extending beyond the level of the posterior margin of orbit; lateral line scales 26−29; four scale rows between dorsal fin origin and lateral line and 2½−3 from the lateral line and pelvic fin origin; 8−11 predorsal scales, sharp truncate lower jaw edge. Body color golden yellow with chocolate tint or chocolate, particularly the lateral line scale and below. Maximum size ∼1000 mm SL.

Distribution. Ganga, Brahmaputra and Surma-Megha drainages, Eastern Himalaya.

Neolissochilus kaladanensis Lalramliana, Lalronunga, Kumar & Singh, 2019
(Fig. 4.117)

Neolissochilus kaladanensis. Lalramliana, Lalronunga, Kumar & Singh, 2019: 56 (type locality: Kaladan River, Kawlchaw Village, Lawntlai District, Mizoram, India).

Diagnosis. Lateral line scales 20−22; lateral transverse scales $^1/_2$ 3/1/2$^1/_2$−3 scales; predorsal scales 8−10; circumferential scales 17−18; circumpeduncular scales 12 and total vertebrae 35;

FIGURE 4.117

Neolissochilus kaladanemsis, 142.0 mm SL, Kaladan River, Mizoram.

gill rakers on the descending arm of the first gill arch 13−14; body coloration: darkish brown dorsum, sides becoming lighter and abdomen yellowish to orange hue and inconspicuous dark brownish spots on cheek, opercle and caudal peduncle and black edged caudal fin. Maximum size ∼150 mm SL.

Distribution. Kaladan River at Kawlchaw, Mizoram, India.

Neolissochilus spinulosus **(McClelland, 1845)**
(Fig. 4.118)

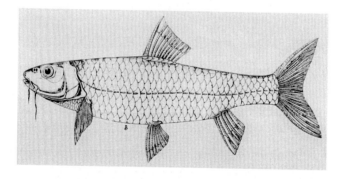

FIGURE 4.118

Neolissochilus spinulosus, McClelland's (1845) pl. 21, fig. 3 reproduced.

Barbus spinulosus McClelland, 1845: 280, pl. 21, fig. 3 (type locality: rivers at foot of Sikkim Mountains, India).

Diagnosis. Body elongated, back slightly arched; eyes in the middle of head; dorsal fin origin midway between tip of snout and caudal-fin base; last simple dorsal-fin ray osseous and straight, slender, shorter than first branched ray; lateral line scales 32; body color olive green above, white below, fins pale, Maximum size ∼120 mm SL.

Distribution. Teesta drainage, northern Bengal, India.

Neolissochilus stevensonii **Day, 1870**
(Fig. 4.119)

FIGURE 4.119

Neolissochilus stevensonii, 189.0 mm SL, Kaladan River, Mizoram.

Barbus (Barbodes) stevensonii Day, 1870b: 100 (type locality: hills near Akyab, Myanmar).
Neolissochilus stevensonii: Rainboth, 1985: 29 (placement under the genus).
Diagnosis. A conspicuous black blotch on caudal-fin base; eye diameter less than snout length (23.4%−26.9% HL); 8−10 predorsal scales; 25−27 lateral-line scales on body; body depth 25.7%−32.3% SL; lateral transverse scales ½4/1/2½; 3½ scales between lateral line and anal-fin origin; three branchiostegal rays; 12 circumpeduncular scales; 3 + 7 gill rakers; and 39−40 vertebrae.
Distribution. Akyab (now Sittwe), Myanmar and Kolo River, Mizoram, both Kaladan drainage.

Neolissochilus stracheyi **(Day, 1871)**
(Fig. 4.120)

FIGURE 4.120

Neolissochilus stracheyi, 96.7 mm SL, Chakpi River, Manipur.

Barbus (Barbodes) stracheyi Day, 1871: 307 (type locality: Akyab and Moulmein, Burma).
Diagnosis. Rostral barbel extends to middle of orbit and maxillary to posterior edge of orbit; scales large, lateral line complete with 23−24 scales; scales between origin of dorsal fin and lateral line 3½ and between it and origin of pelvic fin 2½; predorsal scales 8−9; body color above lateral line and dorsal half of head dark brown with golden yellowish tinge, ventral half silvery with golden yellow tinge. Maximum size ~600 mm SL.
Distribution. Chindwin-Irrawaddy drainage, Manipur, India and Myanmar, Akyab, Myanmar.

Genus *Oreichthys* Smith, 1933
Oreichthys Smith, 1933: 63 (type species: *Oreichthys parvus* Smith, 1933). Gender: masculine.
Body deep short, compressed, abdomen rounded; snout obtuse; no barbels; interorbital region, cheeks, and opercle with rows of minute pores; lateral line scales larger than others; gill rakers absent; eyes large; dorsal fin origin above distal half of pectoral fin, no spinous ray in dorsal fin; lateral line complete or incomplete; caudal fin forked.

Oreichthys andrewi **Knight, 2014**
(Fig. 4.121)

FIGURE 4.121

Oreichthys andrewi ~30.0 mm SL, Dibru River, Tinsukia, Assam.

Photo courtesy: M. Beta.

Oreichthys andrewi Knight, 2014: 5358, figs. 1, 2A (type locality: River Dibru at Guijan Gnat, Tinsukia District, Assam, India).

Diagnosis. Dorsal profile arched, hump at nape; lateral line complete with 30 + 1 scales; transverse scales ½4/1/2½; predorsal scales 9; dorsal height equals body depth; dorsal fin with two simple and 8½ branched rays; pelvic fin reaching anal fin origin; faint humeral spot covering fifth and sixth lateral line scale; black spots on distal ends of dorsal and anal fin; no spot at caudal fin peduncle or base; pelvic fin bright yellow. Maximum size ~39 mm SL.

Distribution. Dibru River, Guijan Ghat, Tinsukia District, Assam.

Oreichthys cosuatis **(Hamilton, 1822)**
Cyprinus cosuatis Hamilton, 1822: 338, 392 (type locality: Kosi River, Uttar Pradesh).

Diagnosis. Head with numerous fine parallel sensory folds on sides; barbels absent; dorsal fin fairly high, its origin slightly ahead of the distance between snout tip and caudal-fin base, four to five scales ahead; predorsal scales 8–9; a dusky caudal spot at caudal-fin base; a faint spot on anal fin. Maximum size ~60 mm SL.

Distribution. Ganga and Brahmaputra drainage, northeast India.

Oreichthys crenuchoides **Schäfer, 2009**
(Fig. 4.122)

FIGURE 4.122

Oreichthys crenucoides, 36.2 mm SL, Dikrong River, Doimukh, Arunachal Pradesh.

Oreichthys crenuchoides Schäfer, 2009: 202, figs. 1−2, 6−10 (type locality: River Jorai, a tributary of Brahmaputra River, near Assam border, Coochbehar District, West Bengal, India).

Diagnosis. *Oreichthys* with a conspicuous black blotch at caudal-fin base; second simple to second branched rays of dorsal fin black distally, the fin in matured males variegated with numerous grayish markings with a blackish stream at distal anterior margin; no black spot on anal fin; cheek with 11−13 rows of pores; lateral line incomplete, 17−19 + 2 scales in lateral series, pored scales 2−5; predorsal scales 7−9. Maximum size ∼300 mm SL.

Distribution. Brahmaputra River drainage in northern West Bengal, Assam, and Arunachal Pradesh.

Genus *Oreinus* McClelland, 1838

Oreinus McClelland, 1838: 943, 946 (type species: *Oreinus guttatus* McClelland, 1838: 946). Gender: masculine.

Diagnosis. Mouth transverse; premaxillaries suspended horizontally so as to be carried by the muscular structure of the snout upward and forward in opening the mouth; lower jaw short and broad as against *Schizothorax* which has intermaxillaries fixed; lower jaw long, narrow and shorter than upper; lips soft and round (Tilak, 1987).

Oreinus molesworthi **Chaudhuri, 1913**
(Fig. 4.123)

FIGURE 4.123

Oreinus molesworthi, 188.0 mm SL, Teesta River, Rangit, Sikkim.

Oreinus molesworthi Chaudhuri, 1913: 247, pl. 7, figs. 2, 2a−b [type locality: Abor Hills, Assam, Yembung (Yombong)].

Diagnosis. Snout broadly rounded, width of mouth nearly 2½ times the head length; edge of rostral flap entire; numerous conical tubercles on snout and rostral flap; a narrow groove in front of the nostrils, anterior nostril fleshy and tubular; rostral barbell longer than half diameter of eye; maxillary concealed in lower lip and is 3/4th of eye; both lips overturned to form sectorial disk; lower lip more broken up; margin of lower jaw mounted with a hard horny covering. Dorsal spine shorter than head length, serrated strongly; lateral line scales more than 100. Maximum size ∼210 mm SL.

Distribution. Teesta and Brahmaputra drainages, Himalayan foothills.

Genus *Osteobrama* Heckel, 1843

Osteobrama Heckel, 1843: 1033 (type species: *Cyprinus cotio* Hamilton, 1822). Gender: feminine.

Diagnosis. Body compressed, fairly deep and short; abdomen edge trenchant wholly or before pelvic fin origin; mouth directed forward and slightly upward; dorsal fin origin slightly behind pelvic fin origin, its last simple ray osseous and serrated posteriorly; pharyngeal teeth in three rows; anal fin long.

Osteobrama belangeri (Valenciennes, 1844)
(Fig. 4.124)

FIGURE 4.124

Osteobrama belangeri, 162.0 mm SL, fish farm near Loktak Lake, Manipur.

Leuciscus belangeri Valenciennes in Cuvier and Valenciennes, 1844a, 1844b: 99 (type locality: Bengal).

Diagnosis. Lateral line scales 72–80; abdominal edge sharp and keeled entirely; length of caudal peduncle longer than its depth; branched anal fin rays 17; predorsal scales 23–25. Maximum size ~380 mm SL.

Distribution. Chindwin and Irrawaddy drainages of India and Myanmar; cultured in farms in Manipur.

Osteobrama cotio (Hamilton, 1822)
(Fig. 4.125)

FIGURE 4.125

Osteobrama cotio, 93.0 mm SL, Jiri River, Barak drainage, Manipur.

Cyprinus cotio Hamilton, 1822: 339, 393 (type locality: Bengal).

Diagnosis. Body trapezoid and considerably compressed; its depth 36%–38% SL, abdominal edge trenchant from behind pelvic-fin base to anal fin but rounded in front of pelvic fin; barbels absent; dorsa spine weak and serrated; lateral line scales 57−70; body color silvery, darkest along the back and sometimes with a silvery stripe; a black blotch before the base of dorsal fin and another on the nape; anal fin with 27−33 branched rays. Maximum size ~120 mm SL.

Distribution. Ganga, Brahmaputra and Barak-Surma-Meghna drainages.

Osteobrama cunma (Day, 1888)
(Fig. 4.126)

FIGURE 4.126

Osteobrama cunma, 87.0 mm SL, Manipur River, Sugnu, Manipur.

Rohtee cunma Day, 1888: 807 (type locality: Moulmein, Burma).

Diagnosis. Body depth 35.5%−38.3% SL; anal fin with 25−26 branched rays; lateral line compete with 44−48 scales; predorsal scales 28; head short, conical, dorsal profile steep with distinct rise from occiput and humped between occiput and dorsal fin origin; dorsal spine osseous weakly serrated; caudal fin forked, lower lobe slightly longer; olive above, becoming silvery on sides, and below, brassy tinge on lateral line and over cheeks and gill covers, fins amber, caudal with a narrow black edge. Maximum size ~120 mm SL.

Distribution. Chindwin and Irrawaddy drainages, India and Myanmar.

Osteobrama feae Vinciguerra, 1890
(Fig. 4.127)

FIGURE 4.127

Osteobrama feae, 187.0 mm SL, Yu River, Tamu, Myanmar.

Osteobrama feae Vinciguerra, 1890: 311, pl. 10, fig. 10 (type locality: Burma: Kokarit).

Diagnosis. Body deep, its depth 44.2%−47.4% SL; abdominal edge keeled; barbels two pairs; dorsal fin origin midway between the tip of snout and the base of caudal fin; dorsal fin spine weak and serrated; lateral line scales 62−65; transverse scales 15/1/12; predorsal scales 38−40; anal fin long, its origin well behind the origin of dorsal fin, 28−30 branched rays; length of caudal peduncle shorter than its height; caudal fin forked. Maximum size ∼200 mm SL.

Distribution. Chindwin drainage, Manipur (India) and Myanmar border, Irrawaddy drainage, Yunnan, China and Myanmar.

Genus *Pethia* Pethiyagoda, Meegaskumbura & Maduwage, 2012

Pethia Pethiyagoda, Meegaskumbura & Maduwage, 2012: 80 (type species: *Barbus nigrofasciatus* Günther, 1868). Gender: feminine.

Diagnosis. Adult size small, usually less than 50 mm SL; rostral barbels absent; maxillary barbels minute if present; last unbranched dorsal-fin ray serrated; three or four unbranched and eight branched dorsal-fin rays, three unbranched and five branched anal-fin rays; gill rakers simple, no predorsal spinous ray; infraorbital three deep, free uroneural absent; four supraneurals; 11−13 precaudal and 13−16 caudal vertebrae; postepiphysial fontanel absent; lateral line usually incomplete, 19−24 scales in lateral series; a black blotch on caudal peduncle and frequently also other black blotches, spots or bars on side of body.

Pethia ater (Linthoingambi & Vishwanath, 2007)
(Fig. 4.128)

FIGURE 4.128

Pethia ater, 53.4 mm SL, Loktak Lake, Manipur.

Puntius ater Linthoingambi & Vishwanath, 2007: 46 (type locality: Iril River, Bamon Kampu, Chindwin drainage, Manipur, India).

Diagnosis. Dorsal-fin edge black, its spine weak, serrated posteriorly with 13−17 serrae, spine length 16.8%−23.3% SL; predorsal scales 10 or 11; lateral line incomplete with 5−11 pored scales, 25−29 scales in lateral-line row; transverse scales ½ 4/1/4 ½; a black longitudinal stripe covering the upper half of the first scale row and lower half of the second scale row above lateral-line row of scales; a black spot extending over 19th and 20th scales of lateral-line row at the level above the posterior end of the anal-fin base. Maximum size ∼60 mm SL.

Distribution. Chindwin River basin, Manipur India.

***Pethia aurea* Knight, 2013**

Pethia aurea Knight, 2013: 174 (type locality: Ponds in South 24−Parganas District, West Bengal, Ganga basin, India)

Diagnosis. Lateral line incomplete, with three to four pored scales; 25−26 + 1 scales in the lateral series, ½5/1/3−3½ scales in transverse line on body; predorsal scales 9; last unbranched dorsal-fin ray slender, serrated, with 19−22 serrae on posterior margin; barbels absent; a black band around the caudal peduncle covering scales 23−25 of the lateral series; body golden yellow, a black blotch beneath the origin of the dorsal fin and a black spot above the origin of the anal fin. Maximum size ∼24 mm SL.

Distribution. Ponds in South 24-Parganas, West Bengal, India.

***Pethia canius* (Hamilton, 1822)**

Cyprinus canius Hamilton, 1822: 320 (type locality: Pond in Cooch Behar District, West Bengal, India).

Diagnosis. Combination of the following characters: lateral line incomplete, with three to four pored scales; 20−21 + 1 scales in the lateral series, ½4/1/2 scales in transverse line on body; predorsal scales 8; last unbranched dorsal-fin ray strong, curved, serrated, with 24−27 large, curved serrae on posterior margin; barbels absent; a broad black band around caudal peduncle covering scales 17−19 of longitudinal series. Three diffuse black blotches on body, first behind opercle, second below dorsal-fin origin, third above anal-fin origin. Black spots at base and root of dorsal, anal, and pelvic fins. Maximum size ∼30 mm SL.

Distribution. West Bengal.

***Pethia conchonius* (Hamilton, 1822)**
(Fig. 4.129)

FIGURE 4.129

Pethia conchonius, 45.0 mm SL, Barak River, Silchar, Assam.

Cyprinus conchonius Hamilton, 1822: 317, 389 (type locality: ponds of northeastern Bengal; Kosi and Ami rivers, India).

Diagnosis. Body deep, its depth 41.0−43% SL; barbels absent; lateral line incomplete, 7−10 pored scales; predorsal scales 9−10; a black blotch just anterior to the caudal peduncle on the sides; last simple ray serrated posteriorly; dorsal, pelvic, and anal fins with black coloration at the margins. Maximum size ∼105 mm SL.

Distribution. Widely distributed in the rivers of India.

Pethia expletiforis **Dishma & Vishwanath, 2013**
(Fig. 4.130)

FIGURE 4.130

Pethia expletiforis, 53.0 mm SL, Ka-ao River, Kaladan drainage, Mizoram.

Pethia expletiforis Dishma & Vishwanath, 2013: 83 (type Ka-ao River near New Serkawr village, Kaladan drainage, Saiha District, Mizoram).

Diagnosis. A complete lateral line with 21−23 pored scales; a black blotch extending over scales 17−19 of the lateral line at level above the posterior end of anal-fin base; absence of a humeral mark; nine predorsal scales; ½4 scales between dorsal-fin origin and lateral line, and 3½ scales between lateral line and pelvic-fin origin; last simple dorsal-fin ray strong and serrated with 14−19 serrae, its length 16.6%−19.7% SL; body depth 37.1%−42.9% SL; and absence of barbels. Maximum size ∼53 mm SL.

Distribution. Ka-ao River, New Serkawr, Saiha District, Mizoram, India.

Pethia gelius **(Hamilton, 1822)**
(Fig. 4.131)

FIGURE 4.131

Pethia gelius, 37.0 mm SL, Brahmaputra River, Goalpara, Assam.

Cyprinus gelius Hamilton, 1822: 320, 390 (type locality: ponds and ditches of northeastern Bengal).

Diagnosis. Lateral line incomplete, with three to four pored scales; 21−22 + 1 scales in lateral series, ½4/1/2½ scales in transverse line on body; predorsal scales 8; last unbranched dorsal-fin ray thick, straight, serrated, with 20−25 serrae on posterior margin; barbels absent; a broad black band around the caudal peduncle covering scales 19−21 in the longitudinal series; three diffuse black blotches on the body, first behind opercle, second below dorsal-fin origin, third above anal-fin origin; black spot at base and origins of dorsal, anal and pelvic fins. Maximum size ∼37 mm SL.

Distribution. Widely distributed in Brahmaputra and Ganga drainages of India, Bangladesh and Nepal.

Pethia guganio (Hamilton, 1822)
(Fig. 4.132)

FIGURE 4.132

Pethia guganio, 50.0 mm SL, Dikrong River, Brahmaputra drainage, Arunachal Pradesh.

Cyprinus guganio Hamilton, 1822: 338, 392 (type locality: Brahmaputra and Yamuna rivers).

Diagnosis. Body with no bars; translucent; barbels absent; last simple dorsal fin ray osseous, strong and serrated posteriorly; lateral incomplete, pored scales 5−6; back with small black dots, sides of body silvery; a small black spot at caudal-fin base. Maximum size ∼60 mm SL.

Distribution. Ganga and Brahmaputra drainages, India and Bangladesh.

Pethia khugae (Linthoingambi & Vishwanath, 2007)
(Fig. 4.133)

FIGURE 4.133

Pethia khugae, 45.5 mm SL, holotype, Khuga River, Chindwin drainage, Manipur.

Puntius khugae Linthoingambi & Vishwanath, 2007: 49 (type locality: Khuga River, Churachandpur District, Chindwin drainage, Manipur, India).

Diagnosis. Dorsal fin edge plain, its spine serrated posteriorly with 10−12 serrae, spine length 16.0%−20.9% SL; predorsal scales 11−12; lateral line incomplete, pored scales 8−11; 28−30 scales in lateral-line row; transverse scales ½5/1/4½; a black blotch on caudal peduncle at level of one scale behind posterior end of anal origin. Maximum size ∼47 mm SL.

Distribution. Khuga River, Chindwin drainage, Manipur, India.

Pethia manipurensis (Menon, Rema Devi & Vishwanath, 2000)
(Fig. 4.134)

FIGURE 4.134

Pethia manipurensis, 29.0 mm SL, Sekmai River, Pallel, Chindwin drainage, Manipur.

Puntius manipurensis Menon, Rema Devi & Vishwanath, 2000: 263 (type locality, Loktak Lake, Chindwin basin, Manipur, India).

Diagnosis. Barbels absent; fins and caudal fin scarlet red; lateral line incomplete with four to five pored scales; two black spots: one on interlace area between third and fourth scale of lateral series of scales, another on the 17th scale; last simple dorsal ray osseous and finely serrated posteriorly. Maximum size ∼45 mm SL.

Distribution. Chindwin basin, Manipur, India.

Pethia meingangbii (Arunkumar & Tombi Singh, 2003)
(Fig. 4.135)

FIGURE 4.135

Pethia meingangbi, 38.8 mm SL, Lokchao River, Moreh, Yu River drainage, Manipur.

Puntius meingangbii Arunkumar & Tombi Singh, 2003: 483 (type locality: Moreh Bazar, Manipur, India, bordering Myanmar).

Puntius bizonatus Vishwanath & Laisram, 2004: 131 (Lokchao River, Moreh).

Diagnosis. Two blue-black bars on the sides of the body, one above the middle of pectoral fin and another on the caudal peduncle above anal fin; caudal fin and flank of lateral side distinctly red, dorsal fin with two to three bars; last simple dorsal fin ray osseous and serrated posteriorly, prenatal scales 12; lateral line income with six pored scales. Maximum size ~40 mm SL.

Distribution. Yu River drainage, Moreh, Manipur, India.

Pethia phutunio (Hamilton, 1822)
(Fig. 4.136)

FIGURE 4.136

Pethia phutunio, 32.0 mm SL, Barak River, Silchar, Assam.

Cyprinus phutunio Hamilton, 1822: 319, 390 (type locality: ponds of north-east Bengal).

Diagnosis. Incomplete lateral line, three to four pored scales, 20—23 scales in lateral series, predorsal scales 9; each scale with blackish base; three black spots on body, one behind the operculum, another above anal fin and one on caudal peduncle, a cross bar on dorsal fin. Maximum size ~30 mm SL.

Distribution. Water bodies of Orissa, West Bengal, Bangladesh and Assam.

Pethia rutila Lalramliana, Knight & Laltlanhlua, 2014
(Fig. 4.137)

Pethia rutila Lalramliana, Knight & Laltlanhlua, 2014a: 367, fig. 2a and b (type locality: Aivapui River, in the vicinity of Phuldungsei Village, Mizoram, India).

Diagnosis. Lateral line complete with 21—22 scales; a black blotch on caudal peduncle extending over 16—18 or 17—19 scale of lateral line, above the origin of the last anal fin ray; an inconspicuous black humeral spot on the scale row below third and fourth lateral line scales, predorsal scale 8—9; lateral transverse scales ½4/1/3½; dorsal fin with two rows of black bars on interradial membrane posterior to last unbranched ray. Maximum size ~45 mm SL.

Distribution. Karnaphuli drainage, Mizoram.

Pethia shalynius (Yazdani & Talukdar, 1975)
(Fig. 4.138)

Puntius shalynius Yazdani & Talukdar, 1975: 218 (type locality, Barapani Lake, Khasi Hills, Meghalaya, India).

FIGURE 4.137

Pethia rutila, 44.6 mm SL, Aivapui River, Kaladan drainage, Mizoram.

Photo courtesy: Lalramliana.

FIGURE 4.138

Pethia shalynius, 36.5 mm SL, Umiam Lake, Khasi Hills, Meghalaya.

Diagnosis. No barbels, lateral line incomplete with seven pored scales, a blue black thin lateral stripe from two to three scales behind the operculum to caudal peduncle, fainter dusky coloration above and below the stripe, two dark spots on caudal peduncle. Maximum size ~45 mm SL.

Distribution. Brahmaputra basin, northeast India.

Pethia stoliczkana **(Day, 1871)**
(Fig. 4.139)

FIGURE 4.139

Pethia stoliczkana, 48.3 mm SL, Yu River, Tamu, Myanmar.

Barbus (Puntius) stoliczkanus Day, 1871: 328 (type locality: Eastern Myanmar).
Puntius stoliczkanus: Linthoingambi & Vishwanath, 2007: 42, fig. 6 (redescription).

Diagnosis. Lateral line almost complete, reaches at least to caudal peduncle, pored scales 19−23, scales in series 21−24, a black bar on dorsal fin, black in preserved specimens, a black blotch on 17th to 19th scale of lateral scale row, two incomplete bars on dorsal fin, one on the membrane between third and fifth branched rays and another one at outer margin. Maximum size ∼40 mm SL.

Distribution. Chindwin basin in Manipur and Myanmar; also reported in Thailand and Laos.

***Pethia ticto* (Hamilton, 1822)**
(Fig. 4.140)

FIGURE 4.140

Pethia ticto, 38.0 mm SL, wetlands of Hajo, North Kamrup District, Brahmaputra basin, Assam.

Cyprinus ticto Hamilton, 1822: 314, 389 (type locality: southern parts of Bengal)
Puntius ticto: Linthoingambi & Vishwanath, 2007: 50, fig. 5 (redescription).

Diagnosis. Last unbranched dorsal fin ray osseous and serrated posteriorly with 15−17 serrae, lateral line incomplete with 6−11 pored scales, 22−26 scales in lateral series, two black spots on lateral line scale row, one on the fourth scale and another on 17th−20th scales, two black bars on dorsal fin. Maximum size ∼90 mm SL.

Distribution. Ganga and Brahmaputra basins in north-east India.

***Pethia yuensis* (Arunkumar & Tombi Singh, 2003)**
(Fig. 4.141)

FIGURE 4.141

Pethia yuensis, 42.0 mm SL, Lokchao River, Moreh, Yu River drainage, Manipur.

Puntius yuensis Arunkumar & Tombi Singh, 2003: 482 (type locality: Maklang River, near Moreh, Yu River drainage, Manipur, India).

Puntius ornatus Vishwanath & Laisram, 2004: 132 (Lokchao River, Moreh, Manipur, India).

Diagnosis. Body yellowish; a wide blue black bar forming a ring around caudal peduncle, overlaid by a dark brown to black blotch on the sides; dorsal spine serrated posteriorly; lateral line incomplete with 5−21 pored scales, 21−24 scales in lateral row; circumpeduncular scales 10−11; predorsal scales 8−9; prenatal scales 14. Maximum size ~60 mm SL.

Distribution. Yu River drainage, Manipur, India and Myanmar.

Genus *Poropuntius* Smith, 1931

Poropuntius Smith, 1931: 14 (type species: *Poropuntius normani* Smith, 1931). Gender: masculine.

Diagnosis. Body moderately elongated, compressed; snout blunt and rounded, its median part covered with rows of large pores, mouth subterminal, lips continuous, postlabial groove of lower lip interrupted in the middle, lower lip separated from lower jaw by a sulcus, which has a horny covering; rostral and maxillary barbels well developed; scales large, posterior-most simple dorsal fin ray osseous and strongly denticulated posteriorly; branched dorsal fin rays with 8½; anal fin with 5½ branched rays; gill membranes joined to isthmus.

Pethiyagoda et al. (2012) did not examine specimens of *Poropuntius*, however, referred it tentatively to *Systomus*. *Poropuntius* differs from *Systomus* in having accessory pores (Fig. 4.142) on the canals of the lateral-line system (accessory pore present in *Poropuntius*: Kottelat et al., 1993; Roberts, 1998; Kottelat, 2001) and also in having strongly serrated (Fig. 4.144) or denticulated weakly serrated (vs weakly serrated) dorsal spine.

FIGURE 4.142

Lateral view of snout of *Poropuntius* showing pores.

Poropuntius burtoni (Mukerji, 1934)
(Fig. 4.143)

FIGURE 4.143

Poropuntius burtoni, 123.5 mm SL, Tizu River, Jessami, Chindwin drainage, Manipur.

Barbus clavatus burtoni Mukerji, 1934: 64 (type locality: Phungin Hka, tributary of Mali Hka River, Myitkyina District, Upper Myanmar).

Poropuntius burtoni: Vishwanath & Kosygin, 2001: 32 (species status and redescription).

Diagnosis. Lateral line scales 34−35; scales between dorsal fin origin and lateral line 6½ and 3½ between it and pelvic fin origin; predorsal scales 12−13; circumpeduncular scales 14; body depth 26.3%−29.8% SL; dorsal-fin height 22.6%−28.5% SL; dorsal spine length shorter than body depth. Maximum size ∼160 mm SL.

Distribution. Chindwin-Irrawaddy drainage, Manipur, India and Myanmar.

Poropuntius clavatus (McClelland, 1845)
(Fig. 4.144)

FIGURE 4.144

Poropuntius clavatus, 168.8 mm SL, Barak River, Manipur.

Barbus clavatus McClelland, 1845: 280, pl. 21 (type locality: Sikkim mountains on the northern frontier of Bengal).

Diagnosis. Lateral line scales 41−42; 7½ scales between dorsal fin origin and lateral line and 4½ scales between it and pelvic fin origin; predorsal scales 16−18; body depth 29.5%−32.9% SL; dorsal fin height 28.9%−31.4% SL; dorsal spine length almost equals body depth. Maximum size ∼170 mm SL.

Distribution. Teesta drainage, Sikkim, Barak-Surma-Meghna and Brahmaputra drainage of northeastern India and Bangladesh.

Poropuntius margarianus (Anderson, 1879)

Barbus margarianus Anderson, 1879: 867, pl. 79, fig. 1 (type locality: "Nampong River, Kakhen Hills," near Bhamo, Irrawaddy basin).

Diagnosis. Scale in lateral series 32; predorsal scales 12; transverse scale rows 6/1/3; in fresh condition, almost entirely white or silvery, with fins nearly colorless; dorsal and caudal fin with a little lack pigment distally; middle of dorsal fin with black pigment on interradial membranes; caudal fin lobes without stripes; gills rakers 3 + 6 on the first gill arch. Maximum size ∼160 mm SL.

Distribution. Irrawaddy drainage, Yunnan, China and Myanmar.

Poropuntius shanensis (Hora & Mukerji, 1934)
(Fig. 4.145)

FIGURE 4.145

Poropuntius shanensis, 116.0 mm SL, Manipur River, Sugnu, Chindwin drainage, Manipur.

Barbus shanensis Hora & Mukerji, 1934b: 362, fig. 3 (type locality: Lawksawk stream, southern Shan states, Myanmar).

Diagnosis. Lateral line scales 37−38; scales between dorsal fin origin to lateral line 6½ and from it to pelvic origin 3½; predorsal scales 15−16; circumpeduncular scales 16; dorsal spine shorter than body depth. Maximum size ∼130 mm SL.

Distribution. Chindwin-Irrawaddy drainage, Manipur, India and Myanmar.

Genus *Puntius* Hamilton, 1822

Puntius Hamilton, 1822: 310, 388 (type species: *Cyprinus sophore* Hamilton, 1822). Gender: masculine.
Puntius: Pethiyagoda et al., 2012: 73 (genus redescription).

Diagnosis. Small to medium sized fishes, adult size usually less than 120 mm SL; rostral barbel absent; maxillary barbel present or absent; dorsal fin with three to four simple and eight branched rays; anal fin with three simple and five branched rays; last simple dorsal-fin ray apically segmented, not serrated posteriorly; lateral line complete, with 22−28 pored scales; free uroneural; postepiphysial fontanel; infraorbital 3 slender; pharyngeal teeth 5 + 3 + 2; caudal peduncle usually with a black spot or blotch.

Puntius chola (Hamilton, 1822)
(Fig. 4.146)

Cyprinus chola Hamilton, 1822: 312, 389 (type locality: northeastern Bengal).

Diagnosis. A short maxillary barbel; a black blotch at caudal-fin base, between 21st and 23rd scales of lateral line row, a spot at the base of first to second rayed dorsal fin; dorsal spine weak and smooth; lateral line complete with 24−28 scales. Maximum size ∼120 mm SL.

Distribution. Widely distributed in India, Bangladesh, Myanmar, Pakistan, Nepal, and Sri Lanka.

FIGURE 4.146

Puntius chola, 54.0 mm SL, Loktak Lake, Manipur.

Puntius sophore (Hamilton, 1822)
(Fig. 4.147)

FIGURE 4.147

Puntius sophore, 59.0 mm SL, Loktak Lake, Manipur.

Cyprinus sophore Hamilton, 1822: 310, 389 (type locality Ganga basin, India).

Diagnosis. A *Puntius* with no barbels, a characteristic round black blotch at caudal-fin base, another blotch on the middle of dorsal-fin base; lateral line complete with 22−27 scales, predorsal scale 8−10; opercle shot with gold; fins golden yellowish in color. Maximum size ∼80 mm SL.

Distribution. India, Pakistan, Bangladesh, Nepal, Sri Lanka, and Myanmar.

Puntius terio (Hamilton, 1822)
(Fig. 4.148)

Cyprinus terio Hamilton, 1822: 313, 389 (type locality: northeast Bengal).

Diagnosis. Lateral line incomplete, potred scales 4−7; 22−24 scales in lateral series, barbel absent, dorsal fin long, last unbranched ray osseous and smooth, a black blotch above anal fin, just anterior to the caudal peduncle, a thin black stripe extending from the blotch to the caudal-fin base. Maximum size ∼75 mm SL.

Distribution. Barak and Brahmaputra drainages in Assam, Manipur, India, and Bangladesh.

Genus *Raiamas* Jordan, 1919

Raiamas Jordan, 1919 *Cyprinus bola* Hamilton, 1822; type by a replacement name for Bola Günther, 1868, preoccupied by Bola Hamilton, 1822). Gender: masculine.

FIGURE 4.148

Puntius terio, 52.0 mm SL, Jiri River, Barak drainage, Manipur.

Diagnosis. Body elongated and compressed; mouth sharply pointed; mouth gap very wide extending beyond the posterior margin of orbit; lower jaw with a symphyseal knob fitting into a median notch of upper jaw; two or three rows of brownish or blackish spots on the sides of the body.

Raiamas bola **(Hamilton, 1822)**
(Fig. 4.149)

FIGURE 4.149

Raiamas bola, 98.0 mm SL; Brahmaputra River near Dhubri, Assam.

Cyprinus bola Hamilton, 1822: 274, 385 (type locality; Brahmaputra River, India).
Diagnosis. Body elongated and compressed, scales small; mouth terminal, obliquely directed upward, its cleft extends beyond posterior margin of orbit, snout sharply pointed; symphyseal knob prominent; both rostral and maxillary barbels rudimentary; two or three rows of brownish or blackish oblong diffused spots or blotches on the sides of the body, upper row more prominent and darker than lower row; mouth gap extending beyond the posterior margin of orbit; pectoral lobe and pelvic axillary scale present; 25 predorsal scales; lateral line complete with 85–95 pored scales; caudal fin hyaline. Maximum size ∼270 mm SL.
Distribution. Ganga and Brahmaputra drainages of Assam, India; Nepal, Bhutan, and Bangladesh.

Raiamas guttatus **(Day, 1870)**
(Fig. 4.150)

FIGURE 4.150

Raiamas guttatus, 113.0 mm SL, Thoubal River, Yairipok, Manipur.

Opsarius guttatus Day, 1870a: 620 (type locality: Irrawaddy River, from Prone to Mandalay, Myanmar).

Diagnosis. Body elongated and moderately compressed laterally; body depth greatest at dorsal-fin origin with about 22.7%−24.0% SL; dark brown black blotches of varying size along the lateral sides of the body; snout elongated and sharply pointed; mouth very wide extending beyond the middle of orbit; a black stripe on upper edge of caudal fin; scales moderately large; lateral line complete with 44−48 scales, transverse scale rows 8½/1/3½; predorsal scales 21, circumpeduncular scales rows 18; dorsal fin with iii + 7 rays, its origin nearer to caudal-fin base than to tip of snout; pectoral fin with i + 14 rays; pelvic fin with i + 8 rays; anal fin with iii + 11 rays; caudal fin forked with longer lower lobe; dark blue blotches of varying size along the sides of the body; lower lobe of caudal fin orange; dorsal fin with a dark stripe. Maximum size ∼300 mm SL.

Distribution. Chindwin drainage, Manipur, India and Myanmar. Thailand, Taiwan, Laos, and Cambodia.

Genus *Rasbora* Bleeker, 1859

Rasbora Bleeker, 1859b: 361, 371 (type species: *Leuciscus cephalotaenia* Bleeker, 1852). Gender: feminine.

Diagnosis. Body elongated and moderately compressed; snout pointed, mouth terminal, directed obliquely upward; prominent symphyseal knob present on the lower jaw with a corresponding notch on the upper jaw that receives the knob; barbels absent; anal fin with 5−6½ branched rays.

Rasbora daniconius (Hamilton, 1822)
(Fig. 4.151)

FIGURE 4.151

Rasbora daniconius, 62.3 mm SL, Brahmaputra River, Goalpara, Assam.

Cyprinus daniconius Hamilton, 1822: 327, 391, pl. 15, fig. 89 (type locality: Southern Bengal, India).

Diagnosis. Body elongated and laterally compressed; a black stripe which extends from tip of snout to caudal-fin base; complete lateral line; presence of a prominent symphyseal knob on lower jaw which fits to corresponding notch on upper jaw; lips strongly developed; predorsal scales 14. Maximum size ∼110 mm SL.

Distribution. Ganga and Brahmaputra drainages; Bangladesh, Nepal, Bhutan and Chindwin drainage, Myanmar.

Rasbora ornata **Vishwanath & Laishram, 2005**
(Fig. 4.152)

FIGURE 4.152

Rasbora ornata, 58.4 mm SL, Chatrickong River, tributary of Yu River, Manipur.

Rasbora ornatus Vishwanath & Laisram, 2005: 429, fig. 1 (type locality: Lokchao River, a tributary of the Yu River, Chindwin drainage, Moreh, Manipur, India).

Diagnosis. Lateral line incomplete, 26−28 in series, 11−20 pored scales; transverse scales ½4/l/2; first dorsal fin ray without a fleshy sheath; pharyngeal teeth 2, 4, 5/5, 4, 2; gill rakers on first gill arch 4−5 + 11; dark brown or black stripe of uniform width from behind operculum to end of median caudal-fin rays. Maximum size ∼78 mm SL.

Distribution. Hill streams draining to Chindwin and Yu rivers, Manipur, India.

Rasbora rasbora **(Hamilton, 1822)**
(Fig. 4.153)

FIGURE 4.153

Rasbora rasbora, 32.2 mm SL, Barak River, Silchar, Assam.

Cyprinus rasbora Hamilton, 1822: 329, 391, pl. 2, fig. 90 (type locality: Bengal, India).

Diagnosis. A thin blackish mid-lateral stripe, prominent posteriorly; barbels absent; lateral line complete with 28−31 scales; dorsal and anal fins blackish; caudal fin with well-defined blackish distal border; predorsal scales 12. Maximum size ∼100 mm SL.

Distribution. Widely distributed in India, Bangladesh, and Pakistan and Myanmar.

Genus *Salmostoma* Swainson, 1839

Salmostoma Swainson, 1839: 184 (type species: *Cyprinus oblongus* Swainson, 1839). Gender: neuter.

Diagnosis. Body elongated and considerably compressed; abdomen keeled; snout pointed, mouth oblique; lower jaw longer with a well-developed symphyseal knob fitting into a well-developed notch on the upper jaw; dorsal fin short, placed considerably behind the middle of the body; pectoral fin long with axillary scale; scales moderately large and easily fall-out.

***Salmostoma bacaila* (Hamilton, 1822)**
(Fig. 4.154)

FIGURE 4.154

Salmostoma bacaila, 83.0 mm SL, Barak River, Silchar, Assam.

Cyprinus bacaila Hamilton, 1822: 265, 384 (type locality: freshwater rivers of Gangetic provinces).

Salmostoma bacaila: Ahmed et al., 2013: 286 (valid).

Diagnosis. Body elongated and compressed laterally; mouth oblique, its cleft slightly extends to anterior margin of orbit; a prominent symphyseal knob present on lower jaw is fitted to upper jaw notch; dorsal fin origin opposite to anal-fin origin, far behind tip of snout; pectoral fin longer than head length not reaching ventral-fin origin; pelvic fin short; caudal fin forked, lower lobe longer; scales moderately large; predorsal scales 45—47 scales; lateral line complete with 118—122 pored scales; in preservative: dorso-lateral side of body gray green, often silvery; fins hyaline. Maximum size ~145 mm SL.

Distribution. Ganga, Brahmaputra, and Meghna drainages. India: Nepal and Pakistan.

***Salmostoma phulo* (Hamilton, 1822)**
(Fig. 4.155)

FIGURE 4.155

Salmostoma phulo, 97.3 mm SL, Brahmaputra River, Guwahati, Assam.

Cyprinus phulo Hamilton, 1822: 262, 384 (type locality: northeastern Bengal).

Diagnosis. Lower jaw length about half of head length; gill rakers 13-6; dorsal-fin origin opposite anal-fin origin; scales small; lateral line scales 99-112; silvery body with a bright lateral stripe. Maximum size ~100 mm SL.

Distribution. Ganga and Brahmaputra drainage of India and Bangladesh.

Salmostoma sladoni (Day, 1870)
(Fig. 4.156)

FIGURE 4.156

Salmostoma sladoni, 65.0 mm SL. Chatrickong River, Chindwin drainage, Manipur.

Chela sladoni Day, 1870a: 622 (type locality: Irrawaddy River, as high as Mandalay, Myanmar).

Diagnosis. Body elongated and compressed, both dorsal and ventral profile equally arched. Abdomen keeled from below pectoral fin to vent; mouth extends from snout tip to obliquely downward to the region of anterior margin of eye, lower jaw longer with distinct symphyseal knob; gill rakers 15−16; dorsal fin origin opposite anal fin origin, pectoral fin long not reaching pelvic fin origin, its outer rays longer than the rest to give a pointed edge; lower lobe longer; lateral line complete with 65−68 scales; predorsal scales 44−45; caudal fin edge black. Maximum size ~100 mm SL.

Distribution. Chindwin River basin, Manipur, India and Myanmar.

Genus *Schizothorax* Heckel, 1838

Schizothorax Heckel, 1838: 11 (type species: *Schizothorax esocinus* Heckel, 1838). Gender: masculine.

Diagnosis. Body subcylindrical, slightly compressed, elongated; head blunt, fleshy, snout smooth and covered with pores or warts; mouth inferior, slightly arched, lips fleshy, continuous; lower jaw with hard horny cartilage; a labial plate on chin forming a sucker; pharyngeal teeth in three rows; dorsal fin origin opposite pelvic fin origin; its last simple ray osseous and serrated posteriorly; barbels two pairs; caudal fin forked; scales small.

? *Schizothorax chivae* Arunkumar & Moyon, 2016
(Fig. 4.157)

FIGURE 4.157

Schizothorax chivae, 156.0 mm SL, Leimakhong, tributary of Imphal River, Manipur.

Schizothorax chivae Arunkumar & Moyon, 2016: 66, fig. 1 (type locality: Chiva River, Khengjang Village, Chandel District, Manipur).

Diagnosis. Snout studded with pores, mouth inferior, lips well developed, papillate, lower lip not lobed, lower jaw with a hard horny covering inside, lateral line scales 83−90; predorsal scales 33−35; barbels long, 12.1%−16.6% HL; dorsal fin origin dorsal spine with 8−13 serrae posteriorly; dorsal fin origin in advance of pelvic origin, 17 scales between dorsal fin origin and lateral line. Maximum size ∼185 mm SL.

Distribution. Hills streams of Chindwin River drainage, Manipur.

Schizothorax progastus (**McClelland, 1839**)
(Fig. 4.158)

FIGURE 4.158

Schizothorax progastus 152.0 mm SL, Teesta River, Rangit, Sikkim.

Oreinus progastus McClelland, 1839: 274, 343, pl. 40, fig. 4.4 (type locality: Mountain streams at Simla, Upper Assam).

Diagnosis. Snout pointed and fleshy; lips thick, pendulous; abdomen prominent between pectoral and pelvic fins; mouth arched, upper jaw longer than the lower, both jaws covered with striated horny layer; lower labial fold uninterrupted but with a median lobe; dorsal spine with 20−25 serrae; lateral line scales 104−114; transverse scales 20−21/1/16−17. Maximum size ∼400 mm SL.

Distribution. Ganga and Brahmaputra drainages in the Himalayan foothills.

Schizothorax richardsonii (**Gray, 1832**)
(Fig. 4.159)

FIGURE 4.159

Schizothorax richardsonii, 166.0 mm SL, Teesta River, Rangit, Sikkim.

Cyprinus richardsonii Gray, 1832: pl. 94, fig. 2 (type locality: Nepal).

Diagnosis. Lateral line scales 85−110; mouth inferior, transverse, slightly arched; hard cartilaginous covering below lower jaw extends between corners of mouth; followed by fleshy and flat lower lip covered with raised papillae, forming a sucker; dorsal spine strong, serrated posteriorly, lower labial fold uninterrupted, no medial lobe. Maximum size ~450 mm SL.

Distribution. Himalayan foothills of India; Sikkim, Bhutan, Nepal, Pakistan, and Afghanistan.

Genus *Securicula* Günther, 1868

Securicula Günther, 1868: 332 (type species: *Cyprinus gora* Hamilton, 1822). Gender: feminine.

Diagnosis. Body strongly compressed and elongated; abdomen sharply keeled, not covered by scales; keel anteriorly supported by extension of pectoral girdle; a muscular mass covered by skin and scutes extends on dorsal face of head to nostrils; lateral line complete with about 120 scales.

Securicula gora **(Hamilton, 1822)**
(Fig. 4.160)

FIGURE 4.160

Securicula gora, 102.0 mm SL, Brahmaputra River, Guwahati, Assam.

Cyprinus gora Hamilton, 1822: 263, 384 (type locality; Brahmaputra River near Goalpara, India).

Diagnosis. As of genus; mouth oblique, almost vertical; a symphyseal knob on lower jaw and a corresponding notch on upper jaw. Scales very small, covering anteriorly upto head above the nostrils; lateral line scales 120−160, dorsal fin origin in advance of anal; anal fin with 13−15 branched rays. Maximum size ~180 mm SL.

Distribution. Northern India upto Orissa and Assam; Bangladesh, Nepal, Pakistan.

Genus *Semiplotus* Bleeker, 1860

Semiplotus Bleeker, 1860: 424 (type species: *Cyprinion semiplotus* McClelland, 1839). Gender: masculine.

Semiplotus: Banarescu & Herzig, 1995: 411 (status discussed, species in eastern Himalaya included under *Semiplotus* Bleker, 1860);

Semiplotus: Vishwanath & Kosygin, 2000b: 92 (revision).

Diagnosis. Body short and compressed, abdomen rounded; snout overhanging mouth, one short pair of maxillary barbels (Fig. 4.161A), mouth inferior, wide, called "sector mouth" characteristic in having exposed cornified mandibular cutting edge and a dentary with broad deflected labial surface (Fig. 4.161B), ascending process of premaxilla reduced, lower jaw with a mandibular symphyseal knob and a synarthrotic dentary joint; dorsal fin with 19½−22½ branched rays, last unbranched ray osseous, smooth or serrated posteriorly origin opposite the middle of pectoral and pelvic fin origins, its base long; branched anal fin rays 6½−8½; lateral line complete with 32−26 scales.

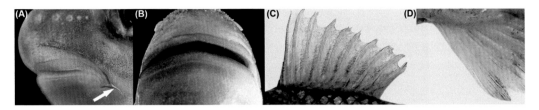

FIGURE 4.161

Figure showing (A) maxillary barbel in *Semiplotus semiplotus*; (B) sector mouth of *Semiplotus*, (C) dorsal fin, and (D) anal fin of *Cyprinion muscatense*.

Photo courtesy: (c, d) L.A. Jawad.

Semiplotus cirrhosus **Chaudhuri, 1919**
(Fig. 4.162)

FIGURE 4.162

Semiplotus cirrhosus, 146.0 mm SL, Tizu River, Jessami, Manipur.

Semiplotus cirrhosus, Chaudhuri, 1919: 280, pl. 22, figs. 3−3a. (type locality: Mountain stream, Putao Plains, Upper Myanmar).

Diagnosis. Body deep, its depth 17.3%−22.1% SL; dorsal fin long, lateral line complete with 32−36 scales, with 20−23 branched rays, last unbranched dorsal fin ray smooth, horny tubercles or open pores on the snout distributed often in two parallel transverse rows, six to eight in each row on the sides. Maximum size ∼185 mm SL.

Distribution. Chindwin drainage in Manipur, India and Myanmar.

Semiplotus modestus **Day, 1870**
(Fig. 4.163)
Semiplotus modestus Day, 1870c: 101 (type locality: hill ranges of Sittwe, Myanmar) (Fig. 4.164).

Diagnosis. Dorsal fin with 20−21 branched rays, its last unbranched ray osseous and serrated (Fig. 4.174) posteriorly with 16−25 serrae; open pores or tubercles on snout tip, often in two transverse rows, five to seven in the upper and three to five in the lower on each side; body immediately behind the operculum from upper angle to lower angle stained with a thin posteriorly curved black bar. Maximum size ∼160 mm SL.

Distribution. Kaladan River in Mizoram, India and Myanmar.

FIGURE 4.163

Semiplotus modestus, 125.5 mm SL, Kaladan River, Kolchaw, Mizoram.

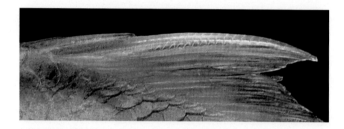

FIGURE 4.164

Dorsal fin of *Semiplotus modestus* showing serrated dorsal spine.

Semiplotus semiplotus (McClelland, 1839)
(Fig. 4.165)

FIGURE 4.165

Semiplotus semiplotus, 146.0 mm SL, Rangit River, Rangit, Sikkim.

Cyprinus semiplotus McClelland, 1839: 346 (type locality: Brahmaputra drainage, Assam).

Diagnosis. Dorsal fin with 23−25 branched rays, last unbranched ray osseous and smooth; snout tip with a transverse row of 10−12 tubercles or open, five to six on each side; lateral line scales 27−33; transverse scales 7/1/4; predorsal scales 11−12; Maximum size ∼400 mm SL.

Distribution. Brahmaputra, Ganga and Teesta drainage in the Himalayan foothills.

Genus *Systomus* McClelland, 1838

Systomus McClelland, 1838: 943 (type species: *Systomus immaculatus* McClelland, 1839). Gender: masculine.

Diagnosis. Medium sized fish, adult size greater than 80 mm SL; maxillary and rostral barbels present; dorsal fin with four simple rays and 8½ branched rays, the last simple ray osseous and strongly serrated; lateral line complete with 27−34 scales; no antrorse predorsal spinous ray; postepiphysial fontanel absent; supraneurals five; infraorbital three, slender; 14−15 abdominal and 17−19 caudal vertebrae; free uroneural present; no bars or blotches on the body.

Systomus sarana (Hamilton, 1822)
(Fig. 4.166)

FIGURE 4.166

Systomus sarana, 133.0 mm SL, Brahmaputra River, Guwahati, Assam.

Cyprinus sarana Hamilton, 1822: 307, 388 (type locality: ponds and rivers of Bengal).

Diagnosis. Barbels two pairs; last simple ray of dorsal fin osseous and finely serrated; lateral line complete with 31−33 scales; 5½ scales between dorsal fin origin and lateral line and 3½ between it and pelvic fin origin; predorsal scales 10−11. Maximum size ∼30 mm SL.

Distribution. Ganga drainage: India, Nepal, and Bangladesh; Brahmaputra drainage: Bhutan, and northeastern India; Pakistan; Myanmar?

Genus *Tariqilabeo* Mirza & Saboohi, 1990

Tariqilabeo Mirza & Saboohi, 1990: 405 (type species: *Labeo macmahoni* Zugmayer). Gender: masculine.

Diagnosis. Body subcylindrical; head small, a roughly pointed snout with a transverse groove and lateral lobes; mouth inferior, lips well developed; lower lip modified into characteristic papillate labial plate; lower jaw with a sharp edge; mental groove long, gap between grooves oval or round; barbels four; dorsal fin with eight branched rays, its origin midway between tip of snout and anal-fin base; anal fin with five branched rays; pectoral with 17 and pelvic with 10 rays; lateral line complete with 35−36 scales; pharyngeal teeth 5, 4, 2/2, 4, 5; caudal fin forked.

Tariqilabeo burmanicus (Hora, 1936)
(Fig. 4.167)

FIGURE 4.167

Tariqilabeo burmanicus, 131.2 mm SL, confluence of Challou and Tizu rivers, Jessami, Manipur.

Crossochilus latius burmanicus Hora, 1936a: 319, 324 (type locality: Chindwin basin in NE India and Myanmar).

Diagnosis. Lateral line complete with 37−38 scales; transverse scales rows between lateral line and dorsal fin origin 5½, between lateral line and pelvic fin origin 4½, between lateral line and anal fin origin 4½; dorsal fin with 8½ branched rays and anal fin with 5½; posteriorly situated anus (34.99%−48.57% of pelvic-anal distance); absence of rostral flap or rostral lobe; presence of thick mid lateral stripe, a black crescentic spot on the fourth to fifth scales of lateral line and a spot at the base of the caudal peduncle. The dark brown broad lateral stripe is found to be very distinct in the smaller specimens, which generally fade with greater length. Maximum size ∼170 mm SL.

Distribution. Chindwin drainage, Manipur and Nagaland.

Tariqilabeo latius (Hamilton, 1822)
(Fig. 4.168)

FIGURE 4.168

Tariqilabeo latius, 74.2 mm SL, Kaldiya River, Jalahghat, Brahmaputra drainage, Assam.

Cyprinus latius Hamilton, 1822: 345, 393 (type locality: Teesta R. near Kalimpong Duars and Silliguri Terai, West Bengal, India).

Diagnosis. Lateral line complete with 40−42 scales; transverse scales rows between lateral line and dorsal fin origin 5½, between lateral line and anal fin origin 5½, dorsal fin with 8½ branched rays and anal fin with 5½; pectoral fin with one simple and 13 branched fin rays; predorsal with 11 scales; a black crescentic spot on the fourth to fifth scales of lateral line; lower lobe of the caudal fin darker than the upper lobe. Maximum size ∼170 mm SL.

Distribution. Ganges, Brahmaputra and Barak-Meghna, Nepal, Bhutan, and Bangladesh.

Genus *Tor* Gray, 1834

Tor Gray, 1834: 96 (type species: *Cyprinus tor* Hamilton, 1822). Gender: masculine.

Diagnosis. Body elongated, moderately compressed; tapers toward caudal peduncle; mouth inferior, lips fleshy, continuous at the angle of mouth; lower lip with a flesh lobe or at least with two notches delimiting usual position of the lobe; post labial groove not interrupted; upper jaw slightly protractile; dorsal fin origin opposite pelvic fin origin, its last simple ray osseous and not serrated posteriorly; scales large; lateral line complete.

***Tor mosal* (Hamilton, 1822)**
(Fig. 4.169)

FIGURE 4.169

Tor mosal, 226.0 mm SL, Barak River, Silchar, Assam.

Cyprinus mosal Hamilton, 1822: 306, 388 (type locality: Nathpur, Kosi River, upper Ganga drainage, Bihar, India).

Diagnosis. Head longer than body depth; lateral line scales 24−26; gill rakers 13 on lower arm of first arch; eye diameter 18.3%−21.9% SL; interorbital space 30.1%−37.5% SL. Maximum size ∼1500 mm SL.

Distribution. Ganga, Brahmaputra, and Surma-Meghna drainage.

***Tor putitora* (Hamilton, 1822)**
(Fig. 4.170)

FIGURE 4.170

Tor putitora, 266.5 mm SL, Barak River, Talenglong, Manipur.

Cyprinus putitora Hamilton, 1822: 303, 388 (type locality: Eastern parts of Bengal).
Barbus progeneius McClelland, 1839: 270 (great rivers in the plains of India).

Diagnosis. Body elongated, compressed; head long, its length longer than body depth; head length 29.5%−32.4% SL; body depth 12.1%−15.6% SL; lateral line scales 29−30; gill rakers 12−13. Body yellowish golden, fins reddish. Maximum size ∼1500 mm SL.

Remarks: Ganga, Brahmaputra, and Surma-Meghna drainages. North and northeastern India, Nepal, Bangladesh.

Tor tor **(Hamilton, 1822)**
(Fig. 4.171)

FIGURE 4.171

Tor tor, 195.0 mm SL, Barak River, Tamenglong District, Manipur. This photograph was wrongly inserted as *Tor progeneius* in Vishwanath et al. (2007).

Cyprinus tor Hamilton, 1822: 305, 338 (type locality: Mahananda = Mahananda River, India).

Diagnosis. Body fairly deep, head length 26.8%−28.8% SL, equals or less than body depth 26.9%−30.6% SL; scales large; mouth inferior, lips fleshy with uninterrupted labial fold; snout smooth; fins reddish yellow; golden and greenish body, silvery below; barbels two pairs, maxillary longer than eye, rostral shorter; dorsal fin with three simple and eight branched rays, last simple ray osseous and smooth; anal fin with three simple and five branched rays; pectoral fin with 18 rays; axillary pelvic scale present; lateral line scales 26; 4−4.5 above and 2.5−3 below; circumferential scales 20, circumpeduncular scales 12. Maximum size ~1500 mm SL.

Distribution. Ganga Brahmaputra, Barak−Surma−Meghna Northern India, Nepal, Bangladesh, northern eastern India.

Tor yingjiangensis **Chen & Yang, 2004**
(Fig. 4.172)

FIGURE 4.172

Tor yingjiangensis, 165.0 mm SL, Tizu River, Jessami, Manipur, Chindwin drainage.

Tor yingjiangensis Chen & Yang, 2004: 186 (type locality: Yingjiang River, Manyun town, Yunnan, China).

Tor yingjiangensis: Qin et al., 2017: 301 (record from Kachin State, Irrawaddy drainage, Myanmar).

Diagnosis. Lateral line scales 24−26; lateral transverse scales 4/1/3; tubercles absent on snout and sides of head; gill rakers 18−20; mouth terminal, postlabial groove continuous, median lobe of lower lip short, its posterior margin truncated; head longer than body depth; rostral and maxillary barbels linger than eye diameter; no color stripe on body. Maximum size ∼180 mm SL.

Distribution. Irrawaddy drainage, Yunnan, China and Chindwin and Irrawaddy drainages, Manipur, India and Myanmar.

Family Psilorhynchidae
Mountain carps

Small sized fishes, not more than 80−90 mm SL; body spindle-shaped, depressed; dorsal profile arched, ventral surface flattened anteriorly; mouth inferior; snout rounded, ventral surface bordered by a deep longitudinal groove on each side; rostral cap dome-shaped, fused with upper lip, often separated by a deep groove; lower lip not continuous with the upper, lower jaw cushion, a flap-like structure composed of two adnate tissue layers covering the lower jaw; globular papillae often present on the cushion; papilliferous skin folds posterolateral to the mouth may or may not be present; jaws with sharp horny edges; barbels absent; gill openings narrow; ventral surface of head flattened; number of unbranched pectoral fin rays at least four; air bladder greatly reduced; paired fins enlarged and horizontal, modified to form adhesive apparatus for burrowing purposes; lateral line complete, with 31−50 scales; pharyngeal bone with one row of four teeth; inhabit freshwater mountain streams (Fig. 4.173).

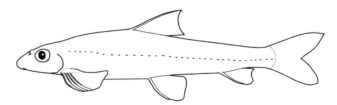

FIGURE 4.173

Outline diagram of Psilorhynchidae.

Genus *Psilorhynchus* McClelland, 1838

Psilorhynchus McClelland, 1838: 944 (type species: *Cyprinus sucatio* Hamilton, 1822). Gender: masculine.

Diagnosis. Small sized fishes, cylindrical body, depressed anteriorly, compressed posteriorly, flattened ventrally; mouth inferior, small, transverse with a projecting snout; barbels absent; paired fins horizontal; simple pectoral fin rays 4−11; chest naked; scales small to moderate; lateral line scales 40−43; pharyngeal teeth uniserial; swim bladder not enclosed in bony capsule; urohyal bone U-shaped (Fig. 4.174).

Psilorhynchus amplicephalus **Arunachalam, Muralidharan & Sivakumar, 2007**
(Fig. 4.175)

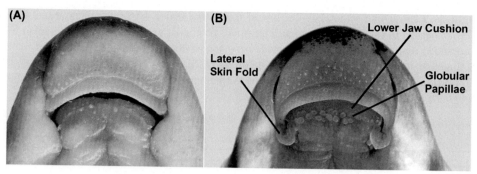

FIGURE 4.174

(A) Lips of *Psilorhynchus rowleyi* showing absence of skin fold; (B) Lip of *Psilorhynchus chakpiensis* showing strongly developed skin folds, lower cushion and globular papillae.

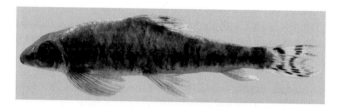

FIGURE 4.175

Psilorhynchus amplicephalus, 61.3 mm SL, Barak River, Silchar, Assam.

Psilorhynchus amplicephalus Arunachalam, Muralidharan & Shivakumar, 2007: 1352, Fig. 1a−b (type locality: Balishwar River of Barak River basin at Malidor Village, Silchar, Assam, India).

Diagnosis. Body deep with six to seven dark brown saddles extending to sides below lateral line and three wavy dark brown bars of similar on caudal fin; dark brownish spots in predorsal region; tubercles present on head along internasal and cheek region; lateral skin fold on mouth; pectoral fin with five unbranched rays; lateral line scales 35−36; abdomen unscaled; papillated posterolateral skin fold present on posterolateral region of mouth. Maximum size ∼62 mm SL.

Distribution. Barak River basin, India.

Psilorhynchus arunachalensis **(Nebeshwar, Bagra & Das, 2007)**
(Fig. 4.176)

FIGURE 4.176

Psilorhynchus arunachalensis, 76.0 mm SL, Dirang River, West Kameng District, Arunachal Pradesh.

Psilorhynchoides arunachalensis Nebeshwar, Bagra & Das, 2007: 1632, pl. 1, figs. 2, 3b, 4b, 5b (West Kameng District, Dirang River at Dirang, Brahmaputra River system, Arunachal Pradesh, India).

Diagnosis. Body elongated with 42−44 lateral line scales; unbranched pectoral fin rays 8−9; postero-lateral skin fold behind the mouth absent; 6−12 dark brown blotches along the lateral line, extending from the region between pectoral and pelvic fin or the middle of dorsal-fin base to caudal fin. Maximum size ~ 100 mm SL.

Distribution. Tributaries of the Brahmaputra in Kameng District, Arunachal Pradesh, India:

Psilorhynchus balitora (Hamilton, 1822)
(Fig. 4.177)

FIGURE 4.177

Psilorhynchus balitora, 53.2 mm SL, Iyei River at Noney, Noney, Manipur, Barak drainage, India.

Cyprinus balitora Hamilton, 1822: 348 (type locality: Northeastern Bengal, India).

Diagnosis. Body short and deep; dorsal fin with eight branched rays; poorly developed posterolateral skin fold around mouth; lateral line scales 32−34; simple pectoral fin rays 6−7; circumferential scales 18; air bladder naked with anterior and posterior chamber. Maximum size ~ 70 mm SL.

Distribution. Northern Bengal, Assam, Ganga-Brahmaputra drainage.

Psilorhynchus bichomensis Shangningam, Kosygin & Gopi, 2019
(Fig. 4.178)

FIGURE 4.178

Psilorhynchus bichomensis, 43 mm SL, holotype, Bichom River, Arunachal Pradesh.

Photo courtesy: B.D. Shangningam.

Psilorhynchus bichomensis Shangningam, Kosygin & Gopi, 2019a: 131 (type locality: Bichom River at Bana (Brahmaputra basin), East Kameng District, Arunachal Pradesh, India.

Diagnosis. Body elongated; lateral line complete with 46 + 2 scales; four scales between dorsal fin origin and lateral line scale three between lateral line scale and pelvic fin origin; predorsal scales 13—14; caudal fin unbranched pectoral-fin rays 9—10; head greatly depressed, covered with fine tubercles; postero-lateral skin fold at the corner of mouth developed; circumpeduncular scales eight. Maximum size ~55 mm SL.

Distribution. Bichom River, tributary of Kameng River, Arunachal Pradesh.

Psilorhynchus chakpiensis **Shangningam & Vishwanath, 2013**
(Fig. 4.179)

FIGURE 4.179

Psilorhynchus chakpiensis, 48.5 mm SL, Chakpi River, Chandel District, Manipur.

Psilorhynchus chakpiensis Shangningam & Vishwanath, 2013a: 383, figs. 1, 2a, 3—4 (type locality: Chakpi River at Chakpikarong, Chandel District, Manipur, India).

Diagnosis. Lateral skin fold on mouth well developed, globular papillae on lower jaw cushion large; unbranched pectoral-fin rays five; caudal peduncle 17.8%—20.5% SL; lateral line scales 30—31; a caudal-fin color pattern consisting of two black bars, one incomplete bar near the base of the upper lobe, and a complete bar across the center of the fin, traversing from the upper to the lower margin of the fin. Maximum size ~50 mm SL.

Distribution. Chakpi River at Chakpikarong, Chandel District, Manipur, India.

Psilorhynchus hamiltoni **Conway, Dittmer, Jezisek & Ng, 2013**
(Fig. 4.180)

FIGURE 4.180

Psilorhynchus hamiltoni, 46.0 mm SL, Rangit River, Sikkim.

Psilorhynchus hamiltoni Conway, Dittmer, Jezisek & Ng, 2013: 232, figs. 23—24 (type locality: Tista River at Tista Barrage, West Bengal, India).

Diagnosis. Lateral stripe well developed; 7—11 round to squarish lateral blotches; six to seven saddles and without contact with lateral blotches; 34—35 lateral line scales, five to six unbranched pectoral fin rays; ventral surface between paired fins with a broad rectangular scale-less patch. Maximum size ~50 mm SL.

Distribution. Teesta River at Teesta Barrage, India.

Psilorhynchus homaloptera (Hora & Mukerji, 1935)
(Fig. 4.181)

FIGURE 4.181

(A) *Psilorhynchus homaloptera*, 44.0 mm SL. Leimatak River; Tamenglong District, Manipur; (B) 83.5 mm SL. Umtrao River at Norbong, Byrnihat, Meghalaya.

Psilorhynchus homaloptera Hora & Mukerji, 1935: 391, pl. 7, figs. 1−6 (type locality: Emilomi, Keleki stream, Brahmaputra drainage system, Nagaland Assam, India).

Psilorhynchus homaloptera: Shangningam and Vishwanath, 2014a: 239 (redescription).

Diagnosis. Body greatly depressed and flattened, depth 13.0%−17.0% SL; eye small, diameter 48%−61% HL; lateral skin fold absent at the corners of mouth, globular papillae weakly developed or absent on lower-jaw cushion; abdominal region naked except one scale anterior to the anus; undivided pectoral fin rays eight; lateral line scales 37−40; transverse scale rows 3½/1/2½; scale rows around caudal peduncle 10, predorsal scales 12−14, scales between anus and anal-fin origin 11−12, and total vertebrae 41. Maximum size ∼60 mm SL.

Distribution. Tributaries of the Brahmaputra and the Surma-Meghna River drainages in northeastern India.

Psilorhynchus kaladanensis Lalramliana, Lalnuntluanga & Lalronunga, 2015
(Fig. 4.182)

FIGURE 4.182

Psilorhynchus kaladanensis, 42.5 mm SL, Tuisi River, Kaladan drainage, Mizoram.

Photo courtesy: Lalramliana.

Psilorhynchus kaladanensis Lalramliana, Lalnuntluana & Lalroununga, 2015a: 173, figs. 1–2 (type locality: Tuisi River, a tributary of Kaladan River in the vicinity of Khopai village, Mizoram, India).

Diagnosis. Lateral skin fold well developed at the corners of mouth; greatly reduced anterior-most branchiostegal ray and a postepiphyseal fontanel smaller than preepiphyseal fontanel; caudal fin with dark brown triangular spot at mid-base, slightly elongated dark mark near base of lower lobe, indistinct v-shaped bar across center, dark brown oblique bar across fin anterior to center; five to six unbranched pectoral fin rays; 30–32 lateral line scales; 32–33 total vertebrae; absence of scales on mid-ventral region between pectoral fins. Maximum size ~50 mm SL.

Distribution. Tuisi River, Kaladan River drainage, Mizoram, India.

Psilorhynchus khopai **Lalramliana, Solo, Lalronunga & Lalnuntluanga, 2014**
(Fig. 4.183)

FIGURE 4.183

Psilorhynchus khopai, 80.2 mm SL, Tuisi River, Kaladan drainage, Mizoram.

© *Lalramliana.*

Psilorhynchus khopai Lalramliana, Solo, Lalronunga & Lalnuntluanga, 2014c: 266, figs. 1–2 (type locality: Tuisi River, a tributary of Kaladan River in the vicinity of Khopai village, Saiha District, Mizoram, India).

Diagnosis. Skin fold in the form of a thin strip at the corner of mouth; body with 9–12 dark brown mid-lateral spots arranged in a row appearing as a lateral stripe; mid-dorsal black spots from behind dorsal fin to caudal fin-base 4–5; unbranched pectoral fin rays 7–9; dorsal and anal fin rays with three unbranched rays each; lateral-line scale rows 39–41. Maximum size ~84 mm SL.

Distribution. Tuisi River, a tributary of the Kaladan in Mizoram, India.

Psilorhynchus konemi **Shangningam & Vishwanath, 2016**
(Fig. 4.184)

FIGURE 4.184

Psilorhynchus konemi, 59.4 mm SL, holotype, Chakpi River at Dujang, Manipur.

Psilorhynchus konemi Shangningam & Vishwanath, 2016: 290, figs. 1−2, (type locality: Chakpi River at Dujang, Chandel District, Manipur, India).

Diagnosis. Deeply arched dorsal profile in front of dorsal fin origin; dorsal-fin origin opposite pelvic-fin origin; unbranched pectoral-fin rays 7−8, skin fold at the corner of mouth absent or weakly developed; with prominent dark brown coloration, the rest pale yellow; principal caudal-fin rays 9 + 8, caudal fin peppered with small dark melanophores at base and an incomplete black v-shaped bar in center of the fin; lateral line scales 39−40; ventral surface between paired fins without scales; body dark olive dorsally and laterally. Maximum size ∼65 mm SL.

Distribution. Chakpi River, Manipur, Chindwin drainage.

Psilorhynchus maculatus **Shangningam & Vishwanath, 2013**
(Fig. 4.185)

FIGURE 4.185

Psilorhynchus maculatus, 61.2 mm SL, holotype, Challou River at Poi Village, Manipur.

Psilorhynchus maculatus Shangningam & Vishwanath, 2013b: 58, figs. 1−2 (type locality: Challou River at Poi Village, Ukhrul District, Manipur, India).

Diagnosis. Body with five to six black blotches on the side behind the insertion of the posterior-most dorsal-fin ray, anterior blotches obscured by deeper background pigmentation; skin fold at the corner of mouth thin; caudal-fin pigmentation pattern consisting of small black blotches on both the upper and lower fin lobes; two rows of black spots in the middle of the dorsal fin, spots only on the branched rays; 35−36 scales on the lateral line; five unbranched pectoral-fin rays; and 36 total vertebrae. Maximum size ∼61 mm SL.

Distribution. Challou River at Poi Village, Ukhrul District, Manipur, Chindwin River drainage, India.

Psilorhynchus microphthalmus **Vishwanath & Manojkumar, 1995**
(Fig. 4.186)

FIGURE 4.186

Psilorhynchus microphthalmus, 60.4 mm SL, Chakpi River at Mombi, Manipur.

Psilorhynchus microphthalmus Vishwanath and Manojkumar, 1995: 249, fig. 1a−d (type locality: Chakpi stream, tributary to Manipur River, at Mombi, 85 km south of Imphal, Manipur, India).

Diagnosis. Eye diameter 22%−24% HL, lateral skin fold absent at the corners of mouth, globular papillae weakly developed or absent on lower-jaw cushion; 39−40 lateral line scales, seven unbranched pectoral fin rays, four black ocelli like marks on occiput and several dark spots on lateral line and mid dorsal line from dorsal fin origin to caudal peduncle. Maximum size ∼60 mm SL.

Distribution. Chakpi stream, a tributary of Manipur River, Chindwin drainage, Manipur, India.

Psilorhynchus nahlongthai Dey, Choudhury, Mazumdar, Thaosen & Sarma, 2020
(Fig. 4.187)

FIGURE 4.187

Psilorhynchus nahlongthai, 54.0 mm SL, Diyung River, Assam.

Psilorhynchus nahlongthai Dey, Choudhury, Mazumdar, Thaosen & Sarma, 2020: 643 [type locality: Diyung River, a tributary of Kopili River (Brahmaputra drainage) near Dehangi, Dima Hasao District, Assam].

Diagnosis. Prominent tubercles densely arranged on head, snout, and rostral cap; lateral skin fold at the corner of mouth well developed; large globular papillae on lower jaw cushion; seven to eight large dark brown blotches of three to four scales broad on the lateral side, contagious to form a sort of a broad stripe, five to six broad saddles, one broad bar in the middle of dorsal fin and two v-shaped bars on the caudal fin, a broad bar interrupted in the middle just behind caudal-fin base; total vertebrae 34 (24 + 10); pectoral fin with five simple rays. Maximum size ∼58 mm SL.

Distribution. Diyung River, tributary of Kopili River, Brahmaputra drainage, Dima Hasao District, Assam.

Psilorhynchus nepalensis Conway & Mayden, 2008
Psilorhynchus nepalensis Conway & Mayden, 2008b: 224, fig. 7 (type locality: Seti River at Khairenitar, Khairentar, Tanahun, Nepal).

Diagnosis. Dorsal-fin rays ii−iii, 9; anal fin ii, 5−6; pelvic fin ii 6−7; pectoral fin v−vi, 10−11; lateral line scales 31−33; transverse scale rows 3½/1/2; predorsal scales 9−11; caudal fin emarginated, upper lobe longer; scales absent from ventral mid-region between pectoral fins. Maximum size ∼50 mm SL.

Distribution. Seti River, Tanahun, Nepal.

Psilorhynchus ngathanu Shangningam & Vishwanath, 2014
(Fig. 4.188)

FIGURE 4.188

Psilorhynchus ngathanu, 59.0 mm SL, holotype, Dutah River at Larong Village, Manipur.

Psilorhynchus ngathanu Shangningam & Vishwanath, 2014b: 28, figs. 1−2 (type locality: Dutah River at Larong village, Chandel District, Chindwin Basin, Manipur, India).

Diagnosis. Papillated skin fold postero-lateral to mouth, absence of scales in mid-ventral region between pectoral fins; presence of two rows of spots on dorsal-fin rays, two black bars on caudal fin, one complete in middle and another incomplete just posterior to caudal-fin base; squarish to rectangular black patches on lateral line, anterior patches obscured by the background pigmentation, patches decrease in size posteriorly; five to six unbranched pectoral-fin rays; and 32−33 lateral-line scales. Maximum size ∼60 mm SL.

Distribution. India: Dutah River, Chindwin drainage, Chandel District, Manipur.

Psilorhynchus nudithoracicus Tilak & Husain, 1980
(Fig. 4.189)

FIGURE 4.189

Psilorhynchus nudithoracicus, 79.0 mm SL, Iyei River, Barak drainage, Manipur.

Psilorhynchus sucatio nudithoracicus Tilak & Husain, 1980: 35, figs. 1−3 (type locality: Bamrauli Canal, near village Bilsanda, Pilibhit District, western Uttar Pradesh, India).

Diagnosis. Large and flap-like skin folds at the lateral corner of the mouth, which extend further posteriorly than the lower jaw cushion and are covered in large "head of cauliflower" papillae, two saddles anterior to that at dorsal-fin origin, irregular dark brown or black markings on dorsal fin, anterior-most lateral blotch equal in size or smaller than more posterior blotches, unbranched pectoral fin ray 4−5 and lateral line scales 32−34. Maximum size ∼85 mm SL.

Distribution. Ganga and Brahmaputra drainages, India; Barak-Meghna drainages in India and Bangladesh.

Psilorhynchus pseudecheneis **Menon & Datta, 1964**

Psilorhynchus pseudecheneis Menon & Datta, 1964: 253, pl. 16, figs. 1−3 [type locality: Dudh Kosi (Koshi) drainage, north of village ghat, Solukhumbu District, Eastern Nepal].

Diagnosis. Unbranched pectoral-fin rays 9−11; lateral line scales 46−48; absence of scales from the dorsal midline between occiput and mid-point between occiput and dorsal-fin origin; presence of highly modified cycloid scales along the ventral surface of the body, forming a series of large flap-like structures between the origins of the pectoral and pelvic fins. Maximum size ∼110 mm SL.

Distribution. Kosi River drainage, eastern Nepal.

Psilorhynchus rowleyi **(Hora & Misra, 1941)**
(Fig. 4.190)

FIGURE 4.190

Psilorhynchus rowleyi, (A) 60.2 mm SL, Chakpi River, Dujang, Manipur; (B) 75.5 mm SL, lectotype. Kora, Myanmar.

Psilorhynchus homaloptera var. *rowleyi* Hora & Misra, 1941: 481, pl. 1, figs. 1−2 (type locality: Upper Chindwin River at Kora, Myanmar).

Psilorhynchus rowleyi: Shangningam et al., 2013: 250, figs. 1, 2 (redescription, lectotype designated).

Diagnosis. Unbranched pectoral-fin rays seven, lateral-line scales 39−42, no papillated postero-lateral skin fold behind mouth; gill rakers absent; dark brown pigments arranged mid-laterally along lateral-line scales, more distinct posteriorly toward caudal-fin base; Pectoral and pelvic-fin bases light yellowish, lined with small dark melanophores; dense cluster of melanophores on fin rays; caudal fin with rows of brown spots forming a faint bar in middle and a deeper bar submarginally at posterior margin. Maximum size ∼76 mm SL.

Distribution. Chakpi River, Manipur, India; Uye River, Kora, Myanmar, Chindwin River drainage.

Psilorhynchus sucatio **(Hamilton, 1822)**
(Fig. 4.191)

Cyprinus sucatio Hamilton, 1822: 347, 393 (type locality: Northern Bengal, India).

Diagnosis. Depressed and pointed snout when viewed from the side; presence of anteriorly directed radii over the anterior field of body scales; rostral cap separated from the upper lip around

FIGURE 4.191

Psilorhynchus sucatio, 55.4 mm SL, stream at Chirang, Brahmaputra drainage, Assam.

the corner of the mouth, the posterior margin of the rostral cap triangular; a complete covering of scales on its ventral surface between paired fins; absence of supraorbital bone; lateral line scales 33−36 and unbranched pectoral fin rays 4−5. Maximum size ∼65 mm SL.

    Distribution. Ganga River drainage in Bangladesh, India and Nepal, Brahmaputra River drainage in Bangladesh and India and Meghna River drainages in Bangladesh.

Family Botiidae
Botiid loaches

Small sized fishes, usually not more than 140.0 mm SL; body compressed; lateral line inconspicuous; barbels in two pairs; suborbital spine; body with oblique barks; popular aquarium fish (Fig. 4.192).

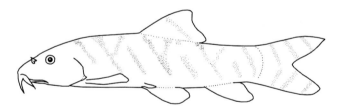

FIGURE 4.192

Outline diagram of Botiidae.

    Key to genera:

    1a. Mental lobe developed into a barbel .... *Botia*
    1b. Mental lobe not developed into a barbel .... *Syncrossus*

Genus *Botia* Gray, 1831

    Botia Gray, 1831: 8 (type species: *Botia almorhae* Gray, 1831). Gender: feminine.

    Diagnosis. Body short, compressed, moderately deep, abdomen rounded; mental lobe developed as a barbel; anterior process of premaxilla entire and not enclosing a cavity; rostral process long with a more or less distinct ridge along inner edge; optic foramen small; fronto-parietal fontanel narrow; a bifid erectile suborbital spine below or in front of eyes; barbels 6—8, four rostral barbels, united at their bases; dorsal fin origin opposite to that of pelvic fin; lateral line present; scales absent on head; air bladder partially enclosed in capsule.

Botia dario (Hamilton, 1822)
(Fig. 4.193)

FIGURE 4.193

Botia dario, (A) 80.0 mm SL, Jiri River, Manipur-Assam Border; (B) 64.0 mm SL, Jiri River, both Barak drainage, Manipur, India.

Cobitis dario Hamilton, 1822: 354, 394. (type locality: Bengal).

Diagnosis. A species distinct in having seven oblique bars extending as saddles, each bar appears forked with two limbs in some, or bars solid, connecting with adjacent bars below in some; caudal fin with two distinct bars and dorsal with one faint bar in the middle. Maximum size ~130 mm SL.

Distribution. Ganga-Brahmaputra drainage, north and northeastern India and Bangladesh.

Botia histrionica Blyth, 1860
(Fig. 4.194)

FIGURE 4.194

Botia histrionica, 103.0 mm SL Chatrikong River, Chindwin drainage, Manipur, India.

Botia histrionica Blyth, 1860: 166 (type locality: Pegu, Tenasserim, Myanmar).

Diagnosis. Characteristic in having five blue black bars continuing as saddles on body against yellowish background and another on head through eyes, one bar in the middle of the dorsal fin and another two across the caudal; bars may bifurcate on the sides and rejoin below leaving a round or oval pale spot in between. Maximum size ∼130 mm SL.

Distribution. Chindwin-Irrawaddy drainage in Manipur, India and Myanmar.

Botia lohachata **Chaudhuri, 1912**
(Fig. 4.195)

FIGURE 4.195

Botia lohachata, (A) 56.3 mm SL. Brahmaputra River at Goalpara, Assam, India; (B) Chaudhuri's (1912), Pl. 40, fig. 2 reproduced.

Botia lohachata Chaudhuri, 1912: 441 (type locality: Gandak River in Saran, Bihar).

Diagnosis. Lateral complete, straight; triangular black mark over snout, a round black blotch in the interorbital space which extend as saddle upto the eye and beyond; a saddle on the occiput and four on the body, often forming loops, one halfway between the occiput and the dorsal fin origin, another from little in front of dorsal fin origin to two-third of dorsal-fin base, another just behind the dorsal-fin base and the last one at the caudal peduncle. The saddles descend to sides as oblique bars, bars in between the bars continued from saddles. The saddles tend to break up into two in larger specimens. Maximum size ∼130 mm SL.

Distribution. Ganga River drainage.

Botia rostrata **Günther, 1868**
(Fig. 4.196)

Botia rostrata Günther, 1868: 367 (type locality: Bengal, India).

Diagnosis. Body with bars of irregular pattern which often forms rings or blotches; snout long, longer than postorbital distance; four pairs of barbels, mental lobe forming a barbel. Maximum size ∼120 mm SL.

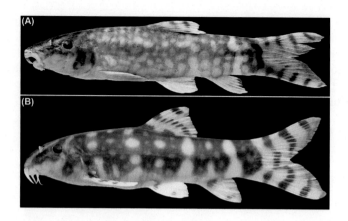

FIGURE 4.196

Botia rostrata, (A) 57.4 mm SL. Dikrong River, Arunachal Pradesh, Brahmaputra drainage; (B) 62.7 mm SL, Brahmaputra River, Goalpara, Assam, India.

Distribution. Ganga-Brahmaputra basin in Assam, West Bengal, Meghalaya, and Bangladesh.

Genus *Syncrossus* Blyth, 1860

Syncrossus Blyth, 1860: 166 (type species: *Syncrossus berdmorei* Blyth, 1860). Gender: masculine.

Diagnosis. Body short, compressed, moderately deep; abdomen rounded; mental lobe with a pair of fleshy papillae at its anterior edge, not developed in a barbel; premaxilla with sickle shaped anterior process, surrounding a cavity between left and right processes, rostral process short without ridge along interior edge; frontoparietal fontanel large, wide; anterior chamber of air bladder partly covered by bony capsule; posterior chamber large; top of supraethmoid narrow; optic foramen large. Suborbital spine not strongly curved backward, bifid. Head naked.

Syncrossus berdmorei (Blyth, 1860)
(Fig. 4.197)

FIGURE 4.197

Syncrossus berdmorei, 74.7 mm SL. Manipur River at Yangkoilok, near Myanmar border, Manipur.

Syncrossus berdmorei Blyth, 1860: 166 (type locality: Tenasserim Provinces, Myanmar).

Diagnosis. Body with about 12–20 rows of horizontal oval spots against the yellowish background on the sides. There are 11–12 oblique blue black bars extending from back to two-third of body below; dorsal with three rows of spots forming three bars and caudal with six to seven bars; mental lobes poorly developed on lower lip; a pair of bifid suborbital spine. Maximum size ~250 mm SL.

Distribution. Chindwin basin in Manipur, India and Irrawaddy basin, Myanmar.

Family Cobitidae
Loaches

Body fusiform, wormlike; mouth subterminal; barbel one to three pairs, rarely absent; erectile spine below or anterior to eye; pharyngeal teeth in one row; cephalic lateral line system conspicuous; caudal fin usually rounded or slightly emarginated; bottom dwellers (Fig. 4.198).

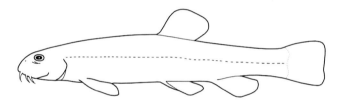

FIGURE 4.198

Outline figure of Cobitidae.

Key to genera:

1a. Barbels absent, lower lip attached to a flap of skin,
    a fringe-like part of upper lip .... *Neoeucirrhichthys*
1b. Barbels in three pairs, lower lip continuous with the upper lip 2

2a. Dorsal-fin origin in posterior half of body, opposite midway between
    pelvic and anal fins .... *Pangio*
2b. Dorsal-fin origin in the middle of the body, nearly opposite
    pelvic-fin origin 3

3a. Scales on vertex and sides of head, lateral line absent .... *Lepidocephalichthys*
3b. Scales absent on vertex and sides of head, lateral line present 4

4a. Dorsal-fin origin in front of pelvic fin origin, caudal fin forked .... *Acantopsis*
4b. Dorsal-fin origin behind the pelvic fin origin, caudal fin truncate .... *Canthophrys*

Genus *Acantopsis* van Hasselt, 1823

Acantopsis van Hasselt, 1823: 133 (type species: *Acantopsis dialuzona* van Hasselt, 1823). Gender: feminine.

Diagnosis. Snout greatly elongated; eye situated in posterior half of head; suborbital spine anterior to eye, not hidden under skin; caudal fin forked or emarginated; branched dorsal-fin rays 8–11½; lower jaw with fringed lip; barbels in three pairs, one of rostral and two of maxillary; dorsal fin with 9–11 branched rays, its origin opposite before pelvic-fin origins; anal fin short with eight to nine rays; scales minute; lateral line complete.

Acantopsis spectabilis (Blyth, 1860)
(Fig. 4.199)

FIGURE 4.199

Acantopsis spectabilis, 152.9 mm SL, Lokchao River, Yu River drainage, Myanmar.

Prostheacanthus spectabilis Blyth, 1860: 167 (type locality: Tenasserim Provinces, Myanmar).

Diagnosis. Very elongated and cylindrical body; 14–17 lateral blotches arranged longitudinally on the flanks and 17 dark brown saddles across the back; a pattern of many small, transverse and horizontal wavy bars and spots arranged longitudinally between saddles and blotches; three rows of dark brown spots on dorsal fin and two on pelvic fin; no black spot at upper base of caudal fin; gill rakers 18 (4 + 14). Maximum size ~ 213 mm SL.

Distribution. Chindwin basin, near Myanmar border, Manipur, India; Irrawaddy River, Ataran River and Salween River basins, Myanmar and Thailand.

Genus *Canthophrys* Swainson, 1938

Canthophrys Swainson, 1938: 364 (type species: *Canthophrys albescens* Swainson, 1839). Gender: feminine.

Diagnosis. Body cylindrical, snout elongated; abdomen rounded; head broad and deep; nostrils close together, open through short tubule; snout with soft warty tubercles; mouth small, inferior, horse shoe shaped; eyes large, superior; lips fleshy, continuous, upper lip thickly papillated protruding over lower lip; a shallow oval depression on occiput; a bifid suborbital spine below eyes; jaws and palate without teeth; barbels in three pairs, one of rostral and two of maxillary; dorsal fin with eight branched rays, no spine, its origin behind the pelvic-fin origin; scales cycloid; lateral line complete; caudal fin truncate.

Canthophrys gongota (Hamilton, 1822)
(Fig. 4.200)

Cobitis gongota Hamilton, 1822: 351, 394 (type locality: North Bengal)

Canthophrys gongota: Kottelat, 1998: 117 (valid).

Diagnosis. Elongated and cylindrical body; eyes large and bulging, covered with skin; a bifid suborbital spine present; dorsal fin with 10 rays, inserted behind base of pelvic fins; head naked; lateral line complete and distinct. Maximum size ~ 130 mm SL.

FIGURE 4.200

Canthophrys gongota, 125.0 mm SL, Barak River, Manipur.

Distribution. India: Ganga and Brahmaputra basin in northern and northeast India, Bangladesh.

Genus *Lepidocephalichthys* Bleeker, 1863

Lepidocephalichthys Bleeker, 1863b: 38, 42 (type species: *Cobitis hasselti* Valenciennes, 1846). Gender: masculine.

Cobitichthys Bleeker, 1858a: 304 (type species: *Cobitis barbatuloides* Bleeker, 1851a, 1851b). Gender: masculine.

Diagnosis. Body elongated, compressed; abdomen rounded; mouth inferior, narrow; lips thick, fleshy, continuous at the angle of mouth; head scaled on cheek and operculum; lower lip deeply divided into posteriorly projecting lobes; eyes lacking free orbital margin; bifid spine under orbit; three pairs of barbels: two rostral and one maxillary; pair of flaps on lower lip; vertebrae 34−38, lateral line absent; dorsal fin with two unbranched and six and half branched rays; pectoral fin with one unbranched and seven branched rays; pelvic fin with one unbranched and six and half branched rays; caudal fin truncated or slightly rounded, with 8 + 8 rays; seventh and eighth pectoral-fin rays fused in adult males.

Lepidocephalichthys alkaia Havird & Page, 2010
(Fig. 4.201)

FIGURE 4.201

Lepidocephalichthys alkaia, 76.0 mm SL, Challou River, tributary of Tizu River, Manipur.

Lepidocephalichthys alkaia Havird & Page, 2010: 140, fig. 4 (type locality: Hpa-Lap stream, of Nam Chim Chaung, of Nan Kwe Chaung, northwest of Myitkyina, Kachin State, Myanmar).

Diagnosis. A mid-lateral dark bluish to brown stripe on caudal fin, extending from base to posterior edge of fin, usually between rays 7 and 10; a truncated or rounded caudal fin; absence of scales on top of head; usually dark stripe, sometimes dark spots on side; dorsal-fin origin opposite slightly posterior to pelvic-fin origin; dark bars on dorsum: 8−10 predorsal and 8−10 postdorsal; prominent adipose crest above caudal peduncle, confluent with the caudal fin. Maximum size ∼80 mm SL.

Distribution. Chindwin drainage near Myanmar border, Manipur, India; Nan Kwe River, Irrawaddy drainage, Kachin Province and Kayin Province, Salween drainage, Myanmar.

Lepidocephalichthys annandalei **Chaudhuri, 1912**
(Fig. 4.202)

FIGURE 4.202

Lepidocephalichthys annandalei, 47.4 mm SL, Dikrong River, Arunachal Pradesh.

Lepidocephalichthys annandalei Chaudhuri, 1912: 442 (type locality: Mahananda River at Siliguri and Tista River, near Jalpaiguri, India).

Diagnosis. Slightly emarginated caudal fin with unique colorations: two v-shaped dark brown marks across, apex toward the posterior tip, two distinct spots, one at base and another at the center of posterior edge; small barbels, rostral pair never reaching anterior nostril; scales on top of head absent or embedded. Maximum size ∼78 mm SL.

Distribution. Slow-flowing streams and oxbows in the Ganga−Brahmaputra and Barak−Surma−Meghna drainages of northeastern India, Bangladesh, and Nepal.

Lepidocephalichthys arunachalensis **(Datta & Barman, 1984)**
(Fig. 4.203)

FIGURE 4.203

Lepidocephalichthys arunachalensis, 81.1 mm SL. Mebo stream, tributary of Siang River, Arunachal Pradesh.

Noemacheilus arunachalensis Datta & Barman, 1984 (type locality: Namdapha River, Namdapha Wildlife Sanctuary, Tirap District, Arunachal Pradesh, India.

Diagnosis. Rounded or truncated caudal fin; body with 12−14 large dark spots on the sides; dorsal-fin origin conspicuously opposite posterior to pelvic-fin origin; caudal fin with dark reticulate bars; a conspicuous thin basicaudal bar. Maximum size ∼82 mm SL.

Distribution. Slow-flowing streams and oxbows in the Ganga−Brahmaputra drainages of Assam, Bangladesh, and Nepal.

Lepidocephalichthys berdmorei **(Blyth, 1860)**
(Fig. 4.204)

FIGURE 4.204

Lepidocephalichthys berdmorei, showing color pattern in caudal fin; (A) 50.2 mm SL, Chakpi River, Manipur; (B) 66.2 mm SL, Yu River, Tamu, Myanmar; (C) 86.2 mm SL, Khuga River, Manipur.

Acanthopis berdmorei Blyth, 1860: 168 (type locality: Tenasserim provinces, Myanmar).

Diagnosis. Caudal fin truncated; scales absent on top of head; anterior rostral barbel long; extending to eye; black spots scattered on the sides extending from dorsal to two-thirds of the body posteriorly; larger spots on lateral series; lamina circularis forming long cylinder like appearance along edge of fin; three to six bars on caudal fin becoming faint posteriorly, number of bars increasing with size. Maximum size ~100 mm SL.

Distribution. Chindwin drainage in Manipur, India; Myanmar, Thailand, Bangladesh, peninsular Malaysia, and Laos.

Lepidocephalichthys goalparensis **Pillai & Yazdani, 1976**
(Fig. 4.205)
Lepidocephalichthys goalparensis Pillai & Yazdani, 1976 (type locality: Assam, India).

Diagnosis. Caudal fin forked with four to five v-shaped bars, apex toward posterior; dorsal fin origin anterior to pelvic fin origin; lamina circularis sometimes forming small dorsally projecting flange, but never large with serrations. Maximum size ~65 mm SL.

Distribution. Slow flowing streams of the Ganga and Brahmaputra drainages of North and northeastern India, Nepal, and Bangladesh.

Lepidocephalichthys guntea **(Hamilton, 1822)**
(Fig. 4.206)

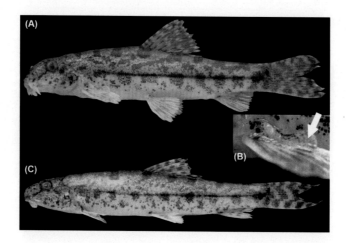

FIGURE 4.205

Lepidocephalichthys goalparensis, (A) 37.0 mm SL, male; (B) pectoral fin of male showing fused 7th and 8th branched rays; (C) 34.0 mm SL, female; Goalpara District, Kaldiya River at Baksa District, Assam.

Cobitis guntea Hamilton, 1822: 353, 394 (type locality: Ganges River, Bengal).

Diagnosis. Rounded or truncated caudal fin; absence of scales on top of head; body with blue-black spots in female and broad dark stripe in male; caudal fin with reticulations, forming wavy bars, an ocellus at caudal-fin base slightly above the center. Maximum size ∼150 mm SL.

Distribution. Ganga and Brahmaputra drainages in northern India, Nepal, and Bangladesh.

Lepidocephalichthys irrorata **Hora, 1921**
(Fig. 4.207)

Lepidocephalichthys irrorata Hora, 1921a: 196 (type locality: Loktak Lake, Manipur, India).

Diagnosis. Small adult size, maturing at 36.4 mm SL; scales on top of head; anal-fin base shorter than or equal to dorsal-fin base; dorsal-fin origin posterior to pelvic-fin origin; dorsum with 8−10 irregular brownish bars on predorsal region and 8−10 on postdorsal region; base of the caudal fin with no spot; reticulated caudal fin with 8−12 elongated brownish spots along upper and 8−11 along lower edges of caudal fin; emarginated caudal fin; first branchial arch with 1 + 9 gill rakers; long and straight intestine; and 36−37 vertebrae; adult male with lamina circularis and an adipose hump behind the dorsal fin (Fig. 4.208A). Maximum size ∼40 mm SL.

Distribution. Sluggish streams of the Chindwin drainage in Manipur, India.

Lepidocephalichthys micropogon **(Blyth, 1860)**
(Fig. 4.208)

Acanthopis micropogon Blyth, 1860: 168 (type locality: Burma: Tenasserim Provinces).

Lepidocephalichthys manipurensis Arunkumar, 2000b: 1097, figs. 4−5 (Lairok Maru, tributary of Lokchao River near Moreh, Chandel District, Manipur, India).

FIGURE 4.206

Lepidocephalichthys guntea, (A) 82.3 mm SL, female, (B) 76.5 mm SL, male, both from Barak River, Silchar, Assam; (C) 82.5 mm SL, Barak River, Maram, Manipur; (D) 74.0 mm SL, female, (E) 75.2 mm SL, male, both from Kolong River, Brahmaputra drainage, Assam; (F) 69.9 mm SL, wetlands around Agartala, Tripura; (G) 67.6 mm SL, Brahmaputra drainage, on the way to Shillong, Meghalaya.

FIGURE 4.207

Lepidocephalichthys irrorata, (A) male 32.0 mm SL, Sekmai River at Sekmaijin, tributary of Imphal River, Manipur, showing fused 7th & 8th P-fin rays and adipose hump; (B) 38.0 mm SL, female, same locality.

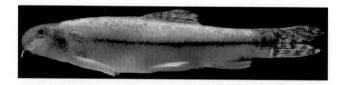

FIGURE 4.208

Lepidocephalichthys micropogon, 67.6 mm SL, Tamu market, Myanmar.

Diagnosis. Head and body much compressed, barbels small, posterior lobe of suborbital spine longer and stronger; 10 oval spots along lateral line, a conspicuous black one at base of caudal fin; caudal fin forked with four wavy v- or w-shaped bars. Maximum size ~70 mm SL.

Distribution. Lokchao River, Chindwin basin in Manipur, India; Salween, Sittang and Irrawaddy drainages of Myanmar.

Genus *Neoeucirrhichthys* Bănărescu & Nalbant, 1968

Neoeucirrhichthys Bănărescu & Nalbant, 1968: 349 (type species: *Neoeucirrhichthys maydelli* Bănărescu and Nalbant, 1968). Gender: masculine.

Diagnosis. Body elongated, head and body strongly compressed; gill openings restricted, extending from just below posttemporal to just below base of first pectoral-fin ray; suborbital spine bifid, with medial point longest and extending to vertical through anterior third of eye; mouth horseshoe-shaped (Fig. 4.222B); rostral groove along anterior margin of upper lip; upper and lower lips slightly papillate, lower lip with fleshy lobe at rictus; lower jaw with a weak symphyseal knob; postlabial groove present along posterior margin of lower lip, groove interrupted medially by narrow frenum between skin of throat and of lower jaw; barbels absent; dorsal-fin origin halfway between snout tip and caudal-fin base; caudal fin emarginated; pectoral fin in males with extension and ramification of distal portion of first branched pectoral fin.

Neoeucirrhichthys maydelli Bănărescu & Nalbant, 1968
(Fig. 4.209)

FIGURE 4.209

Neoeucirrhichthys maydelli, 36.0 mm SL, (A) lateral view, Dudhnoi River, tributary of the Brahmaputra, Goalpara District, Assam; (B) mouth of *N. maydelli*.

Photo courtesy: H. Chowdhury.

Neoeucirrhichthys maydelli Bănărescu & Nalbant, 1968: 349, figs. 14−15 (type locality: Janali River at Raimona, Goalpara District, Brahmaputra drainage, northeastern India).

Diagnosis. Body dark yellow, eight to nine dark brown spots on dorsal midline, spots interspersed with fainter melanophores forming irregular reticulations extending to sides of body, series of 11−12 spots forming bars on sides, ovoid dark spot at dorsal half of caudal-fin base, fins hyaline with dark brown bars. Maximum size ∼40 mm SL.

Distribution. Brahmaputra River drainage in Assam; Brahmaputra and Surma and Meghna drainages in Bangladesh.

Genus *Pangio* Blyth, 1860

Pangio Blyth, 1860: 169 (type species: *Cobitis cinnamomea* McClelland, 1839: 304). Gender: feminine.

Diagnosis. Slender to anguilliform body, compressed to very compressed, dorsal fin origin conspicuously behind the pelvic origin; vertebrae 331−52 + 12−20 = 45−71.

Pangio apoda Britz & Maclaine, 2007

Pangio elongata Britz & Maclaine, 2007: 28 (type locality: Tista River at Tista Barrage, Brahmaputra drainage, West Bengal, India).

Diagnosis. Absence of pelvic girdle and pelvic fin, 38−39 vertebrae, anteriorly placed dorsal fin; absence of nasal barbels and plain brown color pattern. Maximum size ∼38 mm SL.

Distribution. Teesta Drainage, West Bengal, India.

Pangio pangia (Hamilton, 1822)
(Fig. 4.210)

FIGURE 4.210

Pangio pangia, 38.3 mm SL, Imphal River at Mayang Imphal, Manipur.

Cobitis pangia Hamilton, 1822: 355, 394 (type locality: India: north-eastern parts of Bengal, India).

Diagnosis. Pale brown coloration, abdominal vertebrae 34—39; absence of nasal barbel; long pectoral fin, its length 8.2%—9.6% SL; body deep, its depth 13.5%—16.3% SL; presence of pelvic fin and pelvic girdle. Maximum size ∼65 mm SL.

Distribution. Northern Bengal, Assam and Manipur, India; Bangladesh, Myanmar.

Family Balitoridae
Stream loaches

Small sized not more than 100 mm SL; head and body greatly depressed, ventral surface flat; leaf-like; paired fins horizontal, greatly expanded and supported by more number of unbranched and branched fin rays, pads on the ventral surface and act as adhesive organs; three pairs of barbels around the mouth; gill openings often restricted; exoccipitals separated from each other by supraoc-cipital; mesocoracoid fused with an enlarged cleithrum (Fig. 4.211).

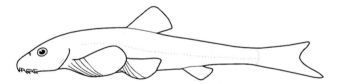

FIGURE 4.211

Outline figure of Balitoridae.

Key to genera:

1a. Gill opening extending to ventral surface 2
1b. space Gill opening not extending to ventral surface .... *Hemimyzon*

2a. Rostral groove in front of mouth, overhung by rostral fold .... *Balitora*
2b. Rostral groove absent, rostral fold absent or poorly developed .... *Homalopteroides*

Genus *Balitora* Gray, 1830

Balitora Gray, 1830: vol. 1, pl. 88 (type species: *Balitora brucei* Gray, 1830). Gender: feminine.

Diagnosis. Body strongly depressed, head and abdomen flattened; mouth small, inferiorly arched, with both jaws covered by horny sheath; rostral flap divided into three lobes, the median one the largest, between rostral barbels; both lips with one or two rows of papillae; lower lip not interrupted; gill opening extending to the ventral surface of head; pelvic fin with two simple rays; 8—10 simple and 10—12 branched pectoral rays; adhesive pads present on ventral surface of the 8—11 anterior most pectoral-fin rays and three to four anterior most pelvic-fin rays; principal caudal-fin rays 10 + 9, the four to five lower ones very closely set, without membranes between them along most of their length.

***Balitora brucei* Gray, 1830**
(Fig. 4.212)

FIGURE 4.212

Balitora brucei, 68.2 mm SL, Irang River, tributary of Barak River, Manipur.

Balitora brucei Gray, 1830: 88 (Priang River near Cherrapunji, Assam, India).
Balitora brucei: Kottelat, 1988: 490 (redescription, neotype designated).
Diagnosis. Head length 19%−22% SL; head width 86.0%−105.0% HL; eye diameter 2.3%−3.3% SL; pectoral fin with 9−10 simple and 10−12 branched rays; caudal peduncle length 16.7%−21.4% SL; belly naked in front of pelvic-fin base; head with numerous unculi of various sizes and irregular shapes; mid-dorsal line with four predorsal, two subdorsal, and five postdorsal dark blotches surrounded by lighter margin. Maximum size ∼105 mm SL.
Distribution. Meghna and Brahmaputra drainages, Assam, Manipur, Meghalaya, Darjeeling, Bhutan, Bangladesh.

***Balitora burmanica* Hora, 1932**
(Fig. 4.213)

FIGURE 4.213

Balitora burmanica, 59.0 mm SL, Chakpi River, Chindwin drainage, Manipur.

Balitora brucei var. *burmanicus* Hora, 1932: 291, pl. 11, fig. 6 (type locality: Burma: Meekalan, Salween drainage).
Balitora burmanica: Kottelat, 1988: 494 (redescription, lectotype designated).
Diagnosis. Head width 73.0%−98.0% HL, eye diameter 3.0%−4.0% SL; interorbital width 37.0%−42.0% HL; body entirely covered by scales except on belly in front of pelvic fin origin; a principal longitudinal keel on each scale and above and below one to two small additional keels; top of head covered by slightly longitudinally elongated unculi; mid-dorsal line with six contiguous regularly shaped large blotches bordered by a light brown area. Maximum size ∼95 mm SL.

Distribution. India: Chindwin basin in Manipur, India and Irrawaddy and Salween basins, Myanmar, and Thailand.

Balitora eddsi **Conway & Mayden, 2010**
Balitora eddsi Conway & Mayden, 2010: 1467, figs. 1, 2a and b, (type locality: Gerwa River, Karnali River basin, Bardiya District, Nepal).
Diagnosis. Elongate body, pointed snout; pectoral fin rays 6−7; pelvic-fin length 12.8%−14.0% SL; posteriorly pointed median lobe between rostral barbels; posteriorly positioned dorsal fin; 66−67 lateral line scales; absence of dark blotches along the dorsal midline. Maximum size ~46 mm SL.
Distribution. Gerwa River, between the towns of Chisapani and Kothiaghat, south-western Nepal.

Genus *Hemimyzon* **Regan, 1911**
Hemimyzon Regan, 1911b: 32 (type species: *Homaloptera formosana* Boulenger, 1894). Gender: masculine.
Diagnosis. Body strongly compressed anteriorly and compressed posteriorly, flattened ventrally; pelvic fins not fused, well separate or contiguous or fused posteriormost rays; last simple anal fin ray not spinous; caudal fin forked; simple pelvic-fin rays 3−7; body widened; gill opening extending to pectoral-fin base, to ventral surface; body covered with small scales and small tubercles; scales absent ventrally; dorsum usually with 7−10 saddles like blotches.

Hemimyzon arunachalensis **(Nath, Dam, Bhutia, Dey & Das, 2007)**
(Fig. 4.214)

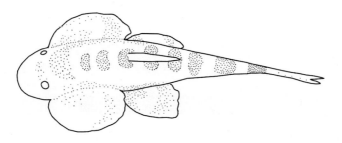

FIGURE 4.214

Hemimyzon arunachalensis.

Outline redrawn from Bhavania arunachalensis Nath P, Dam D, Bhutia PT, Dey SC, Das DN: A new fish species of the genus Bhavania (Homalopteridae: Homalopterinae) from river Noadhing drainage, Arunachal Pradesh, India. Rec Zool Surv India 107(3):71−78, 2007, figs. 1 and 6.

Bhavania arunachalensis Nath, Dam, Bhutia, Dey & Das, 2007: 72, pl. 1 (figs. 1−2; pl. 2, figs. 4−6) (type locality: Noa dihing drainage, near Namsai, about 30 km from Tezu, Arunachal Pradesh, India).
Diagnosis. Head with pores-like outgrowths, rough and arranged in rows around orbit; lateral line complete with 70−75 pores, slightly curved at the origin of ventral fin; pectoral fin origin

opposite to orbit, extends beyond origin of ventral fin; eight saddle-shaped blotches on the back; longitudinal band from base to the tip of lower caudal lobe; caudal fin emarginated; lower lobe longer than upper; dorsal fin with two simple and seven branched rays; pectoral fin with eight simple and 11 branched rays; pelvic fin with three simple and eight branched rays; anal with three and six. Maximum size ~110 mm SL.

Distribution. Noa Dehing Drainage, Tezu, Arunachal Pradesh, India.

Hemimyzon indicus **Lalramliana, Solo, Lalronunga & Lalnuntluanga, 2018**
(Fig. 4.215)

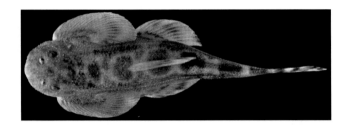

FIGURE 4.215

Hemimyzon indicus, 62.5 mm SL, Kaladan River at Kawlchaw, Mizoram.

Hemimyzon indicus Lalramliana, Solo, Lalronunga & Lalnuntluanga, 2018c: 109, figs. 1−2 (type locality: Kaladan River in the vicinity of Lungbun village, Siaha District, Mizoram, India).

Diagnosis. A *Hemimyzon* species with a wide body; with two simple pelvic-fin rays and 65−67 lateral-line scales; pectoral fin tip reaching to or slightly beyond pelvic-fin origin; pelvic fin widely separated, its fin tip reaching anus; body width at pelvic-fin origin 27.1%−31.2% SL; dorsum with seven to nine dark brown saddles; body with 8−12 irregular dark brown blotches alongside; and pectoral fin with 9−12 unbranched and 9−11 branched rays. Maximum size ~66 mm SL.

Distribution. Kaladan River in the vicinity of Lungbun Village, Siaha District, Mizoram, India.

Genus *Homalopteroides* Fowler, 1905
Homalopteroides Fowler, 1905: 476 (type species: *Homaloptera wassinkii* Bleeker, 1853). Gender: masculine.

Diagnosis. Scales large; no deep groove extending around the corners of the mouth; lips not papillate; body subcylindrical, ventral surface flatted; head depressed but not flattened; snout long and pointed; mouth small, inferior, slightly arched, fringed by thick plain lips, continuous at angle; labial groove widely interrupted; barbels short and stout; eyes moderately large, dorso-lateral situated in the posterior half of the head, not visible from the ventral surface.

Homalopteroides rupicola **(Prashad & Mukerji, 1929)**
(Fig. 4.216)
Chopraia rupicola Prashad & Mukerji, 1929: 188, pl. 8, fig. 3 (type locality: small streams around Kamaing, Myitkyina District, Myanmar).

FIGURE 4.216

Homalopteroides rupicola, 38.6 mm SL, Yu River, Tamu, Myanmar.

Diagnosis. Pectoral fin extending to pelvic fins; dorsal fin origin behind vertically through pelvic fin origin; lateral line scales 42–45; black dots on body arranged in the form of five bars. Maximum size ~40 mm SL.

Distribution. Lokchao River, Manipur, India; Myitkyina, upper Irrawaddy, Myanmar.

Family Nemacheilidae
River loaches

Body elongated, rounded, or compressed; mouth subterminal; lips fleshy, usually with grooves and forming cushions and pads; two pairs of rostral and one pair of maxillary barbels; no spine under or before eyes; single unbranched ray in pectoral and pelvic fins; adipose-like fin present in some; scales present or absent; prepalatine present; body usually with blue or black saddles or bars or both, bars sometimes reticulated (Fig. 4.217).

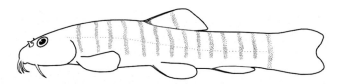

FIGURE 4.217

Outline figure of Nemacheilidae.

Key to genera:

1a. Vent position nearer to pelvic origin than to anal fin origin .... *Aborichthys*
1b. Vent position nearer to anal fin origin 2

2a. Dorsal fin with 8½–18½ branched rays, a black blotch
at upper extremity of caudal-fin base .... *Acanthocobitis*
2b. Dorsal fin with 8½–9½ branched dorsal fin rays, a basicaudal bar,
entire or interrupted 3

3a. Lower lip without median interruption, caudal fin forked with
upper lobe longer .... *Nemacheilus*
3b. Lower lip with median interruption, caudal fin lobes equal 4

4a. Lips hypertrophied, an open oral cavity between the lips .... *Neonoemacheilus*
4b. Lips not hypertrophied, no open oral cavity between the lips 5

5a. Pectoral fin of males not modified 6
5a. Pectoral fin in males modified 7

6a. Mouth moderately arched, lower lip with median interruption,
not forming two lateral triangular pads with deep furrows .... *Schistura*
6b. Mouth strongly arched, lower lip with median interruption
forming two lateral broadly triangular pads with deep furrows .... *Physoschistura*

7a. Pectoral fin of male slanted upward, first branched ray wider than the rest,
thickened with unculiferous pads .... *Mustura*
7b. Pectoral fin in male slanted upward and expanded laterally,
lower lip with median interruption, forming an acute angle,
caudal fin emarginated .... *Rhyacoschistura*

Genus *Aborichthys* Chaudhuri, 1913

Aborichthys Chaudhuri, 1913: 244 (type species: *Aborichthys kempi* Chaudhuri, 1913). Gender: masculine.

Diagnosis. Body greatly elongated and compressed; vent placed far ahead of anal fin origin; nostril with barbel-like outgrowths; preorbital spine absent; lips thick, papillated, continuous at the angle of mouth; lower lip interrupted in middle; jaws and palate without teeth; barbels six; dorsal fin origin slightly behind opposite pelvic fin; paired fins horizontal and provided with adhesive pads on ventral aspect; caudal fin lunate; lateral line incomplete.

Note: Examination of specimens of *Aborichthys* shows variability of the color pattern of caudal fins in different sizes of a species. A. Darshan and L. Tamang (pers. comm.) found the individuals with variegated color pattern to be females and those with dark blue or brown regular submarginals bands to be males. In order to clarify the doubt and to avoid misidentification of a species, a detailed examination of specimens of different age groups and sexes are required. Molecular characterization of the species would also certainly help.

Aborichthys boutanensis (McClelland & Griffith, 1842)
(Fig. 4.218)
Cobitis boutanensis McClelland & Griffith in McClelland, 1842: 586 [type locality: Boutan (Bhutan), on Mishmee Mountains].
Aborichthys cataracta Arunachalam et al., 2014: 34 (Tributary stream of Ranga River, Upper Subansiri District, Arunachal Pradesh, India).
Aborichthys verticauda Arunachalam et al., 2014: 37 (Tributary of Ranga River, Lower Subansiri District, Arunachal Pradesh, India).

FIGURE 4.218

Aborichthys boutanensis, (A) 58.5 mm SL; (B) 82.9 mm SL, both from Dikrong River at Doimukh, Arunachal Pradesh.

Aborichthys boutanensis: Thoni and Hart, 2015: 292 (valid).

Diagnosis. Body with 21−26 narrow oblique blue black to brown bars on the pale yellowish background, bars may unite dorsally and then bifurcate to form saddles. Caudal-fin base with large black spot dorsally which may extend as a short bar below; caudal fin rounded and with graying bars which may be indistinct or regularly arranged in two bars, one prominent bar submarginally and another which is fainter in the middle; dorsal fin with two grayish bars. Maximum size ∼83 mm SL.

Distribution. Brahmaputra drainage in India and Bhutan.

Aborichthys elongatus **Hora, 1921**
(Fig. 4.219)

FIGURE 4.219

Aborichthys elongatus: (A) paratype, 74.0 mm SL; (B) Hora's (1921c) fig in. page 735 reproduced.

Aborichthys elongatus Hora, 1921c: 735, (type locality: Reang River, Darjeeling District, West Bengal, India).

Diagnosis. Body greatly elongated and compressed, dorsal and ventral profiles straight; dorsal fin origin behind pelvic fin origin, its origin equidistant from tip of snout to base of caudal-fin base; body with alternate black and yellowish orange bars forming almost complete rings;

melanophores on rays; caudal fin with whitish margin, black ocellus at upper corner of caudal-fin base. Maximum size ~75 mm SL.

Distribution. Teesta and Brahmaputra drainage, northeast India.

Aborichthys garoensis **Hora, 1925**
(Fig. 4.220)

FIGURE 4.220

Aborichthys garoensis: (A) holotype, 89.5 mm SL; (B) Hora's (1925) Text fig. 1a reproduced.

Aborichthys garoensis Hora, 1925: 233 (type locality: Garo Hills, Meghalaya, India).

Diagnosis. Vent halfway between tip of snout and base of caudal-fin, or slightly nearer to snout-tip; body with 30−35 blue black fork-shaped slightly oblique bars; caudal fin with subterminal blue-black bar, terminal hyaline, a black ocellus at the upper corner of its base; body depth 10.0%−10.5% SL. Maximum size ~90 mm SL.

Distribution. Brahmaputra basin in Meghalaya and Arunachal Pradesh.

Aborichthys iphipaniensis **Kosygin, Gurumayum, Singh, Chowdhury, 2019.**
(Fig. 4.221)

FIGURE 4.221

Aborichthys iphipaniensis, 127.1 mm SL, Iphipani River at Roing, Arunachal Pradesh.

Photo courtesy: S.D. Gurumayum.

Aborichthys iphipaniensis Kosygin, Gurumayum, Singh, Chowdhury, 2019: 70 (Iphipani River at Roing, Lower Divang Valley, Arunachal Pradesh).

Diagnosis. Vent closer to snout tip than caudal-fin base (distance between vent and caudal-fin base 52.0%−56.2% SL); long caudal peduncle (21.5%−23.3% SL), shallow caudal peduncle (9.2%−10.5% SL), shallow body (8.9%−9.9% SL), short predorsal (42.4%−44.4% SL), small eye

(10.9%−14.4% HL), prepectoral (15.1%−17.9% SL, lateral line incomplete reaching pelvic fin origin, 33−35 almost uniform bars on lateral side of the body, and obliquely truncate caudal fin. Maximum size ∼126 mm SL.

Distribution. Iphipani River, Brahmaputra drainage, Lower Divang Valley, Arunachal Pradesh.

Aborichthys kailashi Shangningam, Kosygin, Sinha, Gurumayum, 2020
(Fig. 4.222)

FIGURE 4.222

Aborichthys kailashi, 103 mm SL, Pange River at Ziro, Lower Subansiri District, Arunachal Pradesh.

Photo courtesy: B.D. Shangningam.

Aborichthys kailashi Shangningam, Kosygin, Sinha, Gurumayum, 2020: 362 (type locality: Pange River, Arolenching, Ziro, Lower Subansiri District, Arunachal Pradesh).

Diagnosis. Body with 28−36 black bars, anterior 12 bars upto pelvic fin origin irregular or forked, remaining posterior bars regular, interspaces wider than the bars; dorsal fin rays with two black bars; basicaudal bar with a darker spot/blotch in the dorsal extremity, caudal fin with two concentric black bars, one in the middle and another at the posterior extremity; caudal peduncle with dorsal and ventral adipose crests; incomplete lateral line incomplete, extending only upto the level of pelvic-fin origin; pelvic fin nearer to pectoral-fin origin than to anal-fin origin. Maximum size ∼113 mm SL.

Distribution. Pange River, Brahmaputra basin of lower Subansiri District, Arunachal Pradesh.

Aborichthys kempi Chaudhuri, 1913
(Fig. 4.223)
Aborichthys kempi Chaudhuri, 1913: 245, pl. 7, figs. 1−1b (type locality: N.E. India: Abor Hills: Egar Stream between Renging and Rotung/Dihang River near Yembung/Sirpo River near Renging).

Diagnosis. Vent distinctly nearer to caudal-fin base than to snout-tip. Body with 15−20 slightly oblique bars, bars absent in the caudal peduncle; caudal fin dusky at its anterior third, pale posteriorly, a darker subterminal bar and a terminal hyaline margin. Maximum size ∼90 mm SL.

Distribution. Brahmaputra basin in Meghalaya and Arunachal Pradesh, India.

Aborichthys pangensis Shangningam, Kosygin, Sinha, Gurumayum, 2020
Aborichthys pangensis Shangningam et al., 2020: 4 (type locality: Pange River at Arolenching, Ziro, Arunachal Pradesh).

Diagnosis. Elongate body with 34−38 black bars, anterior 10 bars thin and irregular up to the level of pelvic fin origin, regular and thicker behind, interspaces wider than the bars; caudal peduncle with dorsal and ventral adipose crests; lateral line incomplete, extending up to a little before pelvic fin origin; caudal fin truncate and with clusters of black spots forming four to five broad irregular bars; a black blotch at the upper extremity of the basicaudal bar. Maximum size ∼62 mm SL.

FIGURE 4.223

Aborichthys kempi: (A) 43.7 mm SL, (B) 88.8 mm SL, (C) Chaudhuri's (1913), pl. 7, fig. 1 reproduced; all from Renging stream, Arunachal Pradesh.

Photo courtesy: (a & b) A. Darshan.

Distribution. Pange River, Brahmaputra drainage, Lower Subansiri District, Arunachal Pradesh.

Aborichthys tikaderi **Barman, 1985**
(Fig. 4.224)

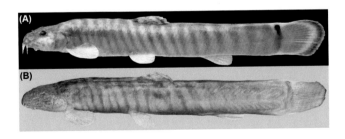

FIGURE 4.224

Aborichthys tikaderi: (A) 82.0 mm SL. (B) 104.0 mm SL. Both from Namdapha wildlife sanctuary, Arunachal Pradesh.

Aborichthys tikaderi Barman, 1985: 680, fig. 1 (type locality: Namdapha Wildlife Sanctuary, Arunachal Pradesh, India.

Diagnosis. Vent nearer to snout-tip than to caudal-fin base, dorsal fin origin nearer to snout-tip than to caudal-fin base; caudal fin plain, darker along posterior edge. Maximum size ~109 mm SL.

Distribution. Brahmaputra drainage, Arunachal Pradesh.

Aborichthys waikhomi **Kosygin, 2012**
(Fig. 4.225)

FIGURE 4.225

Aborichthys waikhomi, 66.5.0 mm SL, Noa-Dihing River, Namdapha, Arunachal Pradesh.

Aborichthys waikhomi Kosygin, 2012: 49, figs. 3, 4 and 5f (type locality: Bulbulia stream near Bulbulia a tributary of Noa-Dihing River, Namdapha, Arunachal Pradesh, India).

Diagnosis. Diagnosed with a combination of characters: vent nearer to caudal-fin base than to tip of snout, dorsal fin origin halfway between snout-tip and caudal-fin base; body with 12−16 slightly oblique bars, interspaces broader at caudal peduncle. Barbels much longer than eye diameter, lateral line incomplete, extending upto slightly ahead of the pelvic fin origin. Caudal fin truncated, with three wavy bars, a black bar at caudal-fin base, an ocellus at its upper corner. Maximum size ~68 mm SL.

Distribution. Noa-Dehing River, Brahmaputra basin in the Namdapha National Park and Tiger Reserve, Changlang District, Arunachal Pradesh, India.

Genus *Acanthocobitis* Peters, 1861

Acanthocobitis Peters, 1861: 712 (type species: *Acanthocobitis longipinnis* Peters, 1861). Gender: feminine.

Acanthocobitis: Singer and Page, 2015: 379 (description of genus).

Diagnosis. Body elongated; lower lip with large papillated pads on each side of a wide medial interruption; the upper lip with two to five rows of papillae and continuous with the lower lip; a conspicuous black ocellus on the upper half of the caudal-fin base; caudal fin pointed; branched dorsal-fin rays 17½−19½ head triangular in lateral view; suborbital flap in male vertical; position of anus halfway between the origins of pelvic and anal fins (Fig. 4.226).

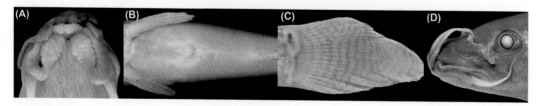

FIGURE 4.226

Acanthocobitis showing (A) Lip structure; (B) anus position, (C) caudal fin, (D) suborbital flap of male.

Acanthocobitis pavonacea **(McClelland, 1839)**
(Fig. 4.227)

Cobitis pavonacea McClelland, 1839: 305, 437, pl. 52, fig. 1 (type locality: Assam, India).

FIGURE 4.227

Acanthocobitis pavonacea, 87.2 mm SL, Tengapani River at Namsai District, Arunachal Pradesh.

Diagnosis. Lower lip with a large papillated pad on either side of a medial interruption; upper lip with two to five rows of papillae and continuous with lower lip; conspicuous black spot with white outline (an ocellus) on upper half of caudal-fin base; caudal fin rounded, slightly pointed in the middle; branched dorsal-fin rays $17^1/_2 - 19^1/_2$; (vs $9^1/_2 - 15^1/_2$; suborbital flap vertically oriented. Maximum size ~156 mm SL.

Distribution. Rivers in northern Bengal, Brahmaputra drainage, northeast India; also collected in Braka drainage in Mizoram.

Genus *Mustura* Kottelat, 2018

Mustura Kottelat, 2018: 2 (type species: *Mustura celata* Kottelat, 2018). Gender: feminine.

Diagnosis. Distinguished from other nemacheilid genera by having: the male pectoral fin slanted upward, first branched ray wider than following rays, branches close together with no membranes in between except near tip (Fig. 4.247); wide, thickened unculiferous pad extending along anterior branched pectoral-fin rays, covered dorsally by small conical tubercles (Fig. 4.228); none

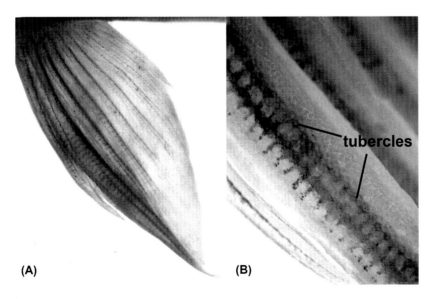

(A)   (B)

FIGURE 4.228

Pectoral fin of male *Mustura* showing (A) expanded first rayed fin; (B) pectoral fin rays showing tubercles.

of these modifications are found in females; air bladder capsules on either side connected by a manubrium, posterior chamber small and circular, or absent; lower lip interrupted medially, two halves forming an acute angle, interrupted lips forming a triangular cushion medially; suborbital flap present with small tubercles on posterior extremity.

Mustura chhimtuipuiensis **(Lalramliana, Lalhlimpuia, Solo & Vanramliana, 2016)**

Physoschistura chhimtuipuiensis Lalramliana, Lalhlimpuia, Solo & Vanramliana, 2016: 193 (type locality: Ngengpui River, a tributary of the Kaladan River, Lunglei District, Mizoram).

Diagnosis. Breast between pectoral-fin bases with deeply embedded and sparsely set scales; branched dorsal-fin rays 8½; branched caudal-fin rays 8 + 7; axillary pelvic lobe present; lateral line incomplete, reaching upto the base of the last anal-fin rays; simple dorsal-fin rays 4; intestine without a loop; body with 11−14 irregularly shaped bars; back with 11−14 saddles, saddles in the caudal peduncle region not continuous with the bars, the rest anterior saddles continuous with the bars; absence of a median notch at the lower jaw; lateral line. Maximum size ∼42 mm SL.

Distribution. Kaladan River drainage, Mizoram.

Mustura chindwinensis **(Lokeshwor & Vishwanath, 2012)**
(Fig. 4.229)

FIGURE 4.229

Mustura chindwinensis, 43.3 mm SL, Lokchao River at Moreh, Manipur, India.

Physoschistura chindwinensis Lokeshwor and Vishwanath, 2012a: 231, figs. 1−4 (type locality: Lokchao River at Moreh, Chindwin basin, Manipur, India).

Diagnosis. A species with 13−17 dark-brown bars on the flank; 11−16 dark-brown saddles on the back; 8 + 8 branched caudal rays; axillary pelvic lobe moderately formed; males with suborbital flap; well-formed free posterior chamber of the air bladder; seven pores in the preoperculo-mandibular canal; a distinct dark brown hexagonal shield like mark in the interorbital region; and lateral line complete. Maximum size ∼45 mm SL.

Distribution. Lokchao River at Moreh (Chindwin basin), Chandel District, Manipur, India.

Mustura dikrongensis **(Lokeshwor and Vishwanath, 2012)**
(Fig. 4.230)

Physoschistura dikrongensis Lokeshwor and Vishwanath, 2012b: 250, figs. 1−4 (type locality: Arunachal Pradesh: from Dikrong River at Doimukh, Brahmaputra basin, India).

Diagnosis. Two v-shaped dark brown bars across caudal fin; 11−15 irregular dark brown bars on body, dorsolateral and dorsal portion of head mottled dark brown; incomplete lateral line with 70−85 pores; four simple and 8½ branched dorsal-fin rays; four simple anal-fin rays; caudal fin

FIGURE 4.230

Mustura dikrongensis, 40.1 mm SL. Dikrong River, Arunachal Pradesh.

with 8 + 7 branched rays; large axillary pelvic-fin lobe; suborbital flap in male; nine pores in preoperculo-mandibular canal. Maximum size ~46 mm SL.

Distribution. Dikrong River at Doimukh, Brahmaputra drainage, Arunachal Pradesh, India.

Mustura harkishorei **(Das & Darshan, 2017)**
(Fig. 4.231)

FIGURE 4.231

Mustura harkishorei: (A) lateral view, 53.4 mm SL, (B) dorsal view, Dibang River, Arunachal Pradesh.

Physoschistura harkishorei Das & Darshan, 2017: 404, fig. 1 (type locality: Dibang River, Lower Dibang Valley District, Brahmaputra basin, Arunachal Pradesh, India).

Diagnosis. Combination of characters: distal filamentous extension of the second branched ray of the pectoral fin; 9−10 brownish vertically elongated spots or blotches along the flank, 8−10 brownish saddles on back, saddles not contiguous with the lateral blotches; lateral line complete; axillary pelvic-fin lobe present; a well-developed free posterior chamber of the air bladder; and caudal fin with 7 + 8 branched rays. Maximum size ~54 mm SL.

Distribution. Dirang River and Lohit River, Brahmaputra drainage, Arunachal Pradesh.

Mustura prashadi **(Hora, 1921)**
(Fig. 4.232)

Nemacheilus prashadi Hora, 1921a: 203, pl. 10, figs. 2, 2a (type locality: Thoubal and Sikmai stream, southern watershed of the Naga Hills, India).

FIGURE 4.232

Mustura prashadi: (A) 58.4 mm SL, Imphal River at Serou, Manipur; (B) 43.6 mm SL, live, Thoubal River, Manipur.

Diagnosis. A large *Physoschistura* with 12−14 irregular light reddish brown colored bars, a few may break up and form blotches, five to six reddish brown irregular saddles on the back, black basicaudal bar wider at its upper extremity possibly connected with last saddle, a black vertically elongated spot at base of last simple and first branched dorsal fin rays, caudal fin with two to three vertical row of spots; lateral line complete, 9 + 8 branched caudal-fin rays, axillary pelvic-fin lobe prominent, males with suborbital flap, and 10 pores in preoperculo-mandibular canal. Maximum size ∼65 mm SL.

Distribution. Streams draining to Yu and of Chindwin rivers, Manipur, India.

Mustura tigrina (Lokeshwor & Vishwanath, 2012)
(Fig. 4.233)

Physoschistura tigrinum Lokeshwor & Vishwanath, 2012c: 97, figs. 1−4 (type locality: Phungrei, Changa River, Chindwin drainage, Ukhrul District, Manipur, India.

Diagnosis. A large species with 12−14 irregular light reddish brown bars, a few of them broken and forming blotches, on side of body, five to six reddish brown irregular saddles on back, arrangement of bars on body with tiger-like appearance; lateral line complete with 90−94 pores, 9 + 8

FIGURE 4.233

Mustura tigrina, 73.6 mm SL, male, Changa River, Chindwin drainage, Manipur.

branched caudal-fin rays, axillary pelvic-fin lobe present but not prominent, males with suborbital flap, and 10 pores in preoperculo-mandibular canal. Maximum size ~74 mm SL.

Distribution. Changa River, Phungrei, tributary of Tizu River, Chindwin drainage. Ukhrul District, Manipur, India.

Mustura tuivaiensis (Lokeshwor, Vishwanath & Shanta, 2012)
(Fig. 4.234)

FIGURE 4.234

Mustura tuivaiensis: (A) 46.0 mm SL, female, Tuivai River at Likhailok, Manipur; (B) in live.

Physoschistura tuivaiensis Lokeshwor, Vishwanath & Shanta, 2012: 6, figs. 1–5 (type locality: Likhailok stream, Brahmaputra River basin, Churachandpur District, Manipur, India).

Diagnosis. A species with 12–14 elongated olivaceous dark blotches on flanks originating from lateral line or slightly above extending below but not reaching belly, 15–17 saddles on back occupying the dorso-lateral portion of flanks sometimes alternate and interrupted with blotches forming a sliding appearance with each other anteriorly, dorsal and dorso-lateral portion of head mottled with dark spots, four simple and 8½ branched dorsal-fin rays, complete lateral line, caudal fin with 8 + 7 branched rays, axillary pelvic lobe well formed, males with suborbital flaps, infraorbital canal with 4 + 8 pores, basicaudal bar black dissociated with an upper small oblique bar and a lower vertical elongated bar. Maximum size ~46 mm SL.

Distribution. Tuivai River and its tributaries, Manipur, Barak drainage, India.

Mustura walongensis (Tamang & Sinha, 2016)
(Fig. 4.235)

Physoschistura walongensis Tamang and Sinha, 2016: 281, figs. 1, 2a, 3a (type locality: small diverted water course of Lohit River at Walong, Brahmaputra River basin, Arunachal Pradesh, India.

Diagnosis. Combination of characters: lateral line incomplete, reaching origin or upto tip of anal-fin; canal pores 71–89; suborbital flap in males rounded, the posterior margin of which

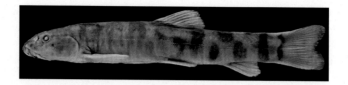

FIGURE 4.235

Mustura walongensis, 53.4 mm SL, Lohit River at Walong, Brahmaputra River basin, Arunachal Pradesh, India.

slightly exceeds the anterior orbital margin, occupying less orbital area; body with 11−18 irregular vertical bars on the flank, more irregular on anterior portion of the body; one to four bars bifurcate ventrally; one to four saddles along the ventro-lateral side of the body; a moderately thick W-shaped basicaudal bar; a prominent axillary pelvic fin lobe; dark brown mottled markings on the head; lips pleated; and forked caudal fin with 10 + 9 principal rays. Maximum size ∼60 mm SL.

Distribution. Lohit River, Walong, Brahmaputra drainage, Anjaw District, Arunachal Pradesh, India.

Genus *Nemacheilus* Bleeker, 1863

Nemacheilus Bleeker, 1863b: 34 (type species: *Cobitis fasciata* Valenciennes, 1846). Gender: masculine.

Diagnosis. Body elongated, lateral line complete, enlarged scales may be present above and below lateral line, acuminate scales may be present above and below lateral line on caudal peduncle, caudal fin forked to deeply forked with prolonger upper lobe, eyes large, mouth small, strongly arched, lips usually thin, usually no median interruption in lower lip, upper jaw with processes dentiformes, lower jaw without median notch, barbels long, males usually have a suborbital flap, thickened pectoral rays two to six with several rows of tubercles and a smaller size.

Nemacheilus corica (Hamilton, 1822)
(Fig. 4.236)

FIGURE 4.236

Nemacheilus corica, ZSIF 9410, Kosi River.

Cobitis corica Hamilton, 1822: 359, 395 (type locality: Kosi River, India).

Diagnosis. A nemacheiline species with a row of 11 rounded to oval blotches on yellowish background along the lateral line, 11 saddles on the dorsum, 7½ branched dorsal-fin rays, forked caudal fin with 8 + 7 branched rays, basicaudal bar dissociated with a median blotch and a small spot near the dorsal extremities, two black spots on dorsal-fin base, an anterior one at bases of simple and first branched dorsal fin rays, the posterior one at the bases of branched rays 4−7, other fins hyaline, basicaudal bar dark brown, dissociated with a median blotch and a small pot near the dorsal extremities of caudal-fin base; males with suborbital flap. Maximum size ∼42 mm SL.

Distribution. Ganga and Brahmaputra river drainages.

Genus *Neonoemacheilus* Zhu & Guo, 1985

Neonoemacheilus Zhu & Guo, 1985: 321 (type species *Nemacheilus labeosus* Kottelat, 1982). Gender: masculine.

Diagnosis. A nemacheiline genus with hypertrophied lips with open oral cavity (Fig. 4.237); nostrils close together, anterior nostril not in the form of barbel; scales imbricated; lateral line complete; caudal fin deeply emarginated; suborbital flap in males.

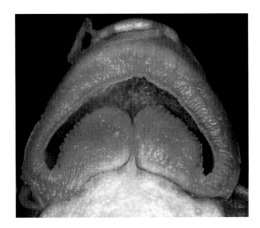

FIGURE 4.237

Mouth of *Neonoemacheilus* showing hypertrophied lips with oral cavity.

Neonoemacheilus assamensis (Menon, 1987)
(Fig. 4.238)

FIGURE 4.238

Neonoemacheilus assamensis, 41.4 mm SL, Jiri River, Barak drainage, Manipur.

Noemacheilus assamensis Menon, 1987: 179 (type locality: Pagladia River, Assam, India).

Diagnosis. Light brown bars, 13—17 (20—21 according to Menon, 1987), extending from dorsum to two-third of flank, bars not reaching ventral surface; branched dorsal fin rays 8½; upper lip thin, moderately hypertrophied with a pad-like structure in the middle of upper lip; snout with four transverse stripes; complete lateral line, forked caudal fin with 9 + 9 branched rays; snout with four transverse brown stripes, basicaudal bars dissociated with a median brown blotch and a brown spot on the dorsal extremities. Maximum size ∼43 mm SL.

Distribution. Barak—Surma—Meghna drainage, Assam and Pagladia River, Brahmaputra basin, Assam.

Neonoemacheilus peguensis (Hora, 1929)
(Fig. 4.239)

FIGURE 4.239

Neonoemacheilus peguensis, 38.6 mm SL, Tamu market, Myanmar.

Nemachilus peguensis Hora, 1929: 321, pl. 14, figs. 1—2 (type locality: Burma: Pegu Yoma range).

Neonoemacheilus morehensis Arunkumar, 2000a, 2000b, 2000c: 44, fig. 1 (description from Lokchao River at Moreh, Indo-Myanmar border, Chindwin drainage, Manipur).

Diagnosis. A species with 20—21 thin bars, bars narrower than interspaces, black basicaudal bar complete, a small dark brown spot at the base of first few dorsal-fin rays, other fins unspotted hyaline; dorsal fin with 9½ branched rays; body covered with scales, belly naked, lateral line complete with 75—88 pores, preoperculo-mandibular canal with 9—10 pores; caudal fin forked with 9 + 8 branched rays, caudal peduncle depth 13.2% SL; anterior nostril pierced on the front side of a flap-like tube; short thumb shaped suborbital flap in males. Maximum size ∼55 mm SL.

Distribution. Khuga River, Manipur, Chindwin basin, India; Irrawaddy basin and Sittang basins, Myanmar.

Genus *Paracanthocobitis* Grant, 2007

Acanthocobitis (*Paracanthocobitis*) Grant, 2007: 3 (type species: *Cobitis zonalternans* Blyth, 1860). Gender: feminine.

Diagnosis. Body elongated, usually spotted; lower lip with papillated pad on either side of the medial interruption, interruption narrow; lips continuous; caudal fin emarginated/truncated caudal fin; 9½—15½ branched dorsal fin rays; rounded head in lateral view; horizontally oriented suborbital flap in males; anus closer to anal fin origin than to pelvic fin origin (Fig. 4.240).

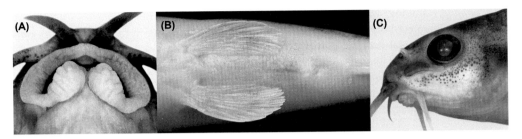

FIGURE 4.240

Paracanthocobitis: (A) showing lip-structure; (B) anus position, and (C) suborbital flap in male.

Paracanthocobitis abutwebi Singer & Page, 2015

Paracanthocobitis abutwebi Singer & Page, 2015: 382 (type locality: Rangapani, Sylhet, Meghna basin, Bangladesh).

Diagnosis. Complete lateral line with 87–103 pores; no axillary pelvic lobe; 11½ dorsal-fin rays; alternating large and small dark spots along lateral line; pattern of dark specks on upper side between small dorsal saddles; four to five dark bands on caudal fin of adult; 13 pectoral-fin rays; 10–13 small dorsal saddles not extending to lateral line and becoming closer together toward head, sometimes with small spots or dashes between saddles; 10–16 dark spots along lateral line, extending onto lower side but not onto venter; speckled pattern on upper side sometimes extending onto lateral line. Round black spot in ocellus near dorsal margin of caudal peduncle; four to five dark bands on caudal fin. Maximum size ~53 mm SL.

Distribution. Karnaphuli, Meghna, and lower Brahmaputra and Ganges river drainages of Bangladesh.

Paracanthocobitis botia (Hamilton, 1822)

(Fig. 4.241)

FIGURE 4.241

Acanthocobitis botia: (A) 57.7 mm SL, Kaldiya River, Barpeta District, Assam; (B) 42.6 mm SL, Manas River, Goalpara, Assam.

Cobitis botia Hamilton, 1822: 350, 394 (type locality: northeastern Bengal).

Diagnosis. Complete lateral line with 83−100 pores; suborbital flap in male; no axillary pelvic lobe; 10½ branched dorsal-fin rays; 12 pectoral-fin rays; eight pelvic-fin rays; small dorsal saddles, narrower than or equal to interspaces, not extending to lateral line; 8−10 oblong to squarish dark blotches along and just below lateral line; a black rounded spot in ocellus near dorsal margin of caudal peduncle; five to seven bars on the caudal fin. Maximum size ∼140 mm SL.

Distribution Brahmaputra and its tributaries in Assam, India.

Paracanthocobitis linypha **Singer & Page, 2015**
(Fig. 4.242)

FIGURE 4.242

Paracanthocobitis linypha, 41.5 mm SL, Yu River, Tamu, Myanmar.

Paracanthocobitis linypha Singer & Page, 2015: 390 (type locality: Kha Wan, Kha Wan stream, Myanmar).

Diagnosis. Body with 10−14 thin dark saddles, often connecting with bars on the sides, shorter bars in between; interspaces wider; ocellus with roundish black spot near dorsal margin of caudal peduncle; caudal fin with five to seven bars; branched dorsal-fin rays 10½; pectoral-fin rays 12; pelvic-fin rays 8; suborbital flap in male; no axillary pelvic lobe; lateral line incomplete with 22−42 pores, extending before the end of addressed pelvic fin. Maximum size ∼43 mm SL.

Distribution Irrawaddy and Sittang basins, Myanmar; Chindwin basin, Manipur, India near Myanmar border.

Paracanthocobitis mackenziei **(Chaudhuri, 1910)**
(Fig. 4.243)

FIGURE 4.243

Acanthocobitis mackenziei, 66.5 mm SL, Ganga River, Patna, Bihar.

Nemachilus mackenziei Chaudhuri, 1910: 183 (type locality: northern India at Cheriyadhang near Kathgodam; Jaulasal, Nainital District, Uttar Pradesh; Jharai and Jamwari Nadi near Siripur, Saran District, Bengal; Purnahia, Champaran District).

Diagnosis. Lateral line complete or almost compete; axillary pelvic lobe present; suborbital flap in males; branched dorsal-fin rays 10½−12½; 11−13 thin dark brown saddles descending to connect with blotches below the lateral line forming a zig-zag pattern; a round black spot below upper margin of caudal peduncle; on caudal fin with five to seven v-shaped bars, their apex pointing posteriorly. Maximum size ∼73 mm SL.

Distribution. Widely distributed in the Gaga River basin of Nepal and India, the Meghna River basin in Bangladesh, the Mahanadi River basin in eastern India, and the upper Indus River basin of northern India and eastern Pakistan.

Paracanthocobitis marmorata Singer, Pfeiffer & Page, 2017
(Fig. 4.244)

FIGURE 4.244

Acanthocobitis marmorata: (A) 42.7 mm SL, male; (B) 39.9 mm SL, female, both from tributary stream of Imphal River, Kalapahar, Manipur; (C) Hora's (1921a) Pl. X, fig. 3 from Manipur valley labeled *Nemachilus zonalternans* reproduced.

Paracanthocobitis marmorata Singer, Pfeiffer & Page, 2017: 103, fig. 11 (type locality: Kangchup hills) (Chindwin drainage, Manipur).
Nemachilus zonalternans: Hora, 1921a: 199, pl X, figs 3, 3a (description from Manipur valley).
Paracanthocobitis tumitensis Arunkumar & Moyon, 2019: 102, figs. 1−3 (description from Tumit River, Chumbang Village, Chandel District, Chindwin drainage, Manipur).

Diagnosis. Lateral line incomplete ending just beyond dorsal-fin insertion, 25—31 pores; branched dorsal-fin rays 9½—10½; no axillary pelvic lobe; no black stripe alongside of body; marmorated pattern of dark lines and blotches between irregularly shaped saddles, dark blotches alongside of body; teardrop-shaped black spot on upper margin of caudal-fin base; eight branched upper caudal-fin rays. Maximum size ~43 mm SL.

Distribution. Kangchup Hills, Chindwin drainage, Manipur, India.

Paracanthocobitis triangula (**Singer, Pfeiffer & Page, 2017**)

Paracanthocobitis triangula Singer, Pfeiffer & Page, 2017: 101, fig. 10 (type locality: Meghna drainage, Bangladesh).

Diagnosis. Incomplete lateral line ending near dorsal-fin insertion, 20—37 pores; 9½—10½ branched dorsal-fin rays; axillary pelvic lobe present; series of black blotches, sometimes overlain with faint dusky stripe alongside of body not obscuring lateral blotches, ending just beyond dorsal-fin insertion; small black triangular blotch in ocellus on upper margin of caudal-fin base; black pigment of the midlateral stripe not extending onto pectoral-fin base; no marmorated pattern between dorsal saddles and lateral blotches; dorsal saddles usually extending ventrally just past faint lateral stripe, usually connecting to lateral blotches; eight branched upper caudal-fin rays. Maximum size ~29 mm SL.

Distribution. Meghna drainage and Jamuna in Brahmaputra drainage.

Genus *Physoschistura* Bănărescu & Nalbant, 1982

Physoschistura Bănărescu & Nalbant, in Singh et al., 1982: 208 (type species: *Nemacheilus brunneanus* Annandale, 1918). Gender: feminine.

Diagnosis. Mouth strongly arched, 1.5—2.0 wider than long, lower lip with median interruption forming two lateral broadly triangular pads with deep furrows, air bladder with a conical free posterior chamber immediately behind capsule, size small, less than 40 mm SL.

Physoschistura elongata Sen & Nalbant, 1982
(Fig. 4.245)

FIGURE 4.245

Physoschistura elongata, 46.2 mm SL, ZSI-ERS, Shillong, Barapani, Meghalaya.

Physoschistura elongata Sen & Nalbant in Singh et al., 1982: 210, figs. 17—20 (type locality: Barapani, near Shillong, Meghalaya, Brahmaputra basin, India).

Diagnosis. A species characteristic in having a slender body (13-2—17.1% SL); eight branched dorsal fin rays; 9—14 bars on body, three to four in front of dorsal and four to six behind; an incomplete lateral line, reaching the region below dorsal fin origin or slightly behind and scales present in the posterior half of body, Maximum size ~48 mm SL.

Distribution. Brahmaputra basin, northeast India.

Genus *Rhyacoschistura* **Kottelat, 2019**

Rhyacoschistura Kottelat, 2019: 152 (type species: *Rhyacoschistura larreci*, 2019). Gender: feminine.

Diagnosis. Pectoral fin in male slanted upward and expanded laterally, the first branched ray rigid and wider, about four times than the following rays, without a membrane between the branches except the tip, anterior branch not further branched, posterior branch branched again covered by small pointed tubercles at maturity; between the branches except near the tip; the capsule of the anterior chamber of the air bladder in two halves, connected by a manubrium; the lower lip with a wide median interruption, the two halves forming an acute angle, not in contact medially, wide and fleshy medially, partly free from the jaw and connected to the isthmus by a frenum; the suborbital flap is present or absent; the body depth is about equal from behind head to caudal-fin base; depth of caudal peduncle 1.1−1.3 times in its length; caudal fin emarginated; and scales very distinct, covering whole body, including predorsal and prepectoral areas.

"Rhyacoschistura" ferruginea (**Lokeshwor & Vishwanath, 2013**)
(Fig. 4.246)

FIGURE 4.246

Rhyacoschistura ferruginea, 41.1 mm SL, rivulet draining to Barak River, Maram, Senapati District, Manipur.

Schistura ferruginea Lokeshwor & Vishwanath, 2013a: 50, figs. 1−4 (type locality: rivulet of Barak River at Don Bosco College campus, Maram, Manipur, India).

Diagnosis. A species in having a combination of characters: 17−19 thin dark brown bars on flank, 12 pectoral-fin rays, eight pelvic-fin rays, 7½ branched dorsal-fin rays; presence of suborbital flap in males and absence of an axillary pelvic lobe. Maximum size ∼47 mm SL.

Distribution. Rivulet in Maram, tributary of the Barak at Maram, Senapati District, Manipur, India.

"Rhyacoschistura" maculosa (**Lalronunga, Lalnuntluanga & Lalramliana, 2013**)
(Fig. 4.247)

FIGURE 4.247

Schistura maculosa, 75.3 mm SL, Pharsih River, Braka drainage, Mizoram.

Photo courtesy: Lalramliana.

Schistura maculosa Lalronunga et al., 2013: 584 (type locality: Pharsih River, a tributary of Tuivai River, Barak drainage, in the vicinity of Kawlbem, Champhai District, Mizoram, India).

Diagnosis. Three to four rows of black spots horizontally across dorsal-fin; five to seven more or less organized rows of black spots on rays vertically across caudal fin; slightly emarginated caudal fin with 8 + 8 branched rays; 20−30 narrow black bars on the body; incomplete lateral line extending up to vertical through pelvic-fin origin; with 26−35 pores; males with a suborbital flap; and intestine looped behind the stomach. Maximum size ∼76 mm SL.

Distribution. Tuingo and Pharsih Rivers, tributaries of Tuivai River, Barak drainage, Mizoram, India.

"Rhyacoschistura" manipurensis **(Chaudhuri, 1912)**
(Fig. 4.248)

FIGURE 4.248

Rhyacoschistura manipurensis, (A) 63.0 mm SL, Chiru Falls; (B) 71.5 mm SL, Sekmai River, Pallel, and (C) 68.0 mm SL, Imphal River, Kanglatongbi, Manipur.

Nemachilus manipurensis Chaudhuri, 1912: 443, pl. 40, fig. 4; pl. 41, fig. 1 (type locality: Chindwin basins of Nagaland and Assam, India).

Diagnosis. A species with thin irregular 17−21 bars on the flank, usually broken, processus dentiformis, 8 + 7 branched caudal fin rays, 7½ branched dorsal fin rays, suborbital flap in males, incomplete lateral line with 35 pores, reaching pelvic fin origin, two oblique rows of spots on dorsal fin and no axillary pelvic lobe. Maximum size ∼70 mm SL.

Distribution. Chindwin basin in Manipur and Nagaland, Manipur.

"Rhyacoschistura" porocephala **(Lokeshwor & Vishwanath, 2013)**
(Fig. 4.249)

FIGURE 4.249

Rhyacoschistura porocephala, 53.5 mm SL, paratype, Mat River, Lunglei District, Mizoram.

Schistura porocephala Lokeshwor & Vishwanath, 2013b: 2, figs. 1−5 (type locality: Stream of Mat River near Thualthu, Lunglei District, Mizoram, India).

Diagnosis. A species having a cephalic lateral line system consisting of prominent pores, lateral line incomplete with 27−38 pores, 20−23 thin olivaceous dark bars on body against yellowish cream background, three simple and 7½ branched dorsal-fin rays, two rows of black spots on dorsal fin, slightly emarginated caudal fin and an elongated suborbital flap. Maximum size ∼53 mm SL.

Distribution. Mat River at Thualthu, Koladyne basin, Lunglei District, Mizoram, India.

Genus *Schistura* McClelland, 1838

Schistura McClelland, 1838: 944, 947 (type species: *Schistura rupecula* McClelland, 1838). Gender: feminine.

Diagnosis. Mouth arched moderately, width 2.0−3.5 of length, lower lip with median interruption, not forming two lateral pads; median notch in lower jaw present or not; body sides usually with more or less regular bars, usually a black bar, two spots or a bar and spot at caudal-fin base; dorsal fin with one or two black marks along its base; caudal fin truncated, emarginated or forked; sexual dimorphism present or not, if present, males usually with suborbital flap.

Schistura aizawlensis Lalramliana, 2012

Schistura aizawlensis Lalramliana, 2012: 98, figs. 1 and 2 (type locality: Mizoram, Muthi River, a tributary of Tuirial River in vicinity of Zemabawk, Aizawl, India).

Diagnosis. A species with five to seven regular broad dark brown bars, typically two predorsal, two postdorsal, two postdorsal; dorsal side of pectoral fin with small tubercules; a black spot at base of simple and first branched rays of dorsal fin, basicaudal bar black, continuous, orange color patch between last bar and caudal-fin base, usually at upper half, two rows of dark spot at posterior half of caudal-fin; very low or no adipose crest on dorsal and ventral side of caudal peduncle; intestine without loop but slightly bent some distance behind the stomach, and males with suborbital flap. Maximum size ∼49 mm SL.

Distribution. India: Mizoram: Aizawl: Muthai River, a tributary of Tuirial River in the vicinity of Zemabawk, Barak-Surma-Meghna basin.

Schistura beavani (Günther, 1868)

Nemachilus beavani Günther, 1868: 350 (type locality: Kosi River, Uttar Pradesh, India).

Diagnosis. A species which has a dorsal fin origin nearer to snout-tip than to caudal-fin base, caudal fin emarginated, body with nine bars, broader than interspaces, a black basicaudal bar, black dots on dorsal and caudal fin. Maximum size ∼77 mm SL.

Distribution. Kosi River basin, Uttar Pradesh.

Schistura chindwinica **(Tilak & Husain, 1990)**
(Fig. 4.250)

FIGURE 4.250

Schistura chindwinica, 48.0 mm (A), 52.0 mm (B) and 56.0 mm (C), all from Leimatak River, tributary of Barak River, Manipur.

Nemacheilus chindwinicus Tilak & Husain, 1990: 51, figs. 1−5 (type locality: Brahmaputra basin, Lanjha stream, Manipur, India).

Diagnosis. A *Schistura* with 8−11 obscure dark gray bars, interspaces gradually wider behind dorsal-fin origin, 8½ branched dorsal-fin rays, a black spot on the bases of simple and first branched dorsal-fin rays, a thick row of spots at mid-length of branched rays; basicaudal bar black extending from the base of lower simple principal ray to about three-fourth of caudal peduncle depth, a black spot at the upper extremity of caudal-fin base; caudal fin with two faint vertical rows of spot at mid length; male with conspicuous suborbital flap, processus dentiformes weak, and no adipose crest. Maximum size ∼65 mm SL.

Distribution. Barak River and its tributaries, Manipur.

Schistura devdevi **(Hora, 1935)**
(Fig. 4.251)

Noemacheilus devdevi Hora, 1935: 54, pl. 3, figs. 5 and 6 (type locality: Eastern Himalayas; small streams below Darjeeling and Sikkim, India).

Diagnosis. A species with 8½ branched dorsal fin rays, incomplete lateral line, terminates above ventral fin lunate caudal fin four to six saddles extending upto lateral line or slightly beyond as bars. Ventral surface scale less; a few broad irregular bars wider than interspaces; bars extending to below lateral line or only upto half of body; dorsal fin with two bars. Maximum size ∼40 mm SL.

Distribution. Teesta and Brahmaputra drainage, Eastern Himalaya.

FIGURE 4.251

Schistura devdevi, 34.2 mm SL, Tezu River, Lohit District, Arunachal Pradesh.

Schistura fasciata **Lokeshwor & Vishwanath, 2011**
(Fig. 4.252)

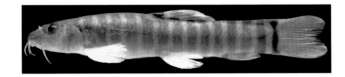

FIGURE 4.252

Schistura fasciata, 51.5 mm SL, holotype, Barak River, Maram Hill, Senapati District, Manipur.

Schistura fasciata Lokeshwor & Vishwanath, 2011: 1515, figs. 1−3 (type locality: Barak River at western side of Maram Hill, Senapati District, Manipur, India).

Diagnosis. A species with a combination of 11−13 dark brown bars on the body against pale yellow background, bars arranged regularly, often fused on mid-dorsal line, bar wider than inter-space; moderately high adipose crest on dorsal and ventral sides of caudal peduncle; lateral line incomplete, reaching vertical to posterior end of anal-fin base; three black spots on dorsal-fin base; 8½ branched dorsal fin rays; large processus dentiformes. Maximum size ∼68 mm SL.

Distribution. Barak River at the western side of Maram Hill, Senapati District, Manipur, India.

Schistura kangjupkhulensis **(Hora, 1921)**
(Fig. 4.253)

FIGURE 4.253

Schistura kangjupkhulensis, 47.6 mm SL, Leimatak stream, Chindwin basin, Manipur.

Nemachilus kangjupkhulensis Hora, 1921a: 202, pl. 10, fig. 4 and 4a (type locality: hill streams of the Manipur valley, Yairipok, Manipur, India).

Diagnosis. A species with processus dentiformes, 7−11 irregular bars on the flank, bars wider than interspace, no suborbital flap in males, incomplete lateral line, reaching about posterior tip of pectoral fin, emarginated caudal fin, black basicaudal bar and axillary pelvic lobe. Maximum size ∼70 mm SL.

Distribution. Chindwin basin, Manipur, India.

Schistura khugae **Vishwanath & Shanta, 2004**
(Fig. 4.254)

FIGURE 4.254

Schistura khugae, 62.0 mm SL, Khuga River, Manipur.

Schistura khugae Vishwanath & Shanta, 2004a: 330 (type locality: Khuga River, Churachandpur District, Manipur, India).

Diagnosis. A species with adipose keel between dorsal and caudal fins; inflated cheeks and swelling body on anterior part in males; upper lip without median incisor; lower lip interrupted in the middle; processus dentiformes with a median notch in lower jaw; eight branched dorsal fin rays; axillary pelvic lobe and incomplete lateral line. Maximum size ∼88 mm SL.

Distribution. Khuga River, Chindwin basin, Manipur, India.

Schistura koladynensis **Lokeshwor & Vishwanath, 2012**
(Fig. 4.255)

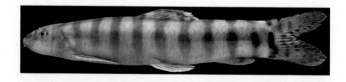

FIGURE 4.255

Schistura koladynensis, 58.6 mm SL, Kaladan River at Kolchaw, Mizoram.

Schistura koladynensis Lokeshwor & Vishwanath, 2012c: 140 (type locality: Mizoram, Lawntlai District, Kaladan River at Kolchaw, India).

Diagnosis. A species with complete lateral line; deeply forked caudal fin; 10−11 dark brown saddles descending on sides to form bars; on the caudal peduncle, the saddles often dissociated from the bars; the black bar on the caudal-fin base confined to the middle and usually not connected with the corresponding saddle; caudal fin with one or two wavy vertical rows of black spots anterior to the fork and another two radiating from the fork, of which the posterior one along the distal margin of the fin and the anterior one at the middle of the lobes. Maximum size ~76 mm SL.

Distribution. Kaladan River, Mizoram, India.

Schistura minuta **Vishwanath & Shanta Kumar, 2006**
Schistura minutus Vishwanath and Shanta Kumar, 2006: 210, figs. 1 and 3B (type locality: Iyei River, Noney, Tamenglong District, Manipur, India).

Diagnosis. A small species (26.0–38.3 mm SL) with processus dentiformes not prominent, 14–18 bars on the body, bars may be paired, incomplete lateral line, a distinct cup shaped band behind the occiput, a spot at dorsal fin origin and basicaudal bar in the form of two spots. Maximum size ~38 mm SL

Distribution. Barak basin in Manipur, India.

Schistura mizoramensis **Lalramliana, Lalronunga, Vanramliana & Lalthanzara, 2014**
(Fig. 4.256)

FIGURE 4.256

Schistura mizoramensis, 47.7 mm SL, Tuirivang River, Barak drainage, Mizoram.

Photo courtesy: Lalramliana.

Schistura mizoramensis Lalramliana, Lalronunga, Vanramliana & Lalthanzara, 2014b: 206, figs. 1 and 2 (type locality: Tuirivang River, tributary of Tuirial River, Barak drainage, Mizoram, northeastern India).

Diagnosis. A species distinct in having 12–16 saddles, extending on the flank as bars to meet the dark brown longitudinal stripe along mid-lateral line, yellowish interspaces slightly reticulate anteriorly and ovoid posteriorly; a complete lateral line and no suborbital flap in males; lateral line complete with 83–95 pores and total vertebrae 35. Maximum size ~56 mm SL.

Distribution. Tuirivang, a tributary of Tuirial River, Barak River drainage, Mizoram, India.

Schistura multifasciata **(Day, 1878)**
Nemacheilus multifasciatus Day, 1878: 617, pl. 153 (type locality: Darjeeling and Assam, India).

Diagnosis. A species with robust body with 14–16 bars as wide as interspaces, truncated caudal fin, 7½ dorsal branched rays and complete lateral line. Maximum size ~98 mm SL.

Distribution. Teesta drainage in Nepal and northern Bengal and Brahmaputra drainage in Assam, India.

Schistura nagaensis **(Menon, 1987)**
(Fig. 4.257)

FIGURE 4.257

Schistura nagaensis, 41.0 mm SL, Challou River, tributary of Tizu River, Nagaland-Manipur border, near Jessami.

Noemacheilus nagaensis Menon, 1987: 117 (type locality: Phodung River, tributary of Tizu River, Brahmaputra basin, Naga Hills, Nagaland, India).

Diagnosis. A species of *Schistura* with 8½ branched dorsal fin rays; 7−14 bars on body; slightly emarginated caudal fin; incomplete lateral line, reaching upto about tip of pectoral fin; and two rows of spots each on caudal and dorsal fins. Maximum size ∼59 mm SL.

Distribution. Tizu river Manipur and Nagaland, Chindwin basin, India.

Schistura papulifera Kottelat, Harries & Proudlove, 2007
(Fig. 4.258)

FIGURE 4.258

Schistura papulifera, 45.1 mm SL, holotype, Jaintia Hills, Krem Umsngat entrance, Meghalaya.

Photo courtesy: M. Kottelat.

Schistura papulifera Kottelat, Harries & Proudlove, 2007: 36, figs. 1−3 (type locality: Krem Umsngat entrance, Jaintia Hills to Synrang Pamiang Cave system, Meghalaya, India).

Diagnosis. A species with pale white body, vestigial eye which is subcutaneous and externally appearing as a small diffuse blackish spot, not communicating with outside by a canal; lower half of head covered by numerous small skin projections and five pores in supratemporal canal of the cephalic lateral line system; 8½ branched dorsal fin rays; incomplete lateral line, extending to level of pelvic fin origin and no axillary pelvic bone. Maximum size ∼49 mm SL.

Distribution. Synrang Pamiang cave system, Jaintia Hills, Meghalaya, India.

Schistura paucireticulata Lokeshwor, Vishwanath & Kosygin, 2013
(Fig. 4.259)

FIGURE 4.259

Schistura paucireticulata, 61.4 mm SL, holotype, Tuirial River near Aizawl, Barak drainage, Mizoram.

Schistura paucireticulata Lokeshwor, Vishwanath & Kosygin, 2013: 582, figs. 1−4 (type locality: Mizoram: Aizawl District: Tuirial River near Aizawl (Barak−Surma−Meghna River system).

Diagnosis. A species with eight to nine bars on body, saddles in front of dorsal fin dividing into two to three small bars forming reticulated appearance as they descend onto side of the body; a black basicaudal bar, dissociated into an upper spot and a lower bar, spotted dorsal fin with 8½ branched rays, complete lateral line, well developed axillary pelvic lobe and a caudal fin with black spots arranged in three to four bars. Maximum size ∼62 mm SL.

Distribution. Tuirial River, tributary of the Barak River, Mizoram, India.

Schistura rebuw **Chowdhuri, Dey, Bharali, Sarma & Vishwanath, 2019**
(Fig. 4.260)

FIGURE 4.260

Schistura rebuw, 69.8 mm SL, Pachai stream at Seppa, draining into Kameng River, Arunachal Pradesh.

Schistura rebuw Chowdhuri, Dey, Bharali, Sarma & Vishwanath,2019: 41 (type locality: Pachai stream at Seppa, draining into Kameng River, Brahmaputra basin, East Kameng District, Arunachal Pradesh, India).

Diagnosis. A light gray body with 10−11 dark brown bars, interspace slightly wider than the bars, predorsal bars broken and/or incomplete, coalescing dorsally in a more or less alternate fashion to form irregular saddles; bars from dorsal fin origin and behind complete, continuing dorsally to form saddles; all bars overlayered on a faint mid-lateral stripe extending from behind the opercle to caudal-fin base; bars on cheek and head reticulated; females with a small suborbital slit and males with a suborbital flap; a compete lateral line; a prominent axillary pelvic lobe; and a deeply emarginated caudal fin. Maximum size ∼71 mm SL.

Distribution. Pachai stream, a feeder of Kameng River at Seppa, Brahmaputra drainage, East Kameng District, Arunachal Pradesh, India.

Schistura reticulata **Vishwanath & Nebeshwar, 2004**
(Fig. 4.261)

FIGURE 4.261

Schistura reticulata: (A) 58.6 mm SL, male; (B) 60.8 mm SL, female, both from Maklang River, Chindwin drainage, Ukhrul District, Manipur.

Schistura reticulata Vishwanath & Nebeshwar, 2004: 324, figs. 1 and 2 (type locality: Maklang River, Ukhrul District, Manipur, India).

Diagnosis. A species with 17−29 bars on body, bars in front of dorsal fin thinner, split and reunite to give reticular appearance; suborbital flap in males, forked caudal fin having two to three v-shaped bars, apex pointing toward base; black basicaudal bar, 8½, dorsal fin branched rays and complete lateral line. Maximum size ~77 mm SL.

Distribution. Maklang, Lokchao and Wanze streams, Chindwin basin, Manipur, India.

Schistura reticulofasciata (Singh & Banarescu, 1982)

Mesonoemacheilus reticulofasciatus Singh & Bănărescu in Singh et al., 1982: 206, figs. 12−16 (type locality: Barani, near Shillong, Meghalaya, India).

Diagnosis. A species with compressed small head; seven or eight branched dorsal fin rays; interrupted lateral line; a net of numerous irregular bar, most of which are vertical and connected by one or two longitudinal stripes; a roundish spot at caudal-fin base. Maximum size ~48 mm SL.

Distribution. Brahmaputra basin, Meghalaya, India.

Schistura rosammae (Sen, 2009)

Aborichthys rosammae Sen, 2009: 15, Fig. 1A and B, pl. I and II (type locality: Pabomukh, Subansiri River, Dhemaji District, Assam, India).

Diagnosis. A species characteristic in having vent situated nearer to caudal-fin base than to snout-tip; body marked with 10−11 bars coalesced on caudal peduncle region; caudal fin truncate with no bars and a black basicaudal bar. Maximum size ~27 mm SL.

Distribution. Brahmaputra basin in Arunachal Pradesh and Assam.

Schistura savona (Hamilton, 1822)

(Fig. 4.262)

Cobitis savona Hamilton, 1822: 357, 394 (type locality: Kosi River at Nathpur, Uttar Pradesh, India.

FIGURE 4.262

Schistura savona, 38.3 mm SL,(A) lateral view, (B) dorsal view, Kameng River, West Kameng District, Arunachal Pradesh.

Diagnosis. A species of *Schistura* with nine bars on the body, bars wider than interspace, 9½ branched dorsal fin rays, deeply emarginated caudal fin, complete basicaudal bar, black bar on dorsal and anal fins and two to three on caudal fin. Maximum size ∼40 mm SL.

Distribution. Kali drainage, Uttar Pradesh to Brahmaputra drainage, Arunachal Pradesh, India.

Schistura scaturigina McClelland, 1839
Schistura scaturigina McClelland, 1839: 308, 443, pl. 53, fig. 6 (type locality: Assam, India).

Diagnosis. Body with 9−12 bars, broader dorsally, narrowing down on sides, extending to two-third of body, scales absent on ventral surface, deeply emarginated or forked caudal fin, caudal fin with black spots forming v-shaped bar. Maximum size ∼95 mm SL.

Distribution. Brahmaputra drainage, northeast India.

Schistura sijuensis (Menon, 1987)
(Fig. 4.263)

FIGURE 4.263

Schistura sijuensis, 63.5 mm SL, Umngi, Balat, Meghna basin. South West Khasi Hills, Meghalaya.

Noemacheilus sijuensis Menon, 1987: 175, pl. 6, fig. 2 (type locality: Siju cave, Meghalaya).

Diagnosis. A species with 8½ branched dorsal fin rays, complete lateral line, 8−10 brown saddles, continuing as bars on the sides, bars anterior to anal fin origin breaking up below lateral line; well-developed processus dentiformes; suborbital flap in males; caudal fin forked; basicaudal bar and two v-shaped bars in the middle of caudal fin. Maximum size ∼65 mm SL.

Distribution. India: Siju cave, Garo Hills, Meghna drainage, South-west Khasi Hills, Meghalaya.

Schistura sikmaiensis **(Hora, 1929)**
(Fig. 4.264)

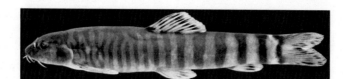

FIGURE 4.264

Schistura sikmaiensis, 64.5 mm SL, Namya River, Indo-Myanmar border, Ukhrul District, Manipur, India.

Nemachilus sikmaiensis Hora, 1921a: 201, pl. 9, fig. 4, pl. 10, fig. 1 and 1a. (Sekmai stream near Pallel, Chindwin basin, Manipur, India. Neotype designated by Kottelat, 1990 from Putao, Myanmar).

Diagnosis. A species characteristic in having a forked caudal fin, 17−21 bars on body; width of bars equals that of interspace; complete lateral line, 8½ branched dorsal fin rays and in the absence of processus dentiformes; caudal fin forked. Maximum size ∼80 mm SL.

Distribution. India: Chindwin basin in Manipur. South west Yunnan, China; Myitkyina and Putao, Myanmar.

Schistura singhi **(Menon, 1987)**
Noemacheilus singhi Menon, 1987: 119, pl. 16, fig. 1 (type locality: Kiphire, Nagaland, India).

Diagnosis. A species diagnosed in having 7½ branched dorsal fin rays; truncated or slightly emarginated caudal fin; 12−13 black bars on body extending from dorsum to below lateral line; bars wider than interspace behind dorsal fin and lateral line incomplete, ending below dorsal fin origin. Maximum size ∼36 mm SL.

Distribution. India: Kiphire, Chindwin drainage, Nagaland.

Schistura syngkai **Choudhury, Mukhim, Dey, Warbah & Sarma, 2019**
(Fig. 4.265)

FIGURE 4.265

Schistura syngkai, 42.1 mm SL, Twahdidoh stream, Surma-Meghna basin, West Khasi Hills District, Meghalaya.

Schistura syngkai Choudhury, Mukhim, Dey, Warbah & Sarma, 2019: 186 (type locality: Twahdidoh stream of Wahblei River, near Seinduli village, Surma-Meghna drainage, West Khasi Hills District, Meghalaya).

Diagnosis. *Schistura* with a color pattern consisting of a prominent dark-brown to black mid-lateral stripe, width of eye diameter or more, overlain with 12−18 ovoid to vertically elongated black blotches; 11−17 dark-brown or greyish saddles, dorsal fin with three oblique bars, one at base and another two in the middle, a dark blotch slightly above the base of the simple rays; caudal fin with two to three bars; lateral line incomplete, extending to the level of the middle of dorsal fin or ahead of pelvic fin origin. Maximum size ∼43 mm SL.

Distribution. Streams and Rivers belonging to the Surma-Meghna drainage, Meghalaya.

Schistura tirapensis (Kottelat, 1990)

Noemacheilus arunachalensis Menon 1987: 129, pl. 16, fig. 10 [type locality: Riwa River at Manpong (Nampong), Arunachal Pradesh, India] (junior primary homonym of *Noemacheilus arunachalensis* Datta and Barman, 1984).

Schistura tirapensis: Kottelat, 1990: 118 (replacement name of *Noemacheilus arunachalensis* Menon, 1987).

Diagnosis. A species with eight branched dorsal fin rays, deeply emarginated caudal fin, incomplete lateral line ending at level of pelvic-fin base, 10−12 bars on body encircling body with yellow interspaces having more or less similar width. Maximum size ∼56 mm SL.

Distribution. Brahmaputra basin, Arunachal Pradesh, India.

Schistura zonata McClelland, 1839
(Fig. 4.266)

FIGURE 4.266

Schistura zonata, 44.2 mm SL, Dibang River, Lower Dibang valley, Arunachal Pradesh.

Schistura zonata McClelland, 1839: 308, 441, pl. 53, fig. 1 (type locality; Upper Assam, India).

Diagnosis. Body with 12−15 bars, interspace wider, extending up to below lateral line, few bars in posterior third of body interrupted, lateral line complete, dorsal fin origin halfway between snout-tip and caudal-fin base. Maximum size ∼50 mm SL.

Distribution. Ganga-Brahmaputra drainage, northern and northeast India.

Family Amblycipitidae
Torrent catfishes

Body slightly compressed; anterior and posterior nostrils close together; anterior nostril situated immediately posterior to the base of the nasal barbel, both lips with double folds, prominent cup like skin flap above the base of pectoral spine; dorsal fin covered by thick skin; adipose fin present, confluent with caudal fin in some species; dorsal-fin base short, spine in fin weak; anal-fin base short, with 9—18 rays; four pairs of barbels; lateral line poorly developed or absent. Pinnate rays along anterior margin of procurrent and medial caudal fin rays (Fig. 4.267).

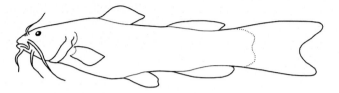

FIGURE 4.267

Outline diagram of Amblycipitidae.

Genus *Amblyceps* Blyth, 1858

Amblyceps Blyth, 1858: 281 (type species: *Amblyceps caecutiens* Blyth, 1858: 281). Gender: masculine.

Diagnosis. Body elongated, subcylindrical. Abodomen rounded. Head small, broad, depressed, covered with thick skin. Snout broadly rounded. Mouth wide, anterior, transverse. Eyes small superior. Lips fleshy. Nostrils close together, separated by nasal barbel. Teeth villiform in bands on jaws. Four pairs of barbels, one each of nasal and maxillary and two of mandibular. A fold of skin above and anterior to pectoral-fin base, and immediately behind gill openings. Dorsal fin inserted above the middle of pectoral fin, with a weak spine. Adipose dorsal fin present. Pectoral fin with a weak spine.

Amblyceps apangi Nath & Dey, 1989
(Fig. 4.268)

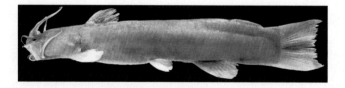

FIGURE 4.268

Amblyceps apangi, 77.8 mm SL, Dikrong River, Doimukh, Arunachal Pradesh.

Amblyceps apangi Nath & Dey, 1989: 2, fig. 2 (type locality: Dikrong River, Arunachal Pradesh, India).

Diagnosis. Skin smooth; jaws equal; rictal fold large, very well developed; pinnate rays on outer margin of principal ray of caudal fin and projections on the median caudal fin rays absent; adipose fin not confluent with caudal fin but very closely placed appearing to be confluent; caudal fin truncated. Maximum size ~150 mm SL.

Distribution. Dikrong River, tributary of the Brahmaputra, Arunachal Pradesh, India.

Amblyceps arunachalensis **Nath & Dey, 1989**
(Fig. 4.269)

FIGURE 4.269

Amblyceps arunachalensis, 82.7 mm SL. Dikrong River, Doimukh, Arunachal Pradesh.

Amblyceps arunachalensis Nath & Dey, 1989: 3, fig. 3 (type locality: Dikrong River, Arunachal Pradesh, India).

Diagnosis. Skin tuberculated; jaws unequal lower longer than upper; rictal fold reduced; pinnate rays on outer margin of procurrent rays and strongly developed projections (Fig. 4.287) on the median principal rays of caudal fin present, i.e., spine-like projections of the rays of the upper lobe facing downwards and those of the rays of the lower lobe facing upwards; caudal fin with upper lobe longer; adipose fin not confluent with caudal fin i.e. widely separated. Maximum size ∼ 130 mm SL.

Distribution. Dikrong River, tributary of the Brahmaputra, Arunachal Pradesh. India (Fig. 4.270).

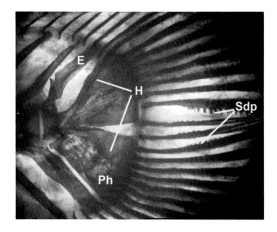

FIGURE 4.270

Caudal complex of *Amblyceps arunachalensis* showing strongly developed projections on the median caudal fin rays; E, epural; H, hypural; Ph; parhypural; Sdp, strongly developed process.

Photo courtesy: A. Darshan.

Amblyceps cerinum **Ng & Wright, 2010**

Amblyceps cerinum Ng & Wright, 2010: 51, figs. 1 and 2 (type locality: Raidak I River at Shipra, near Buxa tiger reserve, West Bengal).

Diagnosis. Truncated caudal fin, long adipose-fin base (32.4%−38.3% SL); slender caudal peduncle (9.2%−11.2% SL); 41−44 post-Weberian vertebrae; lateral line incomplete, terminating at a level just behind the dorsal fin origin, posterior end of adipose fin separated from procurrent caudal fin rays by a distinct notch. Maximum size ~97 mm SL.

Distribution. Brahmaputra River drainage in northern West Bengal, India.

Amblyceps laticeps (McClelland, 1842)
(Fig. 4.271)

FIGURE 4.271

Amblyceps laticeps, 34.79 mm SL, Langkhar Stream, Chirang, District, Assam.

Olyra laticeps McClelland, 1842: 588, pl. 21, fig. 2 (type locality: Khasi Hills, India).

Diagnosis. A species diagnosed in having a narrow head (9.5%−11.1% SL); a strongly projecting lower jaw and a very slender, elongated body (body depth at anus 7.6%−11.1% SL); 41−43 vertebrae and an emarginated caudal fin. Maximum size ~42 mm SL.

Distribution. Brahmaputra basin, northeast India.

Amblyceps mangois (Hamilton, 1822)
(Fig. 4.272)

FIGURE 4.272

Amblyceps mangois, 50.0 mm SL, Jiri River, Barak drainage, Manipur.

Pimelodus mangois Hamilton, 1822: 199, 379 (type locality: Tanks of N. Behar, Kosi River, India).

Diagnosis. Skin smooth; jaws equal; lateral line absent; caudal fin deeply forked, adipose fin not confluent with caudal fin and with rounded posterior margin; number of vertebrae 34−36. Barbels four pairs, compressed throughout their length, nasal when adpressed extends upto dorsal end of head, maxillary extends upto middle of pectoral-fin base, outer mandibular beyond posterior end of pectoral-fin base, inner mandibular extends upto isthmus; median rays of caudal fin with strongly developed projections. Maximum size ~125 mm SL.

Distribution. Brahmaputra drainage, all along the foothills of Himalayas, Bangladesh, and Nepal.

***Amblyceps torrentis* Linthoingambi & Vishwanath, 2008**
(Fig. 4.273)

FIGURE 4.273

Amblyceps torrentis, 93.6 mm SL, Challou River, Chingai village, Ukhrul District, Manipur.

Amblyceps torrentis Linthoingambi & Vishwanath, 2008: 168, Figs. 1−2a, 3 (type locality: Laniye River, Chindwin drainage, Jessami village, Manipur-Nagaland state border, Ukhrul District, Manipur, India).

Diagnosis. Lips with single fold, absence of pinnate rays on procurrent rays of caudal fin, supraoccipital-spine short and not pointed, mesethmoid cornua forming v-shaped median cleft, equal jaw length and smooth dorsal fin spine and total vertebrae 26 + 21. Maximum size ∼96 mm SL.

Distribution. Challou, Wanze, Momo and Laniye River, Ukhrul District, Manipur, India.

***Amblyceps tuberculatum* Linthoingambi & Vishwanath, 2008**
(Fig. 4.274)

FIGURE 4.274

Amblyceps tuberculatum, 90.5 mm SL, Chatrickong River, Kamjong District, Manipur.

Amblyceps tuberculatum Linthoingambi & Vishwanath, 2008: 170, figs. 5−7 (type locality: Lokchao River, Chindwin drainage, Moreh town at Indo-Myanmar border, Chandel District, Manipur, India).

Diagnosis. A species with the combination of characters: lips with double folds, presence of pinnate rays along anterior margin of procurrent rays of upper and lowermost principal caudal fin rays, supraoccipital spine pointed and mesethmoid cornua expand laterally, adipose fin not contiguous with the caudal fin. Maximum size ∼97 mm SL.

Distribution. Chindwin drainage, Manipur, India.

***Amblyceps waikhomi* Darshan, Kachari, Dutta, Ganguly & Das, 2016**
(Fig. 4.275)

FIGURE 4.275

Amblyceps waikhomi, 44.7 mm SL, Nongkon stream, tributary of Noa Dehing River, Arunachal Pradesh.

Amblyceps waikhomi Darshan, Kachari, Dutta, Ganguly & Das, 2016: 2, figs. 1 and 4c (type locality: Nongkon stream at Nongkon village draining into the Noa Dehing River, Brahmaputra River basin, Namsai District, Arunachal Pradesh, India).

Diagnosis. The species is distinct in having the following combination of characters: absence of strongly developed projections on the proximal lepidotrichia of the median caudal-fin rays and in having a longer, wider, and deeper head, posteriorly smooth pectoral spine, a weakly projecting unequal jaw which has a longer lower jaw and deeply forked caudal fin. Maximum size ~45 mm SL.

Distribution. Nongkon stream draining to Noa Dehing River, Brahmaputra basin, Namsai District, Arunachal Pradesh, India.

Family Akysidae
Stream catfishes

Small sized fishes of maximum 50—60 mm length; dorsal fin with a strong spine and a short base, usually four or five soft rays; body with unculiferous tubercles arranged in longitudinal rows, a median mid-dorsal row and usually four lateral rows; dorsal fin with usually five soft rays; adipose fin present and moderate; pectoral fin with strong spine, anterior margin with notch visible dorsally and usually serrated posteriorly. Gill openings relatively narrow; eyes small; barbels four pairs (Fig. 4.276).

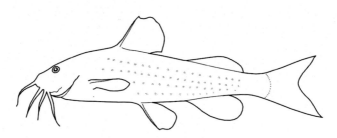

FIGURE 4.276

Outline diagram of Akysidae.

Genus *Akysis* **Bleeker, 1858**

Akysis Bleeker, 1858a: 204, 234 (type species: *Pimelodus variegatus* Bleeker, 1846). Gender: masculine.

Diagnosis. Body elongated, rounded; head anteriorly depressed; mouth subinferior; lips fleshy, papillated; teeth villiform in bands on jaws; barbels in four pairs, one pair each of nasal, maxillary and two of mandibular; skin usually covered with tubercles or granules which are sometimes arranged in longitudinal rows; rayed dorsal with a strong spine, inserted above pectoral fin; pectoral fin with an internally serrated spine; gill opening extending dorsally well above pectoral-fin base; caudal fin forked; lateral line complete.

Akysis manipurensis **(Arunkumar, 2000)**
(Fig. 4.277)

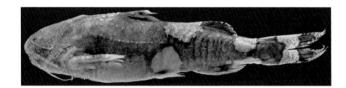

FIGURE 4.277

Akysis manipurensis, 47.0 mm SL, Lokchao River, Moreh, near Myanmar border, Manipur.

Laguvia manipurensis Arunkumar, 2000c: 194, fig. 1 (type locality Lairaok Maru stream near Moreh, Yu River drainage, Manipur, India).

Diagnosis. Characterized by the presence of minute tubercles arranged irregularly in front and in one or two longitudinal rows behind dorsal fin, sparsely on head; a row of three or four tubercles on the outer margin of operculum, a few of them arranged haphazardly behind pectoral fin; dorsal fin with a smooth spine grooved longitudinally on the anterior side, five branched anal fin rays, deeply forked caudal fin with pointed tips; three fontanels on head; premaxillary villiform teeth in two patches which is not exposed when the mouth is closed and 3 + 6 gill-rakers on the first branchial arch; head width 22.4%−24.5% SL; nasal 68.9%−89.4% HL. Maximum size ∼62 mm SL.

Distribution. India: Lokchao River near Moreh and Khuga River, all Chindwin drainage in Manipur, India.

Akysis prashadi **Hora, 1936**
(Fig. 4.278)

Akysis variegatus variegatus Prashad and Mukerji, 1929: 180 (Indawgyi Lake and along west shore near Lonton village, Myitkyina District, Burma). Secondary junior homonym of *Pimelodus variegatus* Bleeker, 1846.

Akysis prashadi Hora, 1936b: 200, figs. 1, 2b and c (type locality: Indawgyi Lake and along west shore near Lonton village, Myitkyina District, Burma). Junior objective synonym of *Akysis variegatus variegatus* Prashad and Mukerji, 1929.

Diagnosis. An *Akysis* species characterized by a stumpy body tapering behind with five to eight longitudinal rows of tubercles on each side and on the head; dorsal, and pectoral spines smooth, strong, and grooved longitudinally on anterior side; caudal fin deeply emarginated, the lower lobe

FIGURE 4.278

Akysis prashadi, 50.9 mm SL, Yu River, Tamu, Myanmar.

being slightly longer. Head width 26.0%−30.1% SL; caudal peduncle length 16.5−18.4 and height 7.7%−9.8% SL; skin with tubercles all over body; lateral line incomplete; caudal fin with lower lobe longer than upper. Maximum size ∼62 mm SL.

Distribution. Lokchao Stream, Chindwin Basin, Manipur, India; Irrawaddy River basin, Myanmar.

Family Sisoridae
Sisorid catfishes

Head and anterior part of body depressed; anterior and posterior nostrils close together, separated by a short nasal barbel; gill membranes attached to isthmus; thoracic adhesive apparatus may be present; barbels four pairs; dorsal and pectoral fins with or without spine; adipose dorsal fin present; anal fin with less than 15 rays; caudal fin forked and separated from anal fin; lateral line complete; small forms inhabiting mountain rapids; first nuchal plate and supraneural bone absent (Fig. 4.279).

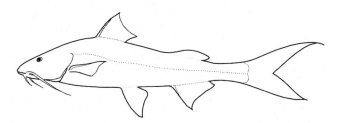

FIGURE 4.279

Outline diagram of Sisoridae.

Key to genera:

1a. Autogenous bony plates present on body,
   adhesive apparatus absent (except *Conta*) 2
1b. Autogenous bony plates absent on body,
   adhesive apparatus on ventral surface of body 11

2a. Adipose fin in the form of a spine, upper lobe of caudal
fin produced into a whip .... *Sisor*
2b. Adipose fin not in the form of a spine, upper lobe of caudal
fin not produced into a whip 3

3a. Superficial bones on head smooth and do not form an armature 4
3b. Superficial bones on head rugose and form an armature 8

4a. Body elongated, caudal fin lobes produced into filament,
elongated adhesive apparatus on thorax .... *Conta*
4b. Body not elongated, no thoracic adhesive apparatus 5

5a. Caudal-fin lobes produced into filament; lower-jaw teeth heterodont .... *Bagarius*
5b. Caudal-fin lobes not produced into filament; lower-jaw teeth not heterodont 6

6a. Pectoral girdle with rugose coracoid process covered with thin skin,
outer and inner mandibular barbels widely separated, not in one line .... *Gogangra*
6b. Pectoral girdle with coracoid process not visible externally, outer and
inner mandibular barbels close together, in one line 7

7a. Head depressed, body terete, maxillary barbel extends to pelvic-fin base .... *Nangra*
7b. Head and body compressed, maxillary barbel extends to pectoral-fin origin .... *Gagata*

8a. Adhesive apparatus absent on thorax and abdomen 9
8b. Adhesive apparatus on ventral surface of body 10

9a. Pectoral-spine serrae on the anterior edge divergent .... *Erethistes*
9b. Pectoral-spine serrae on the anterior edge directed toward tip of spine distally
and toward the body proximally .... *Erethistoides*

10a. Adhesive apparatus on thorax and abdomen 11
10b. Adhesive apparatus in the form of pleats on lips, maxillary barbels,
and paired fins.. 13

11a. Adhesive apparatus on thorax 12
11b. Adhesive apparatus on thorax extending to abdomen .... *Pseudolaguvia*

12a. Thoracic adhesive apparatus rhomboidal or chevron-shaped
formed by pleats of skin .... *Glyptothorax*
12b. Thoracic adhesive apparatus consisting of a series of transverse lamina
separated by grooves, sulcae .... *Pseudecheneis*

13a. Postlabial groove absent .... *Parachiloglanis*
13b. Postlabial groove present 14

14a. Postlabial groove interrupted .... *Creteuchiloglanis*
14b. Postlabial groove continuous 15

15a. Paired fins not greatly expanded, pectoral fin not reaching pelvic fin origin 16
15b. Paired fins greatly expanded, pectoral fin reaching pelvic fin origin .... *Oreoglanis*

16a. Tooth patches in upper jaw separate .... *Exostoma*
16b. Tooth plate in upper jaw juxtaposed .... *Meyersglanis*

Genus *Bagarius* Bleeker, 1853

Bagarius Bleeker, 1853: 121 (type species: *Pimelodus bagarius* Hamilton, 1822). Gender: masculine.
Bagarius: Roberts, 1983: 436 (revision).
Diagnosis. Body elongated, ventrally flattened; head depressed, covered with keratinized skin; mouth wide, terminal; premaxillary bone of each side composed of two ossifications; lower jaw dentition heterodont; two or three outer rows of numerous close-set conical teeth; one or two inner rows of less numerous; widely separated and much larger conical teeth; barbels four pairs, one each of nasal and maxillary and two of mandibular; gill membranes free from each other and also from isthmus; caudal peduncle small and elongated; rayed dorsal fin with a strong smooth spine, inserted above base of pectoral-fin; adipose dorsal-fin long; pectoral-fin with an internally serrated spine; caudal-fin deeply forked, lobes usually produced into filaments.

Bagarius bagarius (**Hamilton, 1822**)
(Fig. 4.280)

FIGURE 4.280

Bagarius bagarius, 146.0 mm SL, Comilla District, Meghna basin, Bangladesh.

Pimelodus bagarius Hamilton, 1822: 186, 378, pl. 7, fig. 62 (type locality: Ganges River, India).
Bagarius bagarius: Roberts, 1983: 437 (revision).
Diagnosis. Pelvic-fin origin well anterior to vertical through base of its last dorsal-fin origin; adipose-fin origin vertical through behind anal-fin origin; no sharp ridge on top of head; no bumps on dorsal mid-line behind dorsal-fin; neural spines distally expanded; body depth 7.6%−8.1% SL; pectoral fin 9−12 rays; eye diameter 10.9%−11.4% HL; adult size up to at about 200 mm SL. Maximum size ~200 mm SL.
Distribution. Ganga and Brahmaputra drainage, India and Bangladesh; Chao Phraya and Mekong basins, Myanmar, Thailand, and Vietnam.

Bagarius yarrelli (Sykes, 1893)
(Fig. 4.281)

FIGURE 4.281

Bagarius yarrelli, 1416.0 mm SL, Brahmaputra River, near Guwahati, Assam.

Bagarius yarrelli Sykes, 1839: 163 (type locality: Mota Mola at Poona, Deccan, India).
Bagarius yarrelli: Roberts 1983: 438 (revision).
Diagnosis. Pelvic fin origin posterior to vertical through the origin of last dorsal fin ray; adipose fin origin vertical through anterior to anal fin origin; skin above neural spines anterior and posterior to the adipose fin not forming distinct ridges; deep body (10.4%−14.3% SL); slender elongated neural spine; circular eyes, its diameter 7.9%−10.1% HL; known to grow upto 1400 mm SL. Maximum size ∼2000 mm SL.
Distribution. Ganga, Brahmaputra, and Chindwin River drainages; Bangladesh, Myanmar, Thailand, Borneo, Java.

Genus *Conta* Hora, 1950
Conta Hora, 1950: 194 (type species: *Pimelodus conta* Hamilton, 1822). Gender: feminine.
Diagnosis. Body elongated and slender; gill openings narrow; a long and narrow thoracic adhesive apparatus; dorsal spine serrated on both anterior and posterior margins; pectoral spine with serrations on anterior margin; upper lip papillated; anal-fin with 9−10 rays; head and body tuberculated; caudal fin deeply forked; its lobes greatly produced.

Conta conta (Hamilton, 1822)
(Fig. 4.282)
Pimelodus conta Hamilton, 1822: 191, 378 (type locality: Mahananda River, northeastern Bengal).
Diagnosis. Anterior edge of pectoral spine with antrose, that is, distally directed serrae; (Fig. 4.283A); caudal peduncle length 21.9%−23.4% SL and depth 3.9%−4.4% SL. Maximum size ∼78 mm SL.
Distribution. Teesta drainage, Northern Bengal and Brahmaputra drainage, Assam.

FIGURE 4.282

Conta conta, 49.0 mm SL, Brahmaputra River, near Goalpara, Assam; inset.

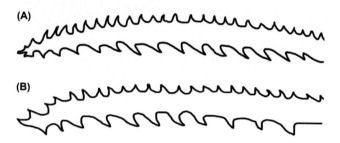

FIGURE 4.283

(A) Sketch of pectoral spine showing antrose serrae on anterior margin; (B) sketch of pectoral spine showing retrose serrae on anterior margin.

Conta pectinata Ng, 2005
(Fig. 4.284)

FIGURE 4.284

Conta pectinata, 106.0 mm SL, Brahmaputra River, Dibrugarh, Assam.

Conta pectinata Ng, 2005b: 24, figs. 1 and 2b (type locality: Dibrugarh, Assam, India).

Diagnosis. Anterior edge of pectoral spine with retrose, that is, anteriorly directed serrae (Fig. 4.301B); caudal peduncle long, its length 24.6%−25.6% SL, its height 2.6%−2.8% SL. Maximum size ∼200 mm SL.

Distribution. Brahmaputra basin in Assam and Meghalaya.

Genus *Creteuchiloglanis* Zhou Li & Thomson, 2011

Creteuchiloglanis Zhou, Li & Thomson, 2011: 227 (type species: *Creteuchiloglanis longipectoralis* Zhou et al., 2011). Gender: masculine.

Diagnosis. Flat ventral surface, ventral surface of maxillary barbel with numerous plicae; unbranched rays of paired fins pad-like and with numerous transverse plicae ventrally (Fig. 4.303); interrupted post-labial fold; extremely broad isthmus, gill openings not extending ventrally onto the undersurface of head; homodont dentition with pointed coniform teeth in both the jaws; tooth patches in upper jaw not extended postero-laterally; branched pectoral-fin rays 14−16 (Fig. 4.285).

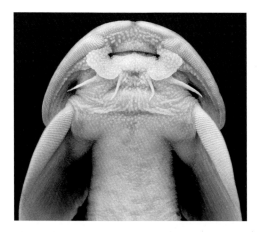

FIGURE 4.285

Ventral surface of mouth and thorax of *Creteuchiloglanis* showing maxillary barbel, interrupted labial fold and plicae on unbranched pectoral fin ray.

Creteuchiloglanis arunachalensis Sinha & Tamang, 2014
(Fig. 4.286)

FIGURE 4.286

Creteuchiloglanis arunachalensis, 87.1 mm SL, holotype, Pange River, Ziro, Arunachal Pradesh.

Photo courtesy: A. Darshan.

Creteuchiloglanis arunachalensis Sinha & Tamang, 2014: 2, figs. 1 and 2a (type locality: Pange River at Aro-Lenching, a tributary of Brahmaputra River, Ziro, Lower Subansiri District, Arunachal Pradesh, India).

Diagnosis. Predorsal length 35.7% SL; body depth 15.1% SL, anal-fin length 12.8% SL; pectoral-fin with one unbranched and 14 branched rays; dorsal surface with light yellow patches, one at base of first dorsal-fin, one each on anterior and posterior parts of the bases of anal-fin, one each on dorsal and posterior parts of caudal-fin base; one bar in the middle and another posterior edge of caudal-fin. Maximum size ~88 mm SL.

Distribution. Pange River, Lower Subansiri District, Arunachal Pradesh, northeast India.

Creteuchiloglanis bumdelingensis Thoni & Gurung, 2018

Creteuchiloglanis bumdelingensis Thoni & Gurung, 2018: 59, figs. 14−15 (type locality: Kuktorgangchhu stream near Tshaling in the Bumdeling Wildlife Sanctuary, Trashiyangtse Dzongkhag, Bhutan).

Diagnosis. Postlabial fold large and extends to or beyond the base of the outer mandibular barbel; caudal peduncle length 6.4%−7.9% SL; branched pectoral-fin rays 15−16; pectoral-fin rays extend beyond pelvic-fin origin; adipose-fin length 29.3%−36.0% SL. Maximum size ~40 mm SL.

Distribution. Tributaries of the Dangmechhu River; Kuktorgangchhu stream near Tshaling, in the Bumdeling Wildlife Sanctuary, Trashiyangtse, Sherichhu River in Mongar, and Gamrichhu River in Trashigang, Bhutan.

Creteuchiloglanis kamengensis (Jayaram, 1966)
(Fig. 4.287)

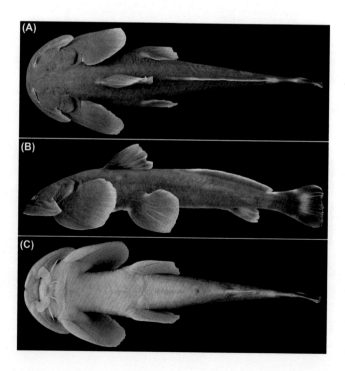

FIGURE 4.287

Creteuchiloglanis kamengensis, 91.0 mm SL, Kameng River, Arunachal Pradesh; (A) dorsal, (B) lateral, and (C) ventral views.

Euchiloglanis kamengensis Jayaram, 1966: 85 (type locality: Norgum River at Kalaktang, Kameng Frontier Division, NEFA (now Arunachal Pradesh, India).

Diagnosis. Predorsal length 34.5%−37.5% SL; body depth 11.1%−11.3% SL; anal-fin length 18.8%−21.9% SL; pectoral-fin with one unbranched and 16 branched rays; caudal-fin with dirty whitish band at about middle; fins fringed white. Maximum size ∼198 mm SL.

Distribution. Kameng and Norgum Rivers, Arunachal Pradesh, India.

Creteuchiloglanis payjab **Darshan, Datta, Kachari, Gogoi, Aran & Das, 2014**
(Fig. 4.288)

FIGURE 4.288

Creteuchiloglanis payjab, 172.2 mm SL, Yomgo River at Mechuka, Arunachal Pradesh.

Photo courtesy: A. Darshan.

Creteuchiloglanis payjab Darshan, Dutta, Kachari, Gogoi, Aran & Das, 2014: 74 (type locality: Arunachal Pradesh state, West Siang District, Yomgo River at Mechuka, a tributary of Siang River, Brahmaputra River basin).

Diagnosis. Predorsal length 30.4%−35.5% SL; pectoral-fin with one unbranched and 14−15 branched rays; body depth 11.4%−13.0% SL; anal-fin length 11.8%−13.4% SL; whitish spot at base of last pectoral-fin ray and another on the inner posterior margin of the skin flap connecting maxillary barbel. Maximum size ∼180 mm SL.

Distribution. Yomgo River at Mechuka, tributary of Siang River, Brahmaputra drainage, Arunachal Pradesh, India.

Creteuchiloglanis tawangensis **Darshan, Abujam, Wamgchu, Kumar, Das & Imotomba, 2019**
(Fig. 4.289)

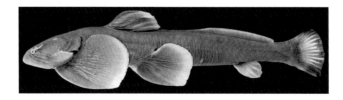

FIGURE 4.289

Creteuchiloglanis tawangensis, 138.7 mm SL, Tawangchu River, Brahmaputra drainage.

Photo courtesy: A. Darshan.

Creteuchiloglanis tawangensis Darshan, Abujam, Wangchu, Kumar, Das and Kumar Imotomba, 2019, Darshan, Abujam, Wangchu, Kumar, Das & Imotomba, 2019c: 18, figs. 1, 4a, 5b (type locality: Tawangchu River at Granger village, headwater of Manas drainage, Brahmaputra River basin, Tawang District, Arunachal Pradesh, India).

Diagnosis. Pinnate rays on the anterior margin of the first branched anal-fin ray (Fig. 4.290); body depth at anus 10.7%−11.1% SL; pectoral fin not reaching pelvic-fin origin; predorsal length

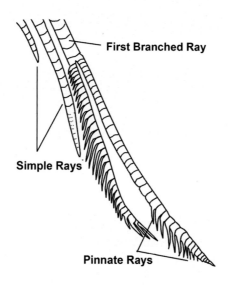

FIGURE 4.290

Sketch of simple and first branched anal-fin ray of *Creteuchiloglanis tawangensis* showing pinnate rays.

Redrawn after A. Darshan et al., 2019.

31.3%−32.5% SL, caudal peduncle depth 6.4%−6.8% SL, adipose-fin base length 33.2%−33.8% SL, pelvic-fin length 19.5%−20.5% SL, 2 + 8 = 10 gill rakers on the first branchial arch, head depth 9.1%−10.0% SL, and eye diameter 5.8%−6.9% HL. Maximum size ∼152 mm SL.

Distribution. Tawangchu River, Tawang District, headwater of Manas River, Brahmaputra drainage, Arunachal Pradesh.

Genus *Erethistes* Müller & Troschel, 1849

Erethistes Müller & Troschel, 1849: 12 (type species: *Erethistes pusillus* Müller and Troschel, 1849). Gender: masculine.

Diagnosis. A sisorid catfish genus with robust body; no thoracic adhesive apparatus; occipital process (OP) reaching nuchal plate; anterior margin of dorsal spine smooth to granulate, posterior margin serrated; anterior edge of pectoral spine flattened, the serrae on anterior edge divergent, and the serrae on posterior edge directed toward the body (antrose); anal fin rays 8−12, papillate upper lip.

Erethistes hara (Hamilton, 1822)
(Fig. 4.291)

FIGURE 4.291

Hara hara, 64.5 mm SL, Barak River at Sekjang Tuifai, Manipur.

Pimelodus hara Hamilton, 1822: 190 (type locality: Nathpur, India).

Hara hara: Ng & Kottelat, 2007: 447 (redescription).

Diagnosis. Anterior edge of pectoral spine finely serrated; supraoccipital spine not reaching nuchal plate; no filamentous extension of the first principal ray of the upper caudal-fin lobe; posterior process of coracoid reaching midway of the distance between bases of pectoral spine and first pelvic-fin ray; adipose-fin base length 11.8%−15.5% SL; caudal peduncle depth 25.8%−28.0% SL; body depth 11.5%−14.3% SL; pectoral spine length 26.9%−33.3% SL; soft pectoral-fin rays seven; total vertebrae 31−34; caudal fin deeply forked. Maximum size ∼65 mm SL.

Distribution. Ganga and Brahmaputra drainages.

Erethistes horai **Misra, 1976**

Hara horai Misra, 1976: 245, Pl. 9, figs. 1−3 (type locality: Terai and Duars, northern Bengal, India).

Diagnosis. Absence of a filamentous extension of the first principal ray of upper caudal-fin lobe; posterior process on coracoid reaching to two-thirds distance between bases of pectoral spine and first pelvic-fin ray; supraoccipital spine not reaching nuchal plate; body depth 15.1%−17.9% SL; dorsal spine length 21.0%−28.4% SL; pectoral spine length 28.0%−33.4% SL; soft pectoral-fin rays seven; adipose-fin base length 10.0%−13.7% SL; caudal peduncle depth 5.2%−6.2% SL; caudal fin length 20.5%−23.5% SL; total vertebrae 31−34. Maximum size ∼70 mm SL.

Distribution. Brahmaputra River drainage.

Erethistes jerdoni **(Day, 1870)**

(Fig. 4.292)

Hara jerdoni Day, 1870d: 39, pl. 4 (type locality: Sylhet District, Bangladesh).

Diagnosis. Supraoccipital spine reaching nuchal plate; posterior process on coracoid reaching to four-fifths distance between bases of pectoral spine and first pelvic-fin ray; pectoral spine length 42.2%−51.7% SL; soft pectoral-fin rays five; total vertebrae 27−29; emarginated caudal fin. Maximum size ∼70 mm SL.

Distribution. Ganga and Brahmaputra River drainages.

Erethistes koladynensis **(Anganthoibi & Vishwanath, 2009)**

(Fig. 4.293)

FIGURE 4.292

Erethistes jerdoni 66.0 mm SL, Surma River drainage, Tripura.

FIGURE 4.293

Erethistes koladynensis, 54.5 mm SL, holotype, Kaladan River at Kolchow, Mizoram.

Hara koladynensis Anganthoibi & Vishwanath, 2009: 466, fig. 1 (type locality: Koladyne River at Kolchaw, Lawntlai District, Mizoram, northeastern India).

Diagnosis. Rough anterior margin of dorsal spine, posterior process on coracoid reaching to two-thirds of the distance between bases of pectoral spine and first pelvic-fin ray; absence of filamentous extension of the first principal ray of upper caudal-fin lobe; head length 36.0%−38.3% SL; body depth 21.9%−25.2% SL; dorsal spine length 20.0%−22.7% SL; pectoral spine length 28.9%−31.0% SL; preanal length 58.5%−60.9% SL; postadipose distance 18.6% SL; caudal peduncle length 18.8%−21.8% SL; caudal-fin length 25.8%−28.0% SL. Maximum size ∼70 mm SL.

Distribution. Kaladan River drainage, Mizoram, India.

Erethistes pusillus **Muller & Troschel, 1849**
(Fig. 4.294)

Erethistes pusillus Müller & Troschel, 1849: 12, pl. 1 (type locality: Assam, India).

Diagnosis. A species of *Erethistes* with a distinctly serrated dorsal spine; head wider than its length; mouth small, teeth villiform in bands on jaws; body with longitudinal rows of tubercles; lateral line complete; caudal fin forked; dorsal fin with one spinelet, one spine and five to six branched

FIGURE 4.294

Erethistes pusillus, 36.2 mm SL, Barak River, Manipur.

rays, the spine serrated posteriorly; pectoral fin with one spine and five branched rays, the spine with 9−10 serrae on posterior margin and divergent serrae on the anterior margin; pelvic fin with one simple and five branched rays; anal with three simple and six branched rays; caudal fin forked with seven branched rays in the upper lobe and six in the lower. Maximum size ∼42 mm SL.

   Distribution. Ganga, Brahmaputra and Barak−Surma−Meghna drainages.

Genus *Erethistoides* Hora, 1950

   Erethistoides Hora, 1950: 190 (type species: *Erethistoides montana* Hora, 1950). Gender: masculine.

   Diagnosis. No thoracic adhesive apparatus; serrations on anterior margin of pectoral spine directed toward tip of spine distally and toward the body proximally (Fig. 4.295); slender body; smooth to granulate anterior margin on dorsal spine; moderate gill openings; papillate upper lip; 9−11 anal-fin rays.

FIGURE 4.295

Pectoral spine of *Erethistoides* showing divergent spine on the anterior margin.

Erethistoides ascita Ng & Edds, 2005

   Erethistoides ascita Ng & Edds, 2005a: 240, figs. 1, 2a, 3a (type locality: Mechi River at Bhadrapur, Jhapa, Nepal).

   Diagnosis. Serrations on the anterior edge of the pectoral spine diverging at the distal quarter, 12−17 proximal-most serrations proximally directed and in having flattened and elongated plaque-like tubercles on the head and body, caudal peduncle length 19.3%−22.5% SL; body depth at anus 8.6%−10.5% SL; absence of a slight median depression on the lower lip margin. Maximum size ∼36 mm SL.

Distribution. Mechi and Kosi River systems, Ganges drainage, Nepal.

Erethistoides cavatura Ng & Edds, 2005

Erethistoides cavatura Ng & Edds, 2005a: 243, figs. 2b, 3b, 5 (type species: Dhungre River at Sauraha, Chitwan, Nepal).

Diagnosis. Serrations on the anterior edge of the pectoral spine more anteriorly directed and diverging in the middle; presence of rounded tubercles on head and body; large nostrils, the length of the narial complex 77%−90% of interorbital width; long caudal peduncle 19.3%−22.5% SL; slender body depth at anus 8.6%−10.5% SL; and absence of slight median depression on the lower lip margin. Maximum size ∼33 mm SL.

Distribution. Lowland inner Tarai valley of south-central Nepal in or near Chitawan National Park, Rapti-Narayani River drainage, Nepal.

Erethistoides infuscatus Ng, 2006

Erethistoides infuscatus Ng, 2006a: 282, figs. 1, 2b, 3a (type locality: East Khasi Hills, Umsing River, Meghalaya, India).

Diagnosis. Serrations on the anterior edge of the pectoral spine distally directed; total vertebrae 31−33; less strongly produced snout with the premaxillary teeth partially exposed when the mouth is closed; presence of an axillary cartilage between the first and second epibranchials and anterior to the infra-pharyngobranchial; brown body with a few indistinct pale patches. Maximum size ∼45 mm SL.

Distribution. Brahmaputra and Meghna River drainages in India and Bangladesh.

Erethistoides montana Hora, 1950
(Fig. 4.296)

FIGURE 4.296

Erethistoides montana, 48.0 mm SL, Brahmaputra River, Dibrugarh, Assam.

Erethistoides montana Hora, 1950: 191, pl. 1, figs. 10−12 (type locality: streamlets near Tangla, Darrang District, Assam, India).

Diagnosis. Head and body greatly depressed, tail long and narrow. Position of mouth, long maxillary barbels and form of caudal fin; skin and bones all along the dorsal and lateral surfaces are covered with denticles while the whole of the ventral surface is covered with small backwardly directed spines; dorsal spine strong, finely serrated externally and pectinated internally; pectoral spine strong and broad, skin of distal end produced into a filiform process along with similar process along with similar structures of some of the other rays, outer border finely serrated and internally strongly pectinate; head and nape with dark band and body with three similar bands one below the first dorsal, one below the adipose dorsal and on caudal-fin base. Maximum size ∼48 mm SL.

Distribution. Brahmaputra drainage, northeastern India.

Erethistoides senkhiensis **Tamang, Chaudhuri & Choudhury, 2008**
Erethistoides senkhiensis Tamang, Chaudhry & Choudhury, 2008: 186 (type locality: Senkhi stream, Arunachal Pradesh, India).

Diagnosis. Three black to light brown cross bars on a dark gray to light brown body background, 29−30 vertebrae, concave caudal fin, 18−29 serrate on the anterior margin of the pectoral sine, snout overhanging presence of the light median depression on the lower lip margin. Maximum size ∼43 mm SL.

Distribution. Confluence of Senkhi and Chimpu streams at Itanagar, Brahmaputra drainage, Arunachal Pradesh, India.

Erethistoides sicula **Ng, 2005**
Erethistoides sicula Ng, 2005c: 2, figs. 1, 2a, 4a (type locality: Schutunga River, tributary of Mansai River, at Ansole, West Bengal, India).

Diagnosis. Dorsally projecting bony splint on the opercle immediately posterior to its articular facet with the hyomandibula; head depth 13.4%−15.1% SL; pectoral spine length 14.6%−28.0% SL; caudal peduncle length 19.6%−22.3% SL. Maximum size ∼39 mm SL.

Distribution. Known from the Mansai River drainage, itself a tributary of the Brahmaputra River in northern West Bengal state in India.

Genus *Exostoma* **Blyth, 1860**
Exostoma Blyth, 1860: 155 (type species: *Exostoma berdmorei* Blyth, 1860). Gender: neuter.

Diagnosis. Body moderately elongated, anteriorly depressed up to pelvic-fin base; nostrils close together and separated by a short flap-like nasal barbel; eyes minute and subcutaneous; mouth ventral and crescentric; lips thick and fleshy, papillated, continuous around mouth to form a sucker; teeth large and oar shaped, upper jaw teeth separated in two distinct patches; no teeth on palate; barbels four pairs; gill membranes confluent with isthmus; thorax with no adhesive apparatus; dorsal fin origin opposite pelvic-fin origin, fin spine absent; adipose dorsal long, may be confluent with caudal-fin; pectoral and pelvic fins with first simple rays ventrally pleated; pectoral-fin with no spine; lateral line complete (Fig. 4.297).

Exostoma barakensis **Vishwanath & Joyshree, 2007**
(Fig. 4.298)
Exostoma barakensis Vishwanath & Joyshree, 2007: 2531 (type locality: Iyei River, Barak River drainage, Noney District, Manipur, India).

Diagnosis. Nostrils midway between tip of snout and anterior margin of eye; wider interorbital space 26.9%−33.1% SL; anal fin rays ii, 4.5−5.0; adipose fin long, uniform deep all along and

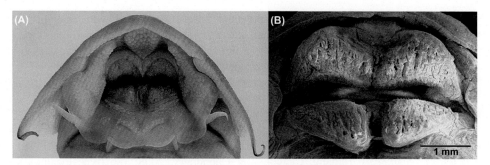

FIGURE 4.297

Ventral view of mouth of *Exostoma labiatum* showing separated upper jaw teeth bands, (A) light microscopy and (B) SEM.

FIGURE 4.298

Exostoma barakensis, 82.0 mm SL, paratype, Iyei River, tributary of Barak River, Manipur.

extend to caudal-fin base, twice the length of dorsal-fin base; caudal fin emarginated, darkish proximally and lower lobe distally dark greyish. Maximum size ∼98 mm SL.

Distribution. Iyei River, tributary of the Barak River in Manipur, Barak−Surma−Meghan drainage, India.

Exostoma dujangensis Shangningam & Kosygin, 2020
Exostoma dujangensis Shangningam & Kosygin, 2020: 545, fig. 1.2 (type locality: Dujang stream flowing into the Chakpi River at Dutuwl near Khubung Khullen village, Chandel District, Manipur).

Diagnosis. Adipose fin short (29.1%−32.3% SL) and confluent with the caudal fin; anastomosing, rounded and curved striae on the anterolateral surfaces of lips and ventral surfaces of the lips and of maxillary barbel; tips of caudal-fin lobes rounded; caudal-peduncle length 17.4%−21.0% SL. caudal-peduncle depth 8.7%−9.4% SL. Maximum size ∼78 mm SL.

Distribution. Tributary streams of Chakpi River, Chindwin basin, Chandel District, Manipur.

Exostoma kottelati Darshan, Vishwanath, Abujam & Das, 2019
(Fig. 4.299)

FIGURE 4.299

Exostoma kottelati, 70.7 mm SL, holotype, Ranga River, Yachuli, Arunachal Pradesh.

Exostoma kottelati Darshan, Vishwanath, Abujam & Das, 2019d: 370 (type locality: tributary of Ranga River at Yazali village, Brahmaputra basin, Lower Subansiri District, Arunachal Pradesh, India).

Diagnosis. Adipose fin distinctly separated from procurrent caudal-fin rays; predorsal length of 38.9%−41.7% SL; adipose-fin base length of 33.4%−36.0% SL; a caudal peduncle length 18.7%−21.1% SL, its depth of 8.8%−9.5% SL; body depth at anus of 12.5%−13.5% SL and elongated and parallel striae on the anterolateral surfaces of the lips; adipose-fin base length 33.4%−36.0% SL; pre-pelvic length 45.6%−47.3% SL; preanal length 73.9%−76.5% SL. Maximum size ∼78 mm SL.

Distribution. Stream at Yazali village, draining into Ranga River, Brahmaputra basin, Lower Subansiri District, Arunachal Pradesh, India.

Exostoma labiatum (McClelland, 1842)
(Fig. 4.300)

FIGURE 4.300

Exostoma labiatum, 53.8 mm SL, Tezu River, Arunachal Pradesh.

Glyptosternon labiatus McClelland, 1842: 588 (type locality: Mishmee Hills, Arunachal Pradesh, India).

Diagnosis. Adipose-fin base length 26.6%−30.5% SL, adipose fin not confluent with caudal fin; caudal peduncle depth with 8.7%−9.9% SL; pectoral fin with 11−12 branched rays; caudal fin emarginated with i,8 + 7,i rays. Maximum size ∼110 mm SL.

Distribution. Upper Brahmaputra River drainage, India; Southwestern China, Nepal.

Exostoma mangdechhuensis Thoni & Gurung, 2018
Exostoma mangdechhuensis Thoni & Gurung, 2018: 56, figs. 12−13 (type locality Nabbey Khola stream at Tingtibi, Zhemgang Dzongkhag, Bhutan).

Diagnosis. A conspicuous anteriorly projected notch on the posterior connection of the adipose fin to the caudal peduncle, head length 22.6%−24.9% SL, preocciput length of 91.9%−102.1% HL, head width 79.7%−88.2% HL, mouth width 33.2%−39.4% HL, caudal peduncle depth 7.7%−

8.4% SL; branched pectoral-fin rays 11 caudal-fin base with a dark blotch. Maximum size ~68 mm SL.

Distribution. Mangdechhu River drainage and its tributaries.

Exostoma sawmteai **Lalramliana, Lalronunga, Lalnuntluanga & Ng, 2015**
(Fig. 4.301)

FIGURE 4.301

Exostoma sawmteai, 65.2 mm SL, Pharsil River, Mizoram.

Photo courtesy: Lalramliana.

Exostoma sawmteai Lalramliana, Lalronunga, Lalnuntluanga & Ng, 2015b: 60, fig. 1 (type locality: Pharsih River, a tributary of Tuivai River, Barak River drainage, in the vicinity of Kawlbem, Champhai District, Mizoram, India).

Diagnosis. Posterior end of the adipose-fin adnate with the dorsal procurrent caudal-fin rays; snout length 48%−55% HL; eye diameter 8%−11% HL; pectoral-pelvic distance 28.6%−32.8% SL; body depth at anus 11.5%−14.1% SL; adipose-fin base length 30.7%−37.2% SL; caudal-peduncle length 18.4%−20.3% SL; caudal peduncle depth 10.1%−11.7% SL; caudal fin lunate; total vertebrae 38−39. Maximum size ~85 mm SL.

Distribution. Pharsih River, a tributary of Tuivai River, Barak River drainage, Mizoram, northeast India.

Exostoma tenuicaudatum **Tamang, Sinha & Gurumayum, 2015**
(Fig. 4.302)

FIGURE 4.302

Exostoma tenuicaudatum, 97.5 mm SL, Siang River, Upper Siang District, Arunachal Pradesh.

© *A. Darshan.*

Exostoma tenuicaudata Tamang, Sinha & Gurumayum−3, 5, 6A (type locality: Siang River, Upper Siang District, Arunachal Pradesh, India).

Diagnosis. Triangular elliptical as per the original description, but figure shows a triangular one plicated adhesive pad medially between the ventral extremity of the snout and the medial furrow of the upper lip, the apex pointing posteriorly; caudal peduncle depth 3.6%−4.7% SL; snout length 60.2%−61.8% SL; predorsal length 34.9%−39.0% SL; adipose-fin base length 30.8%−32.9% SL; dorsal-fin base length of 7.9%−10.8% SL; anal-fin base length 3.9%−5.3% SL; pectoral-fin rays 11−12; adipose-fin not confluent with dorsal-fin base. Maximum size ∼100 mm SL.

Distribution. Tributary of the Siang River, Brahmaputra drainage, Upper Siang District, Arunachal Pradesh, northeastern India.

Genus *Gagata* Bleeker, 1858

Gagata Bleeker, 1858a: 204 (type species: *Pimelodus gagata* Hamilton, 1822). Gender: masculine.
Gagata: Roberts & Ferraris, 1998: 317 (review of the genus).

Diagnosis. Head and body compressed; median longitudinal groove on head extending only up to base of OP; eyes large, lateral; mouth inferior, relatively small and narrow; mesethmoid bone highly modified, strongly curved downward in front of snout; bone laterally compressed, its ventro-medial projection is attached by ascending process of premaxilla by soft tissue; premaxilla with dorsomedial ascending process; jaw teeth finely conical, in few rows, teeth absent in upper jaw in some species; palate toothless; branchiostegal membrane broadly joined to isthmus; outer and inner mandibular barbels in one line (Fig. 4.320A); dorsal fin with six branched rays; caudal fin with 17 principal rays; pectoral girdle with coracoid process covered with thick skin and not visible externally (Fig. 4.303).

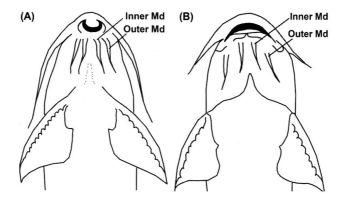

FIGURE 4.303

Figure of ventral surfaces of heads of (A) *Gagata* and (B) *Gogangra* showing arrangements of inner (Inner Md) and outer mandibular (Outer Md) barbels.

Gagata cenia (Hamilton, 1822)

(Fig. 4.304)

Pimelodus cenia Hamilton, 1822: 174, 376, pl. 31, fig. 57 (type locality: rivers of North Bengal).

Gagata cenia: Roberts and Ferraris, 1998: 321, figs. 1, 2 (review).

Diagnosis. A small species reaching to about 100 mm SL; dorsum of body with dark saddles extending ventrally only to lateral line; caudal fin with transverse black bar across peduncle and round

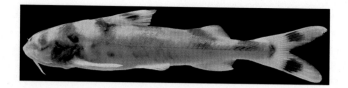

FIGURE 4.304

Gagata cenia, 71.5 mm SL, Jiri River, Barak drainage, Manipur.

or square black spot on middle of each lobe; dorsal-fin with black spot on distal part of anterior rays; upper jaw toothless; lower jaw with few small conical teeth in pocket or depression near symphysis; fourth ceratobranchial without teeth; snout tip acutely pointed in lateral profile, tip separated from rest of snout by distinct notch; distal portions of fins not marked with black. Maximum size ~ 150 mm SL.

Distribution. Ganga, Mahanadi, and Indus River basins, India, Pakistan, and Nepal.

Gagata dolichonema He, 1996
(Fig. 4.305)

FIGURE 4.305

Gagata dolichonema, 68.3 mm SL, Imphal River, Mayang Imphal, Manipur.

Gagata dolichonema He, 1996: 380, fig. 1 (type locality: Yunnan Province, China).
Gagata gasawyuh Roberts and Ferraris, 1998: 325 (Tenasserim River, Myanmar).
Diagnosis. Five oblique saddles: one on head crossing eye, second across dorsal-fin origin, third at the posterior half of dorsal-fin base crossing the lateral line but not reaching pelvic fin, forth at the anterior extent of adipose fin extending obliquely toward anal-fin base but not reaching anal fin and fifth at the base of caudal fin extending ventrally to below lateral line; caudal fin with a dark continuous subterminal lunate mark. Maximum size ~ 130 mm SL.

Distribution. Chindwin-Irrawaddy drainage in Manipur, India, southern China, and Myanmar.

Gagata gagata (Hamilton, 1822)
Pimelodus gagata Hamilton, 1822:197, 379, pl. 39, fig. 65 (type locality: fresh water rivers and estuaries of Bengal).
Gagata gagata: Roberts & Ferraris, 1998: 323, figs. 4 and 5 (review).
Diagnosis. Large species of *Gagata* reaching upto 100 mm SL; head and body silvery with no other markings; fins except caudal black distally; dorsal spine expending past adipose-fin origin when

adpressed; premaxilla with four rows of teeth; total vertebrae 38−39; pectoral-fin with a spine and nine soft rays; anal fin with five to six simple and 11 branched rays. Maximum size ∼ 300 mm SL.

Distribution. Ganga basin in India and Bangladesh.

Gagata sexualis **Tilak, 1970**

Gagata sexualis Tilak, 1970: 207, figs. 1−6 (Chotanagpur, southern Bihar, India).

Gagata sexualis: Roberts & Ferraris, 1998: 332 (review).

Diagnosis. A small species of *Gagata* of maximum size of about 50 mm SL; filamentous extension of dorsal-fin spine in males reaching adipose-fin base; snout bluntly rounded in lateral profile without notch anteriorly; upper jaw toothless; lower jaw with single row of teeth. Maximum size ∼ 55 mm SL.

Distribution. Ganga and Brahmaputra basins.

Genus *Glyptothorax* Blyth, 1860

Glyptothorax Blyth, 1860: 154 (type species: *Glyptosternon striatus* McClelland, 1842). Gender: masculine.

Diagnosis. Body elongated, moderately depressed; head and snout greatly depressed; skin covered with minute to coarse tubercles; genus characteristic in having rhomboidal or nearly rhomboidal or chevron-shaped thoracic adhesive apparatus with or without a central depression, formed by folds of skin (Fig. 4.306 and 4.323); cubito-humeral and scapular processes not prominent; mouth inferior, transverse, and narrow; gill membranes united and also with isthmus; pectoral fin spine serrated internally; dorsal fin with smooth spine; paired fins horizontal; cadual fin deeply forked.

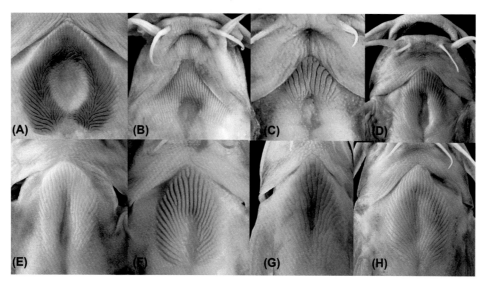

FIGURE 4.306

Shapes of thoracic adhesive apparatus of *Glyptothorax*: (A) deep central depression, (B) unculiferous ridges extending to gular region, (C) chevron shaped with striae only to anterior side, (D) rhomboid with elongated central depression, (E) rhomboid, (F) lanceolate, (G) rhomboid with shallow central depression, apparatus completely enclosed posteriorly, (H) rhomboid with slight or no depression.

Glyptothorax ater **Anganthoibi & Vishwanath, 2011**
(Fig. 4.307)

FIGURE 4.307

Glyptothorax ater, 126.0 mm SL, holotype, Kaladan River at Kolchaw, Mizoram.

Glyptothorax ater Anganthoibi & Vishwanath, 2011: 324 (type locality: Kaladan River at Kolchaw, Lawntlai District, Mizoram, India).

Diagnosis. Dark brown body with two horizontal light creamish stripes, one each along mid-dorsal line and lateral lines; two pale creamish ovoid spots on either side of the dorsal-fin origin connected by thin creamish oblique stripes forming spectacle like mark; caudal fin with pale yellow spots in region of tips of procurrent rays and pale yellow bands at tips of lobes; greatly arched pre-dorsal profile; rhomboidal thoracic adhesive apparatus with a conspicuous central depression, longitudinally elongated and with a constriction at mid-length; ventral surfaces of the pectoral spine and two to four outer rays of the pelvic fin pleated. Maximum size ~126 mm SL.

Distribution. Kaladan River, Mizoram, India.

Glyptothorax botius **(Hamilton, 1822)**
(Fig. 4.308)

FIGURE 4.308

Glyptothorax botius, 87.2 mm SL, Barak River at Silchar, Assam.

Pimelodus botius Hamilton, 1822: 192, 378 (type locality: Ganga River).

Glyptothorax botius: Ng, 2005d: 2, figs. 1 and 2a (redescription).

Diagnosis. Rounded snout; large, prominent tubercles on the head and body; a thoracic adhesive apparatus without a median depression; very slender body with dark brown saddles; differs from *Glyptothorax telchitta* in its broader folds in thoracic adhesive apparatus, longer adipose-fin base and slender caudal peduncle. Maximum size ~90 mm SL.

Distribution. Ganga River drainage.

Glyptothorax burmanicus Prashad & Mukerji, 1929
(Fig. 4.309)

FIGURE 4.309

Glyptothorax burmanicus, 114.0 mm SL, Manipur River at Yankoilok, near Myanmar border.

Glyptothorax burmanicus Prashad & Mukerji, 1929: 184 (type locality: Sankha, a large hill-stream, midway between Kamaing and Mogaung, Myitkyina District, Upper Myanmar).

Diagnosis. Distinct in having a heart shaped thoracic adhesive apparatus with a wide shallow depression at the center (Fig. 4.310A); body with black patches; caudal fin with a submarginal black band forming a v-shaped band by extending to the base of middle rays. Maximum size ~260 mm SL.

Distribution. Chindwin drainage in Manipur, India and Irrawaddy drainage in Myitkyina, Upper Myanmar.

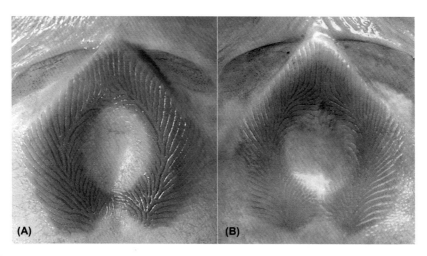

FIGURE 4.310

Thoracic adhesive apparartus of (A) *Glyptothorax burmanicus*, (B) *G. cavia*.

Glyptothorax caudimaculatus Anganthoibi & Vishwanath, 2011
(Fig. 4.311)

Glyptothorax caudimaculatus Anganthoibi & Vishwanath, 2011: 326 (type locality: Koladyne River at Kolchaw, Lawntlai District, Mizoram, India).

FIGURE 4.311

Glyptothorax caudimaculatus, 59.8 mm SL, holotype, Koladyne River at Kolchaw, Mizoram.

Diagnosis. Rhomboidal-shaped thoracic adhesive apparatus with its unculiferous ridges extending anteriorly onto the gular region; central depression on thoracic adhesive apparatus opening posteriorly in an inverted v-shaped form (Fig. 4.329); sparsely granulated skin; papillated lips; long nasal barbel, its length being 35.2%−43.3% HL; acutely pointed snout; ventral surface of paired fin rays nonplaited; and posteriorly serrated pectoral-fin spine. Body light yellowish gray; caudal-fin base with broad oval blue-black spot; dorsal, pectoral, and pelvic-fins with yellowish base followed by dark brown middle band and distally hyaline margin; anal-fin base yellowish with thin dark brown band near the tip; distal margin hyaline; adipose-fin yellowish gray with hyaline distal margin; caudal fin with alternate basal yellowish and middle v-shaped dark brown band with hyaline tip. Maximum size ∼60 mm SL.

Distribution. Kaladan River, Mizoram, India.

Glyptothorax cavia (Hamilton, 1822)
(Fig. 4.312)

FIGURE 4.312

Glyptothorax cavia, 123.4 mm SL, Barak River, Vanchengphai, Manipur.

Pimelodus cavia Hamilton, 1822: 188, 378 (type locality: rivers of Northern Bengal).

Diagnosis. Combination of characters: body depth at dorsal-fin origin 25.7%−26.0% SL; head high, its depth at nape 68.3%−69.5% HL and at eye 51.8%−54.3% HL; teeth band on lower jaw divided only by narrow partition that is not projecting, thoracic adhesive apparatus with a deep central depression (Fig. 4.327B); skin with dark brown spots, sparsely granulated; dorsal and anal-fins with two black stripes: one at base, one submarginal separated by a white one in the middle, fin edges white; adipose fin brown with white edges; paired fins brown with white edges; caudal-fin with a black submarginal band which becomes marginal on the middle rays. Maximum size ∼270 mm SL.

Distribution. Ganga and Brahmaputra drainages, India and Bangladesh.

Glyptothorax chimtuipuiensis **Anganthoibi & Vishwanath, 2010**
(Fig. 4.313)

FIGURE 4.313

Glyptothorax chimtuipuiensis, 57.8 mm SL, holotype, Kaladan River at Kolchaw, Mizoram.

Glyptothorax chimtuipuiensis Anganthoibi & Vishwanath, 2010a: 57 (type locality: Kaladan River at Kolchaw, Lawntlai District, Mizoram, India).

Diagnosis. Short and stout species with granulated skin; the dorsal profile greatly arched anterior to the adipose-fin; dorsal spine short and smooth, its length 5.1%−8.9% SL; pectoral-fin length is 16.8%−21.9% SL, ventral surface of its first simple ray plaited; adipose-fin base length 22.1%−27.3% SL; thoracic adhesive apparatus chevron shaped, wider than long, the median ridges of the apparatus perpendicular to its base, slightly diverging laterally, its base concave, open caudally, with a shallow depression at its posterior end followed by a small fold of skin (Fig. 4.314). Maximum size ∼60 mm SL.

Distribution. Kaladan River at Kolchaw, Lawntlai District, Mizoram, India.

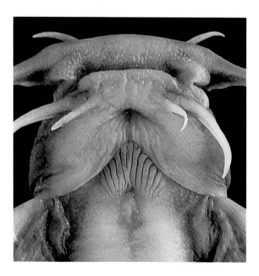

FIGURE 4.314

Thoracic adhesive apparatus of *Glyptothorax chimtuipuiensis*.

Glyptothorax churamanii **Rameshori & Vishwanath, 2012**
(Fig. 4.315)

FIGURE 4.315

Glyptothorax churamanii, 85.5 mm SL, Kaladan River at Kolchaw, Mizoram.

Glyptothorax churamanii Rameshori & Vishwanath, 2012a: 80 (type locality: Kaladan River at Kolchaw, Lawntlai District, Mizoram, India).

Diagnosis. An oblong thoracic adhesive apparatus with an inverted v-shaped median depression on the posterior half, opening caudally, ridges of the apparatus not reaching the gular region; OP not reaching the anterior nuchal-plate element; caudal fin with diffused black submarginal bands on each lobe, a slightly longer ventral lobe than the dorsal lobe; sparsely tuberculated skin; pleated ventral surfaces of pectoral spine and upto two outer rays of pelvic fins, plicae on pectoral spine arranged in rows (Fig. 4.316), continuous distally and dissociated in a series of three to five

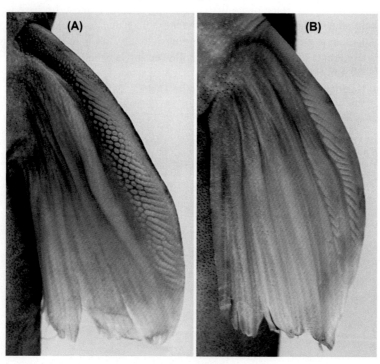

FIGURE 4.316

Ventral surfaces of pectoral (A) and pelvic (B) fins showing plicae.

hexagonal-shaped spots in each row; 3 + 10 gill rakers on the first branchial arch; length of dorsal-fin base 12.0%−13.9% SL; length of caudal fin 25.1%−27.6% SL. Maximum size ∼70 mm SL.
Distribution. Kaladan River at Kolchaw, Mizoram state, India.

Glyptothorax dikrongensis **Tamang & Chaudhry, 2011**
(Fig. 4.317)

FIGURE 4.317

Glyptothorax dikrongensis, 66.4 mm SL, Dikrong River, Arunachal Pradesh.

Glyptothorax dikrongensis Tamang & Chaudhry, 2011: 2 (type locality: Dikrong River at Doimukh, near Khola Camp RCC Bridge, Midpu, Arunachal Pradesh, India).
Diagnosis. A patch of unculiferous striae on the posterior region of the lower lip, in between the inner mandibular-barbel base; unculiferous striae of the thoracic adhesive apparatus extending anteriorly onto the gular region equal distance between the posterior end of the pectoral-fin base to the pelvic-fin origin and between the pelvic fin and the anal-fin origin; pelvic-fin origin anterior to posterior end of the dorsal-fin base; a moderately divergent, chevron-shaped thoracic adhesive apparatus with a somewhat broad anterior apex at the isthmus; weakly developed posterior skin flap of the outer and inner mandibular barbels; posterior margin of pectoral-fin spine with eight serrations; pale yellow body mottled with light and dark brown marks and spots dorsally; dorsal-fin base dusky to light brown, subdistal transverse band brown; caudal fin with dark vertical brown band at base, pale yellow vertical band followed by brown to black melanophores and hyaline tip. Maximum size ∼90 mm SL.
Distribution. Dikrong River, Arunachal Pradesh, northeastern India.

Glyptothorax giudikyensis **Kosygin, Singh & Gurumayum, 2020**
Glyptothorax giudikyensis Kosygin, Singh & Gurumayum, 2020: 2, figs. 1 and 2 (type locality: Giudiky stream near Langpram Village, Barak drainage, Tamenglong District, Manipur).
Diagnosis. Glyptothorax giudikyensis is diagnosed in having a chevron shaped thoracic adhesive apparatus with an inverted V-shaped central depression widely open posteriorly; a furrow running along the entire length of the ventral surface of the pectoral spine; dorsal-fin spine serrated posteriorly; pale yellowish narrow stripes, one along the mid-dorsal line and another along the lateral line; tuberculated skin and on the rays of the paired fins; vertebrae 22−23 + 18−20. Maximum size ∼145 mm SL.
Distribution. Barak drainage, Tamenglong District, Manipur.

Glyptothorax gopii **Kosygin, Das, Singh & Roy Choudhury, 2019**

Glyptothorax gopii Kosygin, Das, Singh & Roy Choudhury, 2019: 569, fig. 1 (type locality: Tuipui River near Champhai Champhai District, Kaladan River drainage, Mizoram, India).

Diagnosis. Anterior nuchal plate element axe-shaped with extensive contact with the posterior nuchal plate; ventral surface of the pectoral-fin spine and outer rays of pelvic fin rays with plicae; thoracic adhesive apparatus elliptical, with lanceolate median depression; body with two pale creamish stripes, one along the mid-dorsal line and another on the lateral line. Maximum size ∼64 mm SL.

Distribution. Tuipui River near Champhai, Champhai District, Kaladan drainage, Mizoram.

Glyptothorax gracilis **(Günther, 1864)**
(Fig. 4.318)

FIGURE 4.318

Glyptothorax gracilis, 110 mm SL, Teesta River, Sikkim.

Glyptosternum gracile Günther, 1864: 186 (type locality: Nepal).

Diagnosis. Granulate skin; thoracic adhesive apparatus longer than broad, no central depression; OP reaching nuchal plate; dorsal-fin height more than body depth; pectoral fin slightly longer than head, ventral surface of its spine with no plaits; pelvic-fin extends considerably beyond anal opening, just reaches anal fin origin; maxillary barbel longer than head; lower lobe of caudal-fin longer than the upper. Maximum size ∼130 mm SL.

Distribution. Nepal and Sikkim.

Glyptothorax granulus **Vishwanath & Linthoingambi, 2007**
(Fig. 4.319)

Glyptothorax granulus Vishwanath & Linthoingambi, 2007: 2620 (Iril River, Chindwin basin, Phungdhar Village at Ukhrul District, Manipur, India).

FIGURE 4.319

Glyptothorax granulus, 76.6 mm SL, Iril River, Manipur.

Diagnosis. Body plain with no longitudinal lines; skin granulated; head rounded, depressed, its depth 60.2%−61.0% SL; thoracic adhesive apparatus well developed, with a central depression which is open caudally; OP separated from dorsal pterigiophore by a considerable distance, its length two times its width; dorsal spine serrated posteriorly on distal part with six antrose serrae; adipose fin well developed, no series of ridges or bumps in front; total vertebrae 35−36. Maximum size ∼90 mm SL.

Distribution. Iril and Lokchao rivers, Chindwin drainage, India.

Glyptothorax igniculus Ng & Kullander, 2013
(Fig. 4.320)

FIGURE 4.320

Glyptothorax igniculus, 72.0 mm SL, Yu River, Tamu.

Glyptothorax igniculus Ng & Kullander, 2013: 553 (Myitta River, 8 km east of Kalaymyo, Sagging Division, Myanmar).

Diagnosis. Thoracic adhesive apparatus with a lanceolate or flame-shaped central depression that is completely enclosed by skin ridges caudally, with a single, nondiverging series of striae running along its edges (Fig. 4.321); dorsal surface of the head with rounded tubercles, uniform body coloration. Maximum size ∼75 mm SL.

Distribution. Tizu River drainage, Manipur, India; Myittha River, a tributary of the Chindwin River in western Myanmar.

Glyptothorax indicus Talwar, 1991
(Fig. 4.322)

Glyptothorax indicus Talwar in Talwar and Jhingran, 1991a, 1991b: 654, fig. 210 (type locality: streams of Terai, northern Bengal).

Diagnosis. Chevron shaped thoracic adhesive apparatus; a unculiferous striae in a patch behind the lower lip and between the bases of inner mandibular barbel (Fig. 4.323), distance between pectoral-fin base and pelvic-fin origin much longer than that of the distance between origins of pelvic and anal fins. Maximum size ∼110 mm SL.

Distribution. Ganga and Brahmaputra drainages.

Glyptothorax jayarami Rameshori & Vishwanath, 2012
(Fig. 4.324)

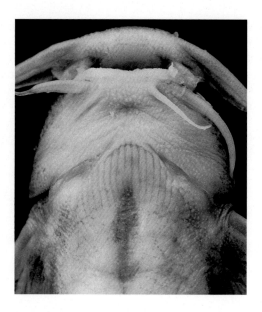

FIGURE 4.321

Thoracic adhesive device of *Glyptothorax igniculus*.

FIGURE 4.322

Glyptothorax indicus, 76.0 mm SL, Brahmaputra River, Goalpara, Assam.

Glyptothorax jayarami Rameshori & Vishwanath, 2012b: 55, figs. 1–3, 4a (type locality: Kaladan River at Kolchaw, Lawntlai District, Mizoram State, northeastern India).

Diagnosis. Elongated, ovoid thoracic adhesive apparatus with an oval central depression, the apparatus not reaching the gular region; the base of the caudal fin with two blackish-brown blotches behind the hypural plate; the OP not in contact with the anterior nuchal-plate element; a long dorsal spine (17.9%–19.5% SL); the ventral surfaces of simple and adjacent branched rays of the pectoral and pelvic fins with well-developed plicae (Fig. 4.325), and first branchial arch with eight to nine gill rakers. Maximum size ∼105 mm SL.

Distribution. Kaladan River at Kolchaw in Mizoram, India.

Glyptothorax kailashi **Kosygin, Singh & Mitra, 2020**
Glyptothorax kailashi Kosygin, Singh & Mitra, 2020: 2, figs. 1, 2a, 3a (type locality: Tuipui River near Champhai, Kaladan River drainage, Champhai District, Mizoram).

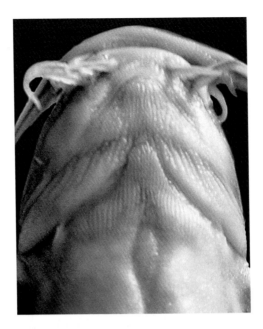

FIGURE 4.323

Thoracic adhesive device of *Glyptothorax indicus*.

FIGURE 4.324

Glyptothorax jayarami, 104.5 mm SL, holotype, Kaladan River at Kolchaw, Mizoram.

    Diagnosis. Thoracic adhesive apparatus elongated and chevron shaped, plicae present all over, closed posteriorly; maxillary barbel longer than head length; pelvic fin reaching anal-fin origin; ventral surfaces of simple rays of paired fin nonplaited; anterior nuchal plate saddle-like with W-shaped extensions; body with pale cream stripes, one each along the mid-dorsal line and another on the lateral line. Max. size ~ 67.3 mm SL.

    Distribution. Tuipui River, Champhai District, Kaladan drainage, Mizoram.

Glyptothorax maceriatus Ng & Lalramliana, 2012
(Fig. 4.326)

Glyptothorax maceriatus Ng & Lalramliana, 2012a: 45, figs. 1 and 2 (type locality: Tlawng River at Sairang, Meghna-Surma River system, Mizoram, India).

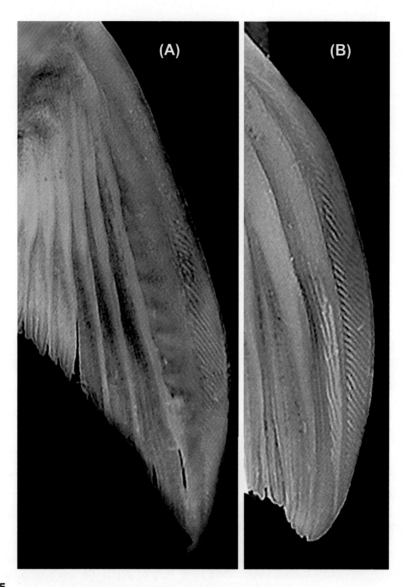

FIGURE 4.325

Ventral views of pectoral (A) and pelvic (B) fins of *Glyptothorax jayarami*, showing plicae.

Diagnosis. Nasal barbel not reaching anterior orbital margin; head length 23.7%−25.3% SL; head depth 12.5%−14.2% SL; body depth at anus 11.3%−13.8% SL; dorsal-spine length 13.0%−17.1% SL; thoracic adhesive apparatus with narrow elliptic central depression that is almost wholly enclosed posteriorly by skin ridges (striae) and with single, nondiverging series of striae running along its edges (Fig. 4.327); width of adhesive apparatus 55.8%−72.1% its length; unculiferous ridges of

FIGURE 4.326

Glyptothorax maceriatus, 72.4 mm SL, Leimatak River, tributary of Barak River, Manipur.

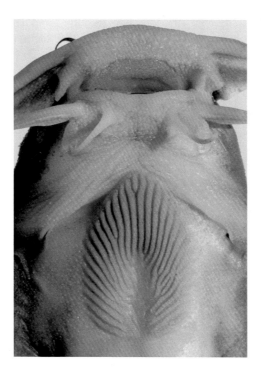

FIGURE 4.327

Thoracic adhesive apparatus of *Glyptothorax maceriatus*.

adhesive apparatus not extending anteriorly onto gular region; absence of striae on first pectoral and pelvic-fin elements; smooth posterior edge of dorsal spine Maximum size ~110 mm SL.

Distribution. Tlawng and Tuirial rivers in Mizoram, northeastern India, Leimatak River, tributary of Barak River in Manipur, Barak−Surma−Meghna River system.

***Glyptothorax manipurensis* Menon, 1955**
(Fig. 4.328)

FIGURE 4.328

Glyptothorax manipurensis, 67.6 mm SL, Barak River, Karong, Manipur.

Glyptothorax manipurensis, Menon, 1955: 23 (type locality: Barak River at Karong, Manipur).

Diagnosis. A species of *Glyptothorax* with a triangular thoracic apparatus, longer than broad, devoid of central pit (Fig. 4.329); OP length three times its width; no ridges or bumps in front of

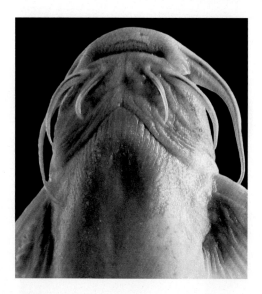

FIGURE 4.329

Thoracic adhesive apparatus of *Glyptothorax manipurensis*.

adipose-fin; dorsal spine serrated with five antrose serrae; pectoral spine with 9−11 serrae; pelvic fin extending upto anus; anal fin short, extending upto vertically through posterior extremity of adipose-fin; caudal-fin lobes equal; a white longitudinal line overlapping lateral line; black spot at dorsal, adipose, and caudal-fin bases present; smooth skin. Maximum size ∼150 mm SL.

Distribution. Barak River, Manipur, India.

***Glyptothorax mibangi* Darshan, Dutta, Kachari, Gogoi & Das, 2015**
(Fig. 4.330)

FIGURE 4.330

Glyptothorax mibangi, 79.0 mm SL, holotype, Tisa River, Tirap District, Arunachal Pradesh.

Glyptothorax mibangi Darshan, Dutta, Kachari, Gogoi & Das, 2015: 116, fig. 1 (type locality: Tisa River near Longding, Arunachal Pradesh, Brahmaputra drainage, India).

Diagnosis. Obtuse leaf-shaped thoracic adhesive apparatus with a spindle shaped median depression, plicae throughout the apparatus including the median depression; simple rays of paired nonplaited ventrally; body depth 10.4%−13.5% SL; caudal peduncle depth 6.8%−8.3% SL; snout length 52.9%−5.8.6% SL; first branchial arch with 2 + 7 rakers. Maximum size ∼80 mm SL.

Distribution. Tisa River, Brahmaputra drainage, Arunachal Pradesh.

Glyptothorax ngapang **Vishwanath & Linthoingambi, 2007**
(Fig. 4.331)

FIGURE 4.331

Glyptothorax ngapang, 82.7 mm SL, holotype, Iril River, Bamonkampu, Imphal, Manipur.

Glyptothorax ngapang Vishwanath & Linthoingambi, 2007: 2619 (Iril River, Chindwin Basin Bamonkampu, Manipur State, India).

Glyptothorax chavomensis Arunkumar & Moyon, 2017: 246, fig. 4 (description from Chavon River, Moyon Khullen, Chandel District, Chindwin drainage, Manipur).

Diagnosis. Combination of characters: head small, its length 22.2%−25.0% SL; gape width 62.6%−71.4% of snout length; eye diameter 20.0%−28.6% snout length; adhesive apparatus width 56.4%−70.0% of its length; its length 14.0%−15.7% SL; adipose dorsal fin well developed and long, its base length 49.1%−65.0% of interdorsal length; caudal peduncle slender, its height 28.0%−34.8% its length; anal fin long, its base length 61.5%−72.0% HL, its height 76.1%−86.9% HL; skin tuberculated, tubercles oval with cornified longitudinal ridges; dorsal spine serrated on distal part only with two to three serrae. Maximum size ∼130 mm SL.

Distribution. Rivers draining to the Imphal and Yu Rivers, Chindwin drainage, Manipur, India.

Glyptothorax pantherinus **Anganthoibi & Vishwanath, 2013**
(Fig. 4.332)

FIGURE 4.332

Glyptothorax pantherinus, 131.2 mm SL, holotype, Noa Dehing River, Changlang District, Arunachal Pradesh.

Glyptothorax pantherinus Anganthoibi & Vishwanath, 2013: 2, figs. 1 and 2 (type locality: Noa Dehing River, Brahmaputra River basin, Deban-Namdapha, Changlang District, Arunachal Pradesh, India).

Diagnosis. Mottled skin, more prominent over the dorsal and caudal region extending to the tip of the caudal fin; a well-developed obtuse leaf-shaped thoracic adhesive apparatus with no central depression; one bean-shaped cream-colored spot on either side of the dorsal-fin origin; ventral surface of the first unbranched rays of the paired fins plaited marginally and distally. Maximum size ~140 mm SL.

Distribution. Noa Dehing River, Brahmaputra River basin, Deban-Namdapha, Changlang District, Arunachal Pradesh, India.

Glyptothorax radiolus **Ng & Lalramliana, 2013**
Glyptothorax radiolus Ng & Lalramliana, 2013: 502, figs. 1 and 2 (type locality: Raidak II River in the vicinity of Buxa Tiger Reserve, Jayanti area, West Bengal, India).

Diagnosis. Prominently plicate ventral surface of pectoral-fin spine and the first pelvic-fin ray; slender body, depth at anus 11.2%−11.4% SL; thoracic adhesive apparatus rhomboid shape consisting of plicae extending from isthmus to level of middle of pectoral-fin base, with wedge-shaped central depression, devoid of skin ridges, apparatus not extending to gular region, absence of mid-lateral stripe and total vertebrae 36; caudal fin deeply forked, lower lobe longer, mid-dorsal stripe. Maximum size ~110 mm SL.

Distribution. Raidak River drainage, a right-bank tributary of the Brahmaputra River, in West Bengal, India.

Glyptothorax rugimentum **Nga & Kottelat, 2008**
(Fig. 4.333)

FIGURE 4.333

Glyptothorax rugimentum, 74.2 mm SL, Yu River, Tamu, Myanmar.

Glyptothorax rugimentum Ng & Kottelat, 2008a: 129 (type locality: Chon Son stream between Kyondaw and Phadaw, Ataran River drainage, Kayin State, Myanmar).

Diagnosis. Thoracic adhesive apparatus rhomboidal which is open posteriorly, a central depression, unculiferous ridges of the apparatus extending anteriorly onto the gular region (Fig. 4.334);

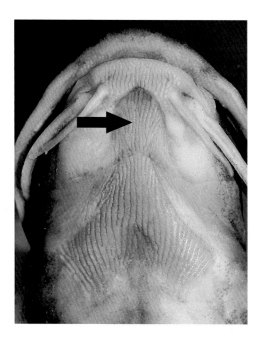

FIGURE 4.334

Thoracic adhesive apparatus of *Glyptothorax rugimentum*, reaching gular region.

short nasal barbel, not reaching anterior margin of orbit; posterior margin of adipose fin steeply sloping; two broad dark brown saddles in front of dorsal fin, two bars: one at the origin of adipose fin and another on the caudal peduncle, two bars on caudal fin, anterior behind cadual-fin base darker and the posterior margin of caudal fin lighter. Maximum size ∼80 mm SL.

Distribution. Chindwin, Ataran, Salween and Sittang River drainages in Myanmar and western Thailand and Chindwin in Myanmar.

Glyptothorax scrobiculus Ng & Lalramliana, 2012
(Fig. 4.335)

Glyptothorax scrobiculus Ng & Lalramliana, 2012b: 2, figs. 1−3, 5 (type locality: Mausam River in the vicinity of NE Khawdungsei village, Tuivai River drainage, Mizoram, India).

Diagnosis. A furrow running along the entire length of the ventral surface of the pectoral spine; unculiferous ridges of thoracic adhesive apparatus rhomboidal, not extending anteriorly onto gular region; depressed area in thoracic adhesive apparatus not wholly enclosed by ridges; chevron-shaped median depression on the posterior half, striae uninterrupted, median striae oriented

FIGURE 4.335

Glyptothorax scrobiculus, 94.0 mm SL, Barak River, Manipur.

anteriorly, lateral ones antero-laterally, smooth posterior edge of dorsal spine; 38−40 vertebrae; absence of both dark saddles on body and dark crescentic mark at base of caudal fin. Maximum size ∼148 mm SL.

Distribution. Mausam and Sur Luite rivers, tributaries of the Tuivai River; the Tuirial River drainage in Mizoram, both draining into the Barak River, Surma-Meghna River system.

Glyptothorax senapatiensis **Premananda, Kosygin & Saidullah, 2015**

Glyptothorax senapatiensis Premananda, Kosygin & Saidullah,2015: 325, figs. 1−3 (Imphal River at Motbung, Senapati District, Chindwin River drainage, Manipur, India).

Diagnosis. Body depth at dorsal-fin origin 21.0%−26.8% SL, caudal peduncle depth 8.4%−9.3% SL; thoracic adhesive apparatus with U-shaped median depression opening caudally; dorsal-fin spine with six to seven serrae posteriorly; pectoral-fin spine with 8−10 serrae posteriorly; nasal barbel extending to middle of orbit; skin densely tuberculated; cream mid-dorsal stripe, extending from behind dorsal-fin base to caudal-fin base; distinctly pale nuchal plate elements. Maximum size ∼60 mm SL.

Distribution. Imphal River at Motbung, Senapati District, Manipur, Chindwin drainage, India.

Glyptothorax striatus **(McClelland, 1842)**
(Fig. 4.336)

FIGURE 4.336

Glyptothorax striatus,102.0 mm SL, Umngi, Balat, Meghna basin, south-west Khasi Hills, Meghalaya.

Glyptosternon striatus McClelland, 1842: 587, pl. 6, figs. 1 and 2 [type locality: Kasyah (Kashya, Khasi) Hills, Meghalaya, India].

Glyptothorax striatus: Ng & Lalramliana, 2013: 506, fig. 4 (redescription).

Diagnosis. Ventral surface of pectoral-fin spine and first pelvic-fin ray with prominent plicate; snout length 51.8%−54.7% HL, wedge-shaped central depression in thoracic adhesive apparatus devoid of skin ridges (Fig. 4.337), pectoral-fin length 18.7%−23.8% SL,

FIGURE 4.337

Ventral surface of anterior part of *Glyptothorax striatus* showing thoracic adhesive apparatus and plicae on pectoral fin.

plicae on ventral surfaces of pectoral-fin spine continuous, dorsal-fin spine length 10.3%−15.7% SL, dorsal-to-adipose distance 24.9%−27.9% SL, body depth at anus 11.0%−14.7% SL, adipose-fin base length 10.7%−13.5% SL, caudal-peduncle length 18.4%−20.7% SL, caudal-peduncle depth 6.8%−8.6% SL; distinct pale mid-lateral stripe on body. Maximum size ∼195 mm SL.

Distribution. Brahmaputra River drainage and the Surma-Meghna River system in India.

Glyptothorax telchitta (Hamilton, 1822)
(Fig. 4.338)

FIGURE 4.338

Glyptothorax telchitta, 88.0 mm SL, Dikrong River, Arunachal Pradesh.

Pimelodus telchitta Hamilton, 1822: 185, 378 (type locality: Hooghly River at Kalna, West Bengal, India).

Glyptothorax telchitta: Ng, 2005d: 8, figs. 2b and 5 (redescription).

Diagnosis. Large, prominent tubercles on the head and body; a thoracic adhesive apparatus without a median depression; pointed snout; absence of saddles on body; caudal peduncle depth 4.7%−5.9% SL. Maximum size ∼100 mm SL.

Distribution. Hoogly and Ganga River drainages.

Glyptothorax ventrolineatus Vishwanath & Linthoingambi, 2006
(Fig. 4.339)

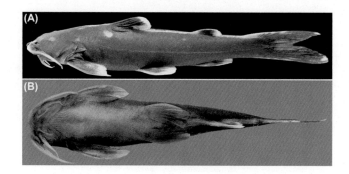

FIGURE 4.339

Glyptothorax ventrolineatus, 78.6 mm SL, (A) lateral view, (B) ventral view, Lokchao River, Moreh, Manipur.

Glyptothorax ventrolineatus Vishwanath & Linthoingambi, 2006: 201, Fig.1 (type locality: Iril River, Ukhrul District, Manipur, India).

Diagnosis. A species of *Glyptothorax* with three creamish stripes on body, one each along mid-dorsal line, lateral lines and mid-ventral line of the body; surface of head, body and adipose dorsal fin granulated; length of nasal barbel twice the internasal length; supra-occipital process not in contact with the nuchal plate, its width 38.3%−44.7% of its length; adipose dorsal-fin base length equals rayed dorsal-fin base length; caudal fin longer than head length.

Distribution. Iril River and Lokchao River, Chindwin drainage, Manipur, India.

Glyptothorax verrucosus Rameshori & Vishwanath, 2012
(Fig. 4.340)

FIGURE 4.340

Glyptothorax verrucosus, 53.6 mm SL, Kaladan River at Kolchaw, Mizoram.

Glyptothorax verrucosus Rameshori & Vishwanath, 2012c: 148, fig. 1 (type locality: Koladyne River at Kolchaw, Lawntlai District, Mizoram, India).

Diagnosis. Densely tuberculated skin on adipose fin and all over the body, tubercles larger, rounded and linearly arranged along the creamish yellow mid-lateral longitudinal stripe; a distinct

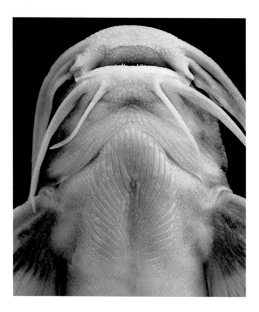

FIGURE 4.341

Thoracic adhesive of Glyptothorax *verrucosus*.

C-shaped dark brown marking on the dorsal fin against a white background, the base of which continues along the entire-fin base; an elliptical thoracic adhesive apparatus with a median depression, not extending to the gular region (Fig. 4.341); an acutely arched predorsal profile; a long nasal barbel reaching the anterior margin of the eye; a long dorsal spine (15.2%−18.7% SL); and poorly developed pectoral fin plicae. Maximum size ∼90 mm SL.

Distribution. Koladyne River (Kaladan) in Mizoram, India.

Genus *Gogangra* Roberts, 2001

Gogangra Roberts, 2001: 83 (type species: *Pimelodus viridescens* Hamilton, 1822). *Gogangra* is the replacement name of *Gangra* Roberts and Ferraris, 1998, since the genus name is preoccupied: *Gangra* Walker in Lepidoptera.

Gangra Roberts & Ferraris, 1998: 333 (type species: *Pimelodus viridescens* Hamilton, 1822).

Diagnosis. Head, mouth, and jaws broad, jaws with several rows of conical teeth; mesethmoid not greatly expanded, its dorsal profile slightly convex, Y-shaped anteriorly; palate toothless; barbels short, slender, and round in cross section; nasal barbel very short, extending posteriorly only to end of posterior naris; maxillary and mental barbels not extending posteriorly beyond head; maxillary barbel membrane absent or greatly reduced; maxillary bone inside maxillary barbel very short; outer and inner mental barbels evenly and widely separated and not in one line (Fig. 4.295B); aired lateral cranial fontanel absent; branchiostegal membranes free from isthmus.

Gogangra laevis Ng, 2005

Gogangra laevis Ng, 2005e: 280 (type locality: Gowain River, and Khal at Gowainghat, Bangladesh).

Diagnosis. The species is distinct in having the anteroventral margin of the opercle, that is, articular facet with the interopercle gently curved, large eye (20.3%−24.8% HL) and also in having a dark, very diffuse midlateral stripe running along lateral myoseptum and a series of darker brown saddles along dorsal and lateral surfaces of head and body, first three to four dorsal-fin rays with evenly distributed brown chromatophores along middle third of fin rays, other parts of dorsal fin hyaline, caudal fin hyaline, with large brown spots on middle third of each lobe. Maximum size ∼80 mm SL.

Distribution. Yamuna and Meghna River drainages, lower Brahmaputra River drainage in Bangladesh.

Gogangra viridescens **(Hamilton, 1822)**
(Fig. 4.342)

FIGURE 4.342

Gogangra viridescens, 82.0 mm SL, Brahmaputra River, Goalpara, Assam.

Pimelodus viridescens Hamilton, 1822: 173, 377 (type locality: Rivers of northern Bengal).

Diagnosis. Distinct in having anteroventral margin of the opercle distinctly notched; head width 19.7%−23.2% SL, eye diameter 16.2%−20.5% HL; inner and outer mental barbels wide separated; origin of inner mental barbel anterior to origin of outer mental barbels; lateral cranial fontanel absent; live specimen with viridescent or silvery supraopercular mark. Maximum size ∼85 mm SL.

Distribution. Ganga and Brahmaputra River drainages.

Genus *Myersglanis* Hora & Silas, 1952

Myersglanis Hora & Silas, 1952: 19 (type species: *Exostoma blythii* Day, 1870e). Gender: feminine.

Diagnosis. Body elongated, flattened ventrally to pelvic-fin base, snout depressed, broadly rounded, eyes small, superior, subcutaneous, anterior half of head, nostrils close together, separated by a flap bearing nasal barbel, eight barbels, maxillary with broad base; mouth inferior, upper jaw longer; lips thick and papillated, lower labial fold continuous (Fig. 4.343); conical pointed or oar shaped teeth on both jaws, continuous teeth band on upper jaw, continuous lower labial fold (Fig. 4.344); no thoracic adhesive apparatus, dorsal fin with no spine; lateral line present; skin smooth; ventral sides of maxillary barbels, labial folds and simple rays of paired fins plicate (Fig. 4.345).

Myersglanis blythii **(Day, 1870)**
(Fig. 4.346)

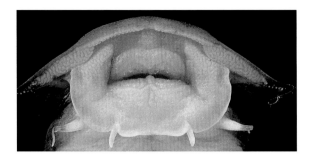

FIGURE 4.343

Mouth of *Meyersglanis jayarami* showing plicae on the barbels. Teeth bands and postlabial groove.

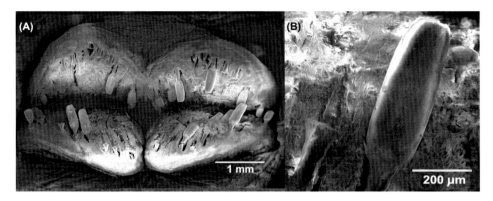

FIGURE 4.344

SEM image of *Meyersglanis jayarami* (A) showing continuous teeth band on upper jaw, (B) oar shaped teeth on the jaws. [*SEM*, Scanning electron microscopy].

Exostoma blythii Day, 1870d: 525 (type locality: Pharping, Nepal).

Diagnosis. Branched pectoral-fin rays 16−17, branched cadual-fin rays 13; anal fin origin nearer to caudal-fin base from pelvic-fin origin; adipose fin low and moderately long. Maximum size ∼70 mm SL.

Distribution. Rivers draining to Ganga, Eastern Nepal; rivers in the Darjeeling Himalaya, India.

Myersglanis jayarami **Vishwanath & Kosygin, 1999**
(Fig. 4.347)

Myersglanis jayarami Vishwanath & Kosygin, 1999: 291, pl. 1, figs. 1 and 2 (type locality: Laniye River at Jessami, Manipur, India).

Diagnosis. Branched pectoral rays 10; branched caudal rays 15−16; anal-fin origin equidistant from pelvic-fin origin and caudal-fin base; and also in having an adipose-fin which is confluent with caudal-fin. Maximum size ∼80 mm SL.

Distribution. India: Laniye River at Jessami, Nagaland-Manipur border, Manipur (Chindwin basin).

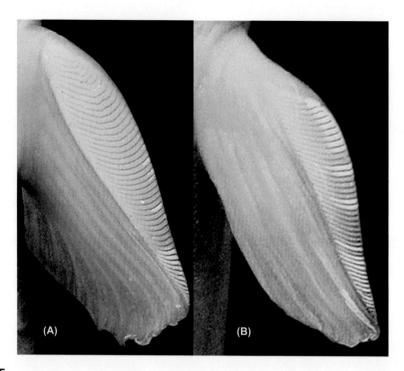

FIGURE 4.345

Ventral views of (A) pectoral and (B) pelvic fins of *Myersglanis jayarami* showing plicae on simple rays.

Genus *Nangra* Day, 1877

Nangra Day, 1877: 493 (type species: *Pimelodus nangra* Hamilton, 1822). Gender: feminine.
Nangra: Roberts & Ferraris, 1998: 335 (review of the genus).

Diagnosis. Body elongated, slender, head depressed, snout slightly to considerably spatulate, with ventral portion flat and more or less markedly projecting anterior jaws; snout supported by broadly expanded plate on the ventral surface of mesethmoid cornua, unlike in other sisorids; eyes small, dorsolateral; mesethmoid large and flattened plate; premaxilla immovably fixed to mesethmoid; jaws well toothed, palatal teeth present on bone tentatively identified as endopterygoid, tooth patch single, elongated and widely separated across midline; teeth few in number but present, branchiostegal rays free from isthmus, barbels long, maxillary extending to at least to pectoral spine tip and usually beyond pelvic-fins, nasal barbel to at least eye; maxillary bone inside maxillary barbel very long, extending posteriorly nearly to end of head or beyond; no adhesive apparatus on thorax.

Nangra assamensis Sen & Biswas, 1994
(Fig. 4.348)

Nangra assamensis Sen & Biswas, 1994: 441, fig. 1; Pl. 1 (type locality: Brahmaputra River at Neematighat, 14 km from Jorhat, Assam, India).

Diagnosis. Nasal barbel short, not extending beyond posterior margin of eye, dorsal-fin typically with six branched rays; branched pectoral fin rays 8−9, of pelvic-fin five, of anal 9−10, of

FIGURE 4.346

Myersglanis blythi, 84.3 mm SL, (A) dorsal, (B) lateral, and (C) ventral views. Bagmati River, near Chandrapur, south of Kathmandu, Nepal.

FIGURE 4.347

Myersglanis jayarami, 78.2 mm SL, Laniye River, Jessami, Manipur.

FIGURE 4.348

Nangra assamensis, 104.5 mm SL, Brahmaputra River, Dibrugarh, Assam.

caudal 16–17; maxillary barbel reaching to middle of anal-fin, outer mandibular to middle of pectoral-fin or little beyond the pectoral-fin. Maximum size ~110 mm SL.

Distribution. Brahmaputra River basin, northeastern India.

Nangra bucculenta **Roberts & Ferraris, 1998**

Nangra bucculenta Roberts & Ferraris, 1998: 336, fig. 14 Ganges River delta, Tangail District, North-central Region, Bangladesh.

Diagnosis. A small species of *Nangra*, largest size 34 mm SL with a terete head, moderately projecting snout and relatively expanded cheeks; nasal barbel extending to eye and maxillary to end of pectoral-fin; maxillary-barbel membrane broadly attached to cheek and maxillary bone extends posteriorly to midway between eye; fins with no filamentous projections. Maximum size ~34 mm SL.

Distribution. Floodplains between tributaries in the Ganges delta in Tangail District of central Bangladesh.

Nangra nangra **(Hamilton, 1822)**
(Fig. 4.349)

FIGURE 4.349

Nangra nangra, 114.2 mm SL, Ganga River, Patna.

Pimelodus nangra Hamilton, 1822: 193, 378, pl. 11, fig. 63 (type locality: Ganges River at Patna, India).

Nangra nangra: Roberts & Ferraris, 1998: 340 (redescription).

Diagnosis. A medium-sized species of *Nangra*, reaching 55.0 mm SL, with a moderately projecting snout; a deep longitudinal groove on dorsum of head, extending to nuchal plate; dorsal-fin typically with eight branched rays; nasal barbel extending posteriorly to end of head or to dorsal-fin origin; maxillary barbel reach vent or end of anal-fin; maxillary bone extending posteriorly almost to end of head; maxillary barbel membrane small, with narrow attachment to cheek; pectoral spine strong with coarse denticulation posteriorly. Maximum size ~120 mm SL.

Distribution. Ganga and Indus River basins.

Nangra ornata **Roberts & Ferraris, 1998**

Nangra ornata Roberts & Ferraris, 1998: 341, fig. 19 (type locality: Gowain River and Khal at Gowainghat, northern Sylhet Province (Surma or Meghna watershed), Bangladesh.

Diagnosis. A small species of *Nangra* reaching maximum size 37.0 mm SL; eye relatively large, eye diameter as long as snout length; nasal barbel extending to just beyond posterior margin of orbit; maxillary barbel reaching to anal-fin origin, barbel with broad membrane; semicircular dark brown or black spot of about the size of eye diameter on middle of caudal-fin base, extending across bases of all principal rays, but not extending onto caudal peduncle; another spot extending across bases of second through fifth branched dorsal-fin rays. Maximum size ~40 mm SL.

Distribution. Surma-Meghna River basin, Bangladesh.

Genus *Oreoglanis* **Smith, 1933**

Oreoglanis Smith, 1933: 70 (type species: *Oreoglanis siamensis* Smith, 1933). Gender: feminine.

Diagnosis. Strongly depressed head and body, greatly enlarged paired fins modified to form an adhesive device along with the mouth and its parts and the maxillary barbels (Fig. 4.350); continu-

FIGURE 4.350

Ventral view of *Oreoglanis majuscula* showing enlarged paired fins, barbels and mouth parts modified as adhesive device.

ous postlabial groove of the lower jaw and an unusual heterodont dentition: pointed teeth in the upper jaw and posterior part of the lower jaw and short, spatulate teeth in the anterior part of the lower jaw.

Oreoglanis majuscula **Linthoingambi & Vishwanath, 2011**
(Fig. 4.351)

FIGURE 4.351

Oreoglanis majuscula, 76.0 mm SL, (A) dorsal view, (B) lateral view, Kameng River, Rupa, Arunachal Pradesh.

Oreoglanis majusculus Linthoingambi & Vishwanath, 2011: 61, figs. 1, 2, 3 (type locality: Kameng River at Rupa, Brahmaputra basin, Arunachal Pradesh, India).

Diagnosis. Ventral surface of pectoral-fin spine and first pelvic-fin ray with prominent plicate; snout length 51.8%−54.7% HL, wedge-shaped central depression in thoracic adhesive apparatus devoid of skin ridges, posterior margin of maxillary barbel with villiform projections (Fig. 4.352); pectoral-fin length 18.7%−23.8% SL, plicae on ventral surfaces of pectoral-fin

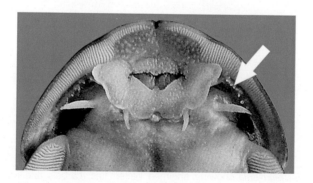

FIGURE 4.352

Lower lip of *Oreoglanis majusculus* showing villiform posterior margin of maxillary barbel.

spine continuous, dorsal-fin spine length 10.3%−15.7% SL, dorsal-to-adipose distance 24.9%−27.9% SL, body depth at anus 11.0%−14.7% SL, adipose-fin base length 10.7%−13.5% SL, caudal-peduncle length 18.4%−20.7% SL, caudal-peduncle depth 6.8%−8.6% SL; posterior margin of the lower lip entire; caudal fin emarginated. Maximum size ∼80 mm SL.

Distribution. Kameng River, Brahmaputra drainage, Arunachal Pradesh, India.

Oreoglanis pangenensis Sinha & Tamang, 2015
(Fig. 4.353)

FIGURE 4.353

Oreoglanis pangenensis, 76.7 mm SL, holotype, Pange River, Lower Subansiri District, Arunachal Pradesh.

Photo courtesy: A. Darshan.

Oreoglanis pangenensis Sinha & Tamang, 2015: 332 (type locality: Pange River, tributary of Brahmaputra River; Aro-Lenching, Ziro Valley, Lower Subansiri District: Arunachal Pradesh).

Diagnosis. Tip of maxillary barbel rounded; one projection on the ventroposterior margin of maxillary barbel; proximal third with lobulate projections and distal two-thirds with laciniate projections; entire lower lip margin; emarginated caudal fin; a dark brown streak on the outer margins of caudal lobes. Maximum size ∼80 mm SL.

Distribution. Pange River, Brahmaputra Drainage, lower Subansiri District, Arunachal Pradesh.

Genus *Parachiloglanis* Wu, He & Chu, 1981

Parachiloglanis Wu, He & Chu,1981: 76, 79 (type species: *Glyptosternum hodgarti* Hora, 1923). Gender: masculine.

Diagnosis. Absence of a postlabial groove; lower lip continuous with isthmus without demarcation; homodont dentition in the lower jaw and heterodont dentition on the lower; extremely posterior vent, occurring just anterior to the anal-fin origin; gill opening extending to ventral surface.

Parachiloglanis benjii Thoni & Gurung, 2018

Parachiloglanis benjii Thoni & Gurung, 2018: 44, figs. 2c, 3−4 (type locality: Mendegangchhu stream, Punakha Dzongkhag, Bhutan).

Diagnosis. Postlabial fold absent; dentition in the upper jaw heterodont and dentition in the lower jaw homodont; posteriorly and widely arched tooth patch with a medial cusp on the upper jaw, adipose fin adnate, black pigmentation on the caudal-fin margin, caudal fin emarginated, absence of a punctate lateral line, head depth 40.0%−43.4% HL, and the absence of a mottled coloration pattern. Maximum size ∼88.2 mm SL.

Distribution. Rivers and streams draining to the Brahmaputra, Bhutan.

Parachiloglanis bhutanensis Thoni & Gurung, 2014

Parachiloglanis bhutanensis Thoni & Gurung, 2014: 307 (type locality: Khalingchhu stream, Trashigang Dzongkhag, Bhutan).

Diagnosis. The species differs is distinct in having 35−40 prominent lateral-lone pores from posterior edge of head to caudal-fin base, large fleshy adipose fin, its height 4.2%−4.6% SL; a deep head, its depth 47%−59% HL; truncated caudal fin, caudal fin with no markings, absence of postlabial groove. Maximum size ∼100 mm SL.

Distribution. Khalingchhu stream and the headwaters of the Dangmechhu River, Brahmaputra drainage.

Parachiloglanis dangmechhuensis **Thoni & Gurung, 2018**

Parachiloglanis dangmechhuensis Thoni & Gurung, 2018: 47, figs. 2a, 6, 7 (type locality: Lingmethangchhu at confluence with Kurichhu, Mongar Dzongkhag, Bhutan).

Diagnosis. Absence of a postlabial groove, lower lip fold, or any other posteriorly produced lip-like structure on the lower jaw, a narrow, semicircular tooth patch on the upper jaw, adipose fin ending with discrete notch, 16 branched pectoral-fin rays, 13−14 branched caudal-fin rays, caudal peduncle length 7.2%−10.6% SL, and the presence of three conspicuous pale spots on the caudal fin. Maximum size ∼70 mm SL.

Distribution. Rivers and streams draining to the Brahmaputra, Bhutan.

Parachiloglanis drukyulensis **Thoni & Gurung, 2018**

Parachiloglanis drukyulensis Thoni & Gurung, 2018: 52, figs. 2d, 8, 9 (type locality: Chaplaychhu stream, Sarpang Dzongkhag, Bhutan).

Diagnosis. Absence of a postlabial fold, of heterodont dentition in the upper jaw and homodont dentition in the lower jaw, having a posterolaterally arched tooth patch on the upper jaw; presence of a mottled pattern covering the entire exposed surface of body, adipose fin adnate with caudal peduncle, the absence of lateral line pores, head depth of 33.5%−47.8% HL, a semilunate caudal-fin. Maximum size ∼74 mm SL.

Distribution. Rivers and streams draining to the Brahmaputra, Bhutan.

Parachiloglanis hodgarti **(Hora, 1923)**
(Fig. 4.354)

FIGURE 4.354

Parachiloglanis hodgarti, 75.0 mm SL, Yomgo River, Upper Siang District, Arunachal Pradesh.

Photo courtesy: A. Darshan.

Glyptosternum hodgarti Hora, 1923: 38, pl. 2, figs. 1−3 (type locality: Pharping, Nepal).

Diagnosis. The species is distinct in the absence or highly reduced lateral line, a shallow head, its depth 36%−45% SL, indented to lunate caudal-fin, presence of postlabial groove on the lower lip, and have elongated, dorso-ventrally flattened bodies, large, semicircular pectoral-fins, inferior mouths, homodont dentition forming a crescent shaped pad on the upper jaw, heterodont dentition on the lower jaw, and a series of adhesive striations on the leading pectoral and pelvic-fin rays. Maximum size ~80 mm SL.

Distribution. Tributaries of the Manas, Mangdechhu, and Dangmechhu Rivers, in southern Bhutan.

Genus *Pseudecheneis* Blyth, 1860

Pseudecheneis Blyth, 1860: 154 (type species: *Glyptosternon sulcatus* McClelland, 1842). Gender: feminine.

Diagnosis. Body elongated, flattened from below head to the region of pelvic-fins; head and snout depressed; OP not reaching nuchal plate; gills narrow, not extending much further ventrally from pectoral-fin origin; gill membranes confluent with isthmus, dorsal spine weak, paired fins horizontal, characteristic in having a thoracic adhesive apparatus consisting of a series of transverse ridges called laminae, separated by grooves, the sulcae (Fig. 4.355).

FIGURE 4.355

Thoracic adhesive apparatus of *Pseudecheneis*, showing transverse ridges and sulcae.

Pseudecheneis crassicauda Ng & Edds, 2005

Pseudecheneis crassicauda Ng & Edds, 2005b: 2, fig. 1 (type locality: Dhankuta District, Nepal).

Diagnosis. Body elongated, body depth at anus 12.9%−14.7% SL; separate pelvic fins separate, not united in the middle; caudal peduncle length 6.0%−6.6% SL; small eye, its diameter 7.5%−8.3%

HL; adipose-fin base length 1.5−2.0 times of anal-fin base; pelvic fins reach base of the first anal-fin ray; presence of pale spots on the body; total vertebrae 38−39. Maximum size ∼130 mm SL.

Distribution. Mewa Khola River, part of the Tamur River, eastern tributary of the Kosi River, which finally flows into the Ganges River in India.

Pseudecheneis eddsi **Ng, 2006**

Pseudecheneis eddsi Ng, 2006a: 51, fig. 4 [type locality: Seti River (Ganges River drainage), Tanahun, Khairenitar, Nepal].

Diagnosis. A prominent bony spur on the anterodorsal surface of the first dorsal fin pterygiophore; pelvic-fins not united; presence of a first dorsal-fin element; pelvic-fin length 18.0%−20.9% SL; caudal peduncle length 3.5%−5.3% SL; adipose-fin base length 19.5%−24.3% SL; total vertebrae 36−39; presence of pale colored patches on the body. Maximum size ∼90 mm SL.

Distribution. Gandaki drainage of central Nepal. Gandaki along with its tributaries become Narayani in Nepal which becomes Gandak in India, a major tributary of the Ganga in India.

Pseudecheneis koladynae **Anganthoibi & Vishwanath, 2010**
(Fig. 4.356)

FIGURE 4.356

Pseudecheneis koladynae, 74.6 mm SL, holotype, Kaladan River, Lawntlai District, Mizoram.

Pseudecheneis koladynae Anganthoibi & Vishwanath, 2010b: 200, figs. 1−3 (type locality: Koladyne River, Lawntlai District, Mizoram, India.

Diagnosis. Combination of characters: three isolated ovoid yellow nuchal patches, one in mid-dorsal line in front of dorsal-fin origin and two on either side of the middle spot, slightly behind; prominent bony spur on the antero-dorsal surface of the first dorsal-fin pterygiophore; short caudal peduncle (length 16.0%−18.8% SL); pelvic fin not reaching the base of the first anal-fin ray; and distinct sexual dimorphism with robust conical papilla in males and flat leaf-like comparatively smaller bilobed papilla with a small mid-ventral lobe in females. Maximum size ∼80 mm SL.

Distribution. Kaladan River, Mizoram State, India.

***Pseudecheneis serracula* Ng & Edds, 2005**
Pseudecheneis serracula Ng & Edds, 2005b: 6, figs. 3, 5 (type locality: Mugu/Bajura, Jhugala, Karnali River, Nepal).
Diagnosis. Body elongated; vertebrae 36−38; pelvic fins not united; strongly elevated neural spines of the last two to three preanal and first six to seven postanal vertebrae; long adipose-fin base (26.8%−30.4% SL), at least 2.0 times length of anal-fin base; presence of pale spots on the body; pelvic fins reaching base of the first anal-fin ray. Maximum size ∼150 mm SL.
Distribution. Karnali drainage, Middle Hills and Tarai and Gandaki drainage, central Nepal, foothills of the Himalayas, Ganga River basin.

***Pseudecheneis sirenica* Vishwanath & Darshan, 2007**
(Fig. 4.357)

FIGURE 4.357

Pseudecheneis sirenica, 109.8 mm SL, holotype, Siren River, Upper Siang District, Arunachal Pradesh.

Pseudecheneis sirenica Vishwanath & Darshan, 2007: 2628, pl. 2; fig. 2 (type locality: Siren River, Upper Siang District, Brahmaputra River drainage, Arunachal Pradesh, India).
Diagnosis. First dorsal-fin pterygiophore with a prominent bony spur on the antero-dorsal surface, pelvic ray reaches anal fin origin, interpelvic gap 1.0−1.5 as wide as eye diameter, rounded caudal peduncle, pectoral fin length with 23.9%−24.9% SL, long anal fin, its length 17.5%−21.7% SL; transverse laminae on the thoracic adhesive apparatus 14−16; caudal peduncle depth 4.1%−4.4% SL and adipose-fin base length 22.7%−23.1% SL. Maximum size ∼110 mm SL.
Distribution. Siren River, Upper Siang District, Brahmaputra River drainage, Arunachal Pradesh, India.

***Pseudecheneis sulcata* (McClelland, 1842)**
(Fig. 4.358)
Glyptosternon sulcatus McClelland, 1842: 587, pl. 6, figs. 1−3 [type locality: Kasyah (Khasi) Hills, Meghalaya, India].
Pseudecheneis sulcata: Ng, 2006a: 47, fig. 1 (identity and redescription).
Diagnosis. Elongated body, its depth 11.8%−16.2% SL; presence of a first dorsal-fin element; absence of a prominent bony spur on the anterodorsal surface of the first dorsal-fin pterygiophore; 12−14 transverse laminae on the thoracic adhesive apparatus, pectoral fin length 121.6%−156.3% HL; separate pelvic fins; adipose-fin base length 17.8%−22.7% SL; caudal peduncle length 25.0%−28.3% SL; total vertebrae 36−39; presence of pale colored patches on the body. Maximum size ∼195 mm SL.
Distribution. Brahmaputra River drainage.

FIGURE 4.358

Pseudecheneis sulcata, 93.7 mm SL, Ranga River, Yachuli, Lower Subansiri District, Arunachal Pradesh.

Pseudecheneis ukhrulensis Vishwanath & Darshan, 2007
(Fig. 4.359)

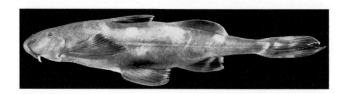

FIGURE 4.359

Pseudecheneis ukhrulensis, 117.7 mm SL, Momo stream, Tusom, Ukhrul District, Manipur.

Pseudecheneis ukhrulensis, Vishwanath & Darshan, 2007: 2627, pl. 1; fig. 1 (type locality: Momo stream, Tusom C.V., Ukhrul District, Manipur, India).

Diagnosis. Combination of characters: first dorsal-fin pterygiophore with a prominent bony spur on the antero-dorsal surface, longest ray of pelvic fin not reaching anal fin origin, interpelvic gap 2.1−2.6 as wide as eye diameter, rounded caudal peduncle, snout length 66.9%−69.0% HL and eye diameter 10.6%−12.1% HL, pectoral fin length 23.9%−24.9% SL, caudal peduncle length 24.9%−26.1% SL and its depth 4.3%−4.9% SL. Maximum size ∼120 mm SL.

Distribution. Streams draining to Tizu River in northern part of Ukhrul District and Chatrickong River in southern part of the District, both the belonging to Chindwin drainage, Manipur, India.

Genus *Pseudolaguvia* Misra, 1976

Pseudolaguvia Misra, 1976: 253 (type species: *Glyptothorax tuberculatus* Prashad and Mukerji, 1929). Gender: feminine.

Diagnosis. Body subcylindrical, moderately elongated, minute tubercles on skin; head depressed; nostrils close together, separated by a flap bearing nasal barbel; eyes small, superior, barbels four short pairs, maxillary and mandibular annulated, gill opening wide, confluent with isthmus; thorax with well-developed adhesive apparatus (Fig. 4.360), longer than wide, almost extending to abdomen, the apparatus with a median depression; shallow median groove on head, reaching base of OP, the later reaching the nuchal plate; cubito-humeral process conspicuous, dorsal spine strong and serrated internally, paired fins not plaited, caudal fin forked.

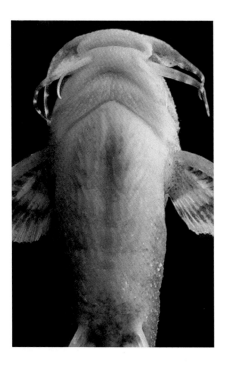

FIGURE 4.360

Thoracic adhesive apparatus in *Pseudolaguvia*.

Pseudolaguvia assula Ng & Conway, 2013

Pseudolaguvia assula Ng & Conway, 2013: 180, figs. 1−2 (type locality: Chitwan Valley, Reu River, near confluence with Rapti River, Nepal).

Diagnosis. Long dorsal-fin spine (20.3%−24.8% SL); large eye (diameter 9%−13% HL); narrow head (21.7%−24.8% SL); long and slender caudal peduncle (length 15.4%−17.0% SL and depth 7.0%−8.3% SL; total vertebrae 28−29; small eye (9%−13% HL); long pectoral fin (27.6%−32.5% SL); no pale mid-dorsal stripe along the body and brown submarginal stripes along the caudal fin lobes; dorsal-fin spine with a smooth anterior margin; four to six serrae on its posterior margin; short caudal peduncle (15.4%−17.0% SL); body depth at anus 13.7−16.2. Maximum size ∼24 mm SL.

Distribution. Rapti River drainage in central Nepal, a tributary of the Narayani River (Gandaki River in India), Ganga River basin.

Pseudolaguvia ferruginea Ng, 2009

(Fig. 4.361)

Pseudolaguvia ferruginea Ng, 2009b: 278, figs. 1−2 (type locality: Raidak I River at Shipra, just outside Beza Tiger Reserve, approximately 8 km toward Barobisha on Siliguri-Guwahati road, West Bengal, India).

FIGURE 4.361

Pseudolaguvia ferruginea, 25.8 mm SL, Rangeet River, Sikkim.

Diagnosis. Slender and elongated body, its depth 10.5%−12.4% SL; long dorsal-fin base, 14.9%−17.3% SL; smooth anterior edge of dorsal spine; long thoracic adhesive apparatus, reaching to midway between bases of last pectoral-fin ray and first pelvic-fin ray; vertebrae 31−32; long dorsal fin to adipose distance, 14.2%−17.3% SL. Maximum size ∼30 mm SL.

Distribution. Raidak River, a tributary of the Sankosh River, Brahmaputra drainage, West Bengal, India.

Pseudolaguvia ferula Ng, 2006
(Fig. 4.362)

FIGURE 4.362

Pseudolaguvia ferula, 26.0 mm SL, Rangeet River, Sikkim.

Pseudolaguvia ferula Ng, 2006b: 60, figs. 1, 2a, 3a, 5a (type locality: Teesta River at Teesta Barrage, West Bengal, India).

Diagnosis. Combination of characters: cylindrical and slightly tapering head and body (body depth 12.1%−13.8% SL); caudal peduncle depth 6.9%−7.8% SL; short adipose-fin base (11.5%−13.0% SL); small anterior fontanel (about one-third the length of the frontals); dorsal spine length 17.3%−18.7% SL; long thoracic adhesive apparatus, pelvic fin not reaching base of the first anal-fin ray; pectoral spines length 17.3%−18.7% SL; less distinct mesethmoid cornua; unculiferous ridges on the thoracic adhesive apparatus joined at their posterior ends; vertebrae 28−30; faint, poorly contrasting cream bands. Maximum size ∼30 mm SL.

Distribution. Teesta River in West Bengal, India.

Pseudolaguvia flavida Ng, 2009
Pseudolaguvia flavida Ng, 2009a: 282, figs. 5 and 6 (type locality: Hooghly, River at Kalna, West Bengal, India).

Diagnosis. Combination of characters: serrated anterior edge of the dorsal spine; body depth at anus 11.0% SL; adipose-fin base length 11.0% SL; snout length 43.0% HL; caudal peduncle depth 6.5% SL; dorsal to adipose distance 16.3% SL; dorsal-fin base length 16.7% SL; pectoral-fin length 22.4% SL. Maximum size ~20 mm SL.

Distribution. Raidak River, a tributary of the Sankosh River, Brahmaputra drainage, West Bengal, India.

Pseudolaguvia foveolata Ng, 2005
(Fig. 4.363)

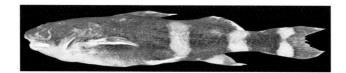

FIGURE 4.363

Pseudolaguvia foveolata, 29.6 mm SL, Teesta River at Rangit, Sikkim.

Pseudolaguvia foveolata Ng, 2005f: 174, figs. 1—2, 4A (type locality: Tista River at Tista barrage, West Bengal, India).

Diagnosis. Combination of characters: short thoracic adhesive apparatus, reaching to middle of pectoral-fin base; a slender and elongated body, body depth at anus: 11.0% SL; a long and slender caudal peduncle, its depth 4.1 times in its length; a long adipose-fin base, 24.0% SL; smooth anterior edge of the dorsal spine; maxillary barbels not reaching base of pectoral spine; adipose fin not reaching the base of the last dorsal-fin ray. Maximum size ~30 mm SL.

Distribution. Teesta River, a tributary of the Brahmaputra River, West Bengal, India.

Pseudolaguvia fucosa Ng, Lalramliana & Lalronunga, 2016

Pseudolaguvia fucosa Lalramliana & Lalronunga, 2016: 547, figs. 1, 2a, 3, 6a (type locality: Mizoram, Lawngtlai District, Tuichawng River in the vicinity of Bandukbanga village, Karnaphuli drainage, India).

Diagnosis. Combination of characters: having a pale, y-shaped marking on the dorsal surface of the head; having serrations on the anterior edge of the dorsal spine; length of dorsal-fin spine 18.0%—21.6% SL; 4—10 serrations on the anterior edge of the dorsal spine; length of pectoral-fin spine 20.7%—26.1% SL; length of adipose-fin base 21.5%—26.3% SL; body depth at anus 12.3%—15.9% SL; caudal peduncle depth 7.5%—9.8% SL; caudal peduncle length 17.0%—20.5% SL; sides of body with pale yellowish patches and irregular bands. Maximum size ~20 mm SL.

Distribution. Headwaters of the Karnaphuli River drainage, Mizoram, India.

Pseudolaguvia inornata Ng, 2005

Pseudolaguvia inornata Ng, 2005g: 36, fig. 1 [type locality: Chittagong District, Koilla Khal (creek), 10 km east of Feni-Chittagong highway on road to Ramgarh, Bangladesh].

Diagnosis. Combination of characters: deep body (13.9%—16.1% SL); adipose-fin base length 13.6%—16.4% SL; dorsal spine 18.7%—21.7% SL; pectoral spine 20.4%—23.3% SL; caudal peduncle length 16.3%−19.0% SL; vertebrae 29−30; thoracic adhesive apparatus reaching beyond the base of the last pectoral-fin ray; uniform brown body with a pale mid-dorsal stripe and without any pale patches and the caudal fin lobes with brown submarginal stripes running along the entire length. Maximum size ∼20 mm SL.

Distribution. Feni River drainage which drains into the Bay of Bengal to the east of the Ganga-Brahmaputra system in Bangladesh.

Pseudolaguvia jiyaensis **Tamang & Sinha, 2014**
(Fig. 4.364)

FIGURE 4.364

Pseudolaguvia jiyaensis, 29.9 mm SL, Jiya stream, Lower Dibang Valley District, Arunachal Pradesh.

Pseudolaguvia jiyaensis Tamang & Sinha, 2014: 45, fig. 5 (type locality: Jiya stream, near Bolik village, Lower Dibang Valley District, Brahmaputra drainage, Arunachal Pradesh, India).

Diagnosis. Combination of characters: thoracic adhesive apparatus almost reaching the pelvic-fin origin; vertebrae 25−27; dorsal-fin spine length 11.9%−15.0% SL; pectoral-fin spine length 16.6%−19.8% SL; pectoral fin length 22.6%−26.0% SL. Maximum size ∼30 mm SL.

Distribution. Jiya stream, Lower Dibang Valley District, Brahmaputra drainage, Arunachal Pradesh, India.

Pseudolaguvia magna **Tamang & Sinha, 2014**

Pseudolaguvia magna Tamang & Sinha, 2014: 38, figs. 1−4 (type locality: Jiya stream, near Bolik village, Lower Dibang Valley District, Brahmaputra drainage, Arunachal Pradesh, India).

Diagnosis. A large species, maximum size 47.0 mm SL; two pale-brown to cream patches on the mid-dorsal region: a rectangular to elliptical patch on the middle of interdorsal region, and an indistinct elliptical patch in between the adipose and caudal fins; a broad rhomboidal thoracic adhesive apparatus; a small pale-brown to cream round spot on the ventro-lateral side of the head, almost perpendicular to the eye. Maximum size ∼47 mm SL.

Distribution.. Jiya stream, Lower Dibang Valley District, Brahmaputra drainage, Arunachal Pradesh, India.

Pseudolaguvia muricata **Ng, 2005**
(Fig. 4.365)

FIGURE 4.365

Pseudolaguvia muricata, 29.0 mm SL, Umngi, Balat, Meghna basin. South West Khasi Hills, Meghalaya.

Pseudolaguvia muricata Ng, 2005g: 41, fig. 3 (type locality: Rangapani Khal (creek), north-northwest of Sylhet-Shillong highway, Bangladesh).

Diagnosis. Dorsal fin length 21.2%−26.7% SL; pectoral-fin length 26.8%−35.7% SL; thoracic adhesive apparatus reaching beyond the base of the last pectoral-fin ray; caudal peduncle length 12.6%−15.7% SL; and fewer vertebrae (28−30); adipose-fin base 12.3%−16.1% SL; caudal-peduncle length 12.6%−15.7% SL; body with light-brown patches. Maximum size ∼30 mm SL.

Distribution. Brahmaputra drainage, India and Bangladesh.

Pseudolaguvia nubila **Ng, Lalramliana, Lalronunga & Lalnuntluanga, 2013**

Pseudolaguvia nubila Ng, Lalramliana, Lalronunga, Lalnuntluanga, 2013: 519, figs. 1−3 (type locality: Sla River, a tributary of the Kaladan River, in the vicinity of Lingpuk Village, Saiha District, Mizoram, India).

Diagnosis. Caudal peduncle depth 9.1%−11.1% SL; head width 19.7%−21.0% SL, dorsal spine length 16.4%−19.3% SL, pectoral-spine length 18.1%−22.0% SL; body depth at anus 13.9%−17.1% SL; body mottled brown with yellowish bands; a weakly projecting snout and the premaxillary teeth barely exposed; eye diameter 10.8%−14.0% HL; body with two irregular creamish color bands: first, on sides of body between dorsal and adipose fins, second, on caudal peduncle. Maximum size ∼30 mm SL.

Distribution. Kaladan River drainage in southern Mizoram.

Pseudolaguvia ribeiroi **(Hora, 1921)**

Laguvia ribeiroi Hora, 1921c: 741, pl. 29, fig. 3 (type locality: Khoila River, tributary of Tista at Jalpaiguri, West Bengal, India).

Diagnosis. Nostrils nearer to tip of snout than to anterior margin of eye; pelvic-fin origin midway between tip of snout and caudal-fin base; dorsal spine finely serrated along the anterior border and also upper one-third of posterior border; adhesive apparatus v-shaped; dorsal-fin origin nearer to snout-tip than to caudal-fin base; pectoral spine serrated externally, eight curved spines internally; two broad yellowish bands on body, anterior between rayed dorsal and adipose-fin and second below posterior half of base of adipose fin. Maximum size ∼95 mm SL.

Distribution. Teesta River system, West Bengal, India.

Pseudolaguvia shawi **(Hora, 1921)**

(Fig. 4.366)

Laguvia shawi Hora, 1921c: 740, pl. 29, fig. 2 (Mahanadi River below Darjeeling, West Bengal, India).

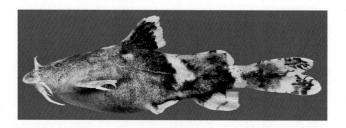

FIGURE 4.366

Pseudolaguvia shawi, 34.0 mm SL, Rangeet River, Sikkim.

Diagnosis. Dorsal fin origin in advance of the pelvic fin origin; nearer to tip of snout than to caudal-fin base; dorsal spine strong, smooth anteriorly and rough posteriorly; pelvic fin extends close to anal fin origin; occipital, cubito-humeral and scapular processes are finely tuberculated and there is a bony nodule covered by skin below the base of dorsal spine. Sides marked with two black bands, anterior below the dorsal-fin base and another below the adipose fin; fins marked with black bands. Maximum size $\sim$40 mm SL.

Distribution. Mahanadi and Sivoke rivers, Darjeeling Himalayas, West Bengal, India.

Pseudolaguvia spicula Ng & Lalramliana, 2010

Pseudolaguvia spicula Ng & Lalramliana, 2010: 62, fig. 1 (type locality: Bawrai River, a tributary of Langkaih River in the vicinity of Zawinuam, Barak drainage, Mizoram, India).

Diagnosis. Caudal peduncle depth 7.9%$-$9.6% SL; dorsal spine 11.6%$-$14.3% SL; pectoral spines 15.7%$-$17.4% SL; indistinct, pale vertical bands on the body; snout length 48.6%$-$51.9% HL; smooth anterior edge of the dorsal-fin spine. Maximum size $\sim$30 mm SL.

Distribution. Upper Barak River drainage in Mizoram, India and the middle Meghna River drainage in Bangladesh.

Pseudolaguvia viriosa Tamang & Sinha, 2012
(Fig. 4.367)

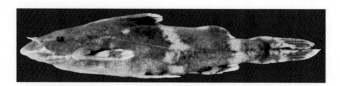

FIGURE 4.367

Pseudolaguvia viriosa, 27.4 mm SL, stream at Tezu, Brahmaputra drainage, Arunachal Pradesh.

Pseudolaguvia viriosa Ng & Tamang, 2012: 82, fig. 1 (type locality: Sille River, Ruskin, about 26 km from Pasighat, Arunachal Pradesh).

Diagnosis. Thoracic adhesive apparatus elliptical in shape with a median depression, extending to midway between the base of last pectoral-fin ray and the pelvic-fin origin; long dorsal and pectoral spines, respectively 23.4%$-$29.0% SL and 26.9%$-$32.9% SL; body depth at anus 16.9%$-$

19.0% SL; caudal peduncle length 14.8%−17.7% SL; caudal peduncle with a creamish blotch or band on sides, elongated creamish patches on dorsal and ventral margins; pelvic and anal fin banded; caudal fin with a large brown band at the distal third, distal tips of the upper and the lower lobe hyaline. Maximum size ∼30 mm SL.

Distribution. Sille River, Upper Siang District, Arunachal Pradesh.

Genus *Sisor* Hamilton, 1822

Sisor Hamilton, 1822: 208, 379 (type species: *Sisor rabdophorus* Hamilton, 1822). Gender: masculine.

Diagnosis. Series of bony plates extending from dorsal fin to base of caudal fin, nuchal plate rugose (Fig. 4.368A); spine in adipose fin; uppermost caudal-fin ray long, more than half length of

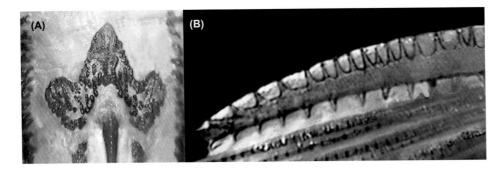

FIGURE 4.368

(A) Nuchal plate of *Sisor barakensis*; (B) pectoral spine of *S. barakensis*.

body; branchiostegal membranes broadly fused to isthmus; outer and inner mental barbels widely separated, with origin of outer barbels anterior to origin of inner barbels; minute teeth in lower jaw, dentition essentially consisting of roughened plate; large serrations on anterior margin of pectoral spine (Fig. 4.368B); well-developed maxillary barbel membrane; palatal teeth absent.

Sisor barakensis Vishwanath & Darshan, 2005
(Fig. 4.369)

FIGURE 4.369

Sisor barakensis, 84.8 mm SL, Barak River, Jiri, Manipur.

Sisor barakensis Vishwanath & Darshan, 2005: 1962, Fig. 1a (type locality: River Barak, Brahmaputra drainage, Jiri, Manipur, India).

Diagnosis. Head depressed, longer than broad, frontal and supra occipital portion of head with ridges converging toward base of OP and remaining portion of head covered with skin; median longitudinal groove on head reaching base of OP, latter not base bone of dorsal fin; gill membranes confluent and with isthmus; dorsal spine length 10.5%−14.3% SL; pectoral spine length 18.5%−20.4% SL; number of serrate on the posterior edge of pectoral spine 15−20; two to four minute pointed premaxillary teeth at the corner of lip little anterior to the base of each slender fleshy process of the upper lip. Maximum size ∼90 mm SL.

Distribution. Barak River, Brahmaputra drainage, Manipur, India.

Sisor chennuah Ng & Lahkar, 2003
(Fig. 4.370)

FIGURE 4.370

Sisor chennuah, 76.6 mm SL, Brahmaputra River, Dibrugarh, Assam.

Sisor chennuah Ng & Lahkar in Ng, 2003: 2876, figs. 2a, 4a, 5 (type locality: Brahmaputra River drainage, Dibrugarh, Assam).

Diagnosis. Combination of characters: body depth at anus 5.6%−5.8% SL; head width 15.7%−15.9% SL; snout length 54.0%−55.9% HL; eye diameter 9.9%−11.4% HL; serrations on posterior edge of pectoral spine 10; nuchal plate width 1.1−1.2 times length. Maximum size ∼98 mm SL.

Distribution. Brahmaputra River Drainage, Assam.

Sisor rabdophorus Hamilton, 1822
(Fig. 4.371)

Sisor rabdophorus Hamilton, 1822: 208, 379 (type locality: Bhagirathi River at crossing point between Kalna, Bardhaman District and Nrisinghapur, Nadia District, West Bengal, India).

Diagnosis. Body depth at vent 5.0%−5.4% SL; eye diameter 9.4%−12.2% HL; body with three autogenous plates; pectoral spine with 12−22 serrae on posterior edge; lateral line ossicles 66−79. Autogenous plates; pectoral spine with 12−22 serrae on posterior edge; lateral line ossicles 66−79. Maximum size ∼180 mm SL.

Distribution. Ganga River drainage.

Sisor rheophilus Ng, 2003

Sisor rheophilus Ng, 2003: 2877, figs. 2c, 4c, 6 (type locality: Kali River, near Muzaffarnagar, Uttar Pradesh State, India).

Diagnosis. Combination of characters: nuchal plate width equal to length; eye diameter 13.2%−16.3% HL; snout length 56.2%−60.3% HL; head width 14.5%−14.8% SL. Maximum size ∼100 mm SL.

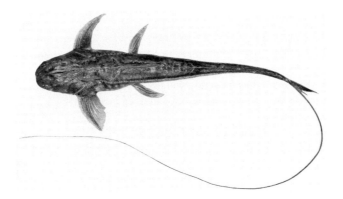

FIGURE 4.371

Sisor rhabdophorus, 92 mm SL, Brahmaputra River at Dhubri, Assam.

Distribution. Middle and upper parts of the Ganges River basin in Bihar and Uttar Pradesh States, India.

Sisor torosus Ng, 2003
Sisor torosus, Ng, 2003: 2878, figs. 2d, 4d, 7 (type locality: Ganga River at Patna, Bihar, India).
Diagnosis. Combination of characters: relatively deep body (body depth at anus 6.0%−7.8% SL); 12−18 serrations on the anterior and posterior edges of the pectoral spine; nuchal plate width 1.2−1.3 times length. Maximum size ~90 mm SL.
Distribution. Middle part of the Ganges River basin in Bihar and Delhi, India.

Family Siluridae
Sheat fishes

Dorsal fin present (absent in some), fewer than seven rays if present, no spines; adipose fin absent; pelvic fin small or absent; anal-fin base elongated; caudal fin forked and separated from anal fin; nasal barbel absent; maxillary barbels elongated, mandibular one or two pairs (Fig. 4.372).

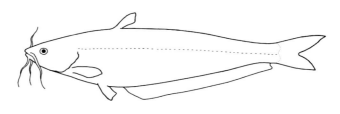

FIGURE 4.372

Outline diagram of Siluridae.

Key to genera:

1a. Dorsal fin rudimentary with one to two soft rays .... *Pterocryptis*
1b. Dorsal fin well developed, short with three to five rays 2

2a. Gape of mouth wide, extending beyond eyes posteriorly .... *Wallago*
2b. Gape of mouth moderate, not extending beyond eyes posteriorly .... *Ompok*

Genus *Ompok* Lacepède, 1803

Ompok Lacepède, 1803: 49 (type species: *Ompok siluroides* Lacepède, 1803). Gender: masculine.

Diagnosis. Body compressed, elongated; abdomen rounded; head small, snout rounded and depressed; mouth superior, its cleft oblique, not extending to the anterior margin of eye; jaws subequal, lower jaw prominent; teeth villiform; barbels two pairs, maxillary and mandibular, mandibular may be rudimentary or absent; dorsal fin origin above the middle of pectoral fin; adipose dorsal absent; anal fin long with more than 50 rays; caudal fin forked; lateral line complete.

Ompok bimaculatus (Bloch, 1794)
(Fig. 4.373)

FIGURE 4.373

Ompok bimaculatus, 272 mm SL, Brahmaputra River, Tezpur, Assam.

Silurus bimaculatus Bloch, 1794: 24 (type locality: Tranquebar, India).

Diagnosis. Body elongated and strongly compressed its depth 19.9%−25.2% SL; eyes moderate, its lower border below level of cleft of mouth. Mouth large and oblique, teeth in villiform bands on jaws, vomerine teeth in two oval patches; two pairs of barbels, maxillary long, extending beyond anal fin origin, mandibular short; pectoral fin spine moderately strong, feebly serrated on its inner edge; pelvic fin with eight rays, not reaching anal fin origin; anal fin rays 55−74 its origin slightly behind the last dorsal fin ray; caudal fin deeply forked, its lobes pointed; body color brown, usually marmorated with conspicuous round black blotch above pectoral-fin base. Maximum size ~450 mm SL.

Distribution. India: throughout; Bangladesh.

Ompok pabda (Hamilton, 1822)
(Fig. 4.374)

Silurus pabda Hamilton, 1822: 150, 374 (type locality: Bengal).

Diagnosis. Snout obtuse, mandibular barbel large, extending upto posterior edge of eye; maxillary barbel longer than head; pelvic fin with eight rays, its origin ahead of opposite dorsal fin

FIGURE 4.374

Ompok pabda, 255.0 mm SL, Brahmaputra River, Guwahati, Assam.

origin, extends to anal fin origin; anal fin with 50–70 rays, its origin opposite slightly behind the last dorsal fin ray. Maximum size ∼270 mm SL.

Distribution. South central Asia: Afghanistan, Pakistan, India, Nepal, Bangladesh, and Myanmar.

Ompok pabo (Hamilton, 1822)
(Fig. 4.375)

FIGURE 4.375

Ompok pabo, 215 mm SL, Brahmaputra River, Goalpara, Assam.

Silurus pabo Hamilton, 1822:153, 375 (type locality: Brahmaputra River towards Assam India).

Diagnosis. Dorsal profile of body gently convex; depth of body 20.7%–22.0% SL; lower border of eye below level of the cleft of mouth; maxillary barbells shorter than head length; nuchal concavity present; dorsal fin small, with one simple and four branched rays; pectoral fin long with one 14 branched rays, extending up to anal-fin base; pelvic fin with one simple and eight branched rays, almost touching the caudal fin, its origin at same vertical level with dorsal fin, pelvic fin rays 9 or 10; caudal peduncle very short, negligible. Lateral line complete, extending to middle of caudal-fin base. Anal fin long with i, 66–67 rays, almost touching caudal fin; integument over anal fin thickened proximally for half of ray length; caudal fin lobes pointed; caudal fin deeply forked with 15 rays, its lobes pointed; body and head creamish and diffusely pigmented. Fin membranes hyaline. Maximum size ∼250 mm SL.

Distribution. Ganga, Brahmaputra and Surma-Meghna basins.

Genus *Pterocryptis* Peters, 1861

Pterocryptis Peters, 1861: 712 (type species: *Pterocryptis gangelica* Peters, 1861). Gender: feminine.

Diagnosis. Body elongated and compressed; head depressed, broader than body; eyes small; mouth subterminal; its gape horizontal and extends just beyond vertically through anterior margin of eye; lower lip papillated; jaws and palate with villiform teeth; vomerine teeth in one or two well separated patches; cheek inflated; dorsal fin small, rudimentary; anal fin long, a notch between anal and caudal fin.

Pterocryptis barakensis **Vishwanath & Nebeshwar, 2006**
(Fig. 4.376)

Pterocryptis barakensis, 128.0 mm SL, Barak River, Vanchengphai, Manipur.

Pterocryptis barakensis Vishwanath & Nebeshwar, 2006: 99 (type locality: Barak River at Vanchengphai, Manipur, India).

Diagnosis. Upper jaw prominent and longer than lower jaw; branchiostegal rays 13; anterior nostrils tubular; dorsal rudimentary with two rays or completely absent; pectoral fin with I,13−14 rays, its spine granulated posteriorly in one or two rows in male, smooth in female; pelvic fin with I,7−8 rays; anal fin with 65−77 rays; caudal fin truncated and with 18 principal rays; vertebrae 54−55; body from head to caudal-fin base with 12−16 transverse rows of sensory pores extending from mid-dorsal region to lateral line; 12 distinct sensory pores, arranged semicircularly on ventral surface of head extending from anterior sides of opercular region on either side to behind lower lip; vomerine teeth band continuous. Maximum size ∼230 mm SL.

Distribution. Barak River, Tamenglong District, Manipur, India.

Pterocryptis berdmorei **(Blyth, 1860)**
(Fig. 4.377)

Pterocryptis berdmorei, 52.0 mm SL, Lokchao River, Indo-Myanmar border, Manipur.

Silurichthys berdmorei Blyth, 1860: 156 (type locality: Tenasserim provinces, Myanmar).

Diagnosis.: Body slender; head depressed; eyes small subcutaneous; teeth short and arranged in villiform bands, vomerine teeth in two oval patches which may be separated or continuous; barbels two pairs: maxillary pair reach pelvic fin; mandibular shorter than head length; dorsal with four branched rays, its origin at vertical level of half of pectoral fin; pectoral with one spine, 11 branched rays, the spine simple short oblong scarcely serrated; pelvic with one simple and nine branched rays, extending beyond anal fin origin; anal long with two simple 60−62 branched rays; caudal cut square or rounded; body leaden purplish below covered all over with minute black spots. Maximum size ∼210 mm SL.

Distribution. Chindwin River drainage, Manipur bordering Myanmar; Tenasserim provinces, Myanmar.

Pterocryptis gangelica **Peters, 1861**

Pterocryptis gangelica Peters, 1861: 712 (type locality: Ganges River, India).

Diagnosis. Dorsal fin rudimentary with two rays, its origin opposite the middle of pectoral fin; adipose dorsal fin absent; pectoral fin with I,13−14 rays, reaching pelvic fin; pectoral spine weakly serrated internally; anal fin long with 67−75 rays, its origin close to pelvic fin; caudal fin rounded with 19 rays. Maximum size ∼165 mm SL.

Distribution. Ganga-Brahmaputra drainage, India.

Pterocryptis indica **(Datta, Barman & Jayaram, 1987)**
(Fig. 4.378)

FIGURE 4.378

Pterocryptis indica, 95.0 mm SL, Namdapha River, Arunachal Pradesh.

Kryptopterus indicus Datta, Barman & Jayaram,1987: 29, fig. 1 (type locality: Hornbill point, Namdapha River, Namdapha Wildlife Sanctuary, Arunachal Pradesh, India).

Diagnosis. Body compressed and elongated, dorsal profile behind head convex and then convex to straight posteriorly; depth of body 18.5% SL, length of head 17.0% SL; maxillary barbel as long as 1½ of head length; mandibular barbel half the length of head; dorsal fin small with one minute ray, its origin opposite the middle of pectoral fin; adipose fin absent; anal fin long with 85−88 rays, a notch between end of anal fin and caudal fin origin; principal caudal fin rounded with 17 rays; pelvic fin with one simple and seven branched rays. Maximum size ∼170 mm SL.

Distribution. Brahmaputra drainage in Arunachal Pradesh.

Pterocryptis subrisa **Ng, Lalramliana & Lalronunga, 2018**

Pterocryptis subrisa Ng, Lalramliana & Lalronunga, 2018: 127 (type locality: Maisa R in the vicinity of Maisa, Saiha District, Mizoram, India).

Diagnosis. Dorsal-fin rays two; head depth 10.6%−11.9% SL; sublabial fold extending posteriorly beyond vertical through posterior orbital margin; anal fin rays 66−75; nearly circular eye, head length 17.6%−19.6% SL; dorsal fin height 2.6%−4.7% SL; pectoral fin length 11.8%−14.0% SL; body depth at anus 14.4%−16.7% SL; caudal peduncle depth 6.8%−8.5% SL; confluent anal and caudal fins separated by a deep notch; 17 principal caudal fin rays and 57 vertebrae. Maximum size ~168 mm SL.

Distribution. Kaladan River drainage in southern Mizoram.

Genus *Wallago* Bleeker, 1851

Wallago Bleeker, 1851a: 265 (type species: *Silurus muelleri* Bleeker, 1846). Gender: masculine.

Diagnosis. Body elongated and compressed; abdomen rounded; head slightly depressed; snout spatulate; mouth gape wide, extending beyond posterior edge of eye, lower jaw longer; teeth villiform in bands on jaws and patches on palate; barbels two pairs, one maxillary extending beyond pelvic fin and mandibular shorter; dorsa fin small, without spine, its origin opposite pectoral fin; adipose dorsal fin absent; anal fin long with more than 70 rays; caudal fin deeply forked, the upper lobe longer; lateral line complete.

Wallago attu (Schneider, 1801)
(Fig. 4.379)

FIGURE 4.379

Wallago attu, 440.0 mm SL, Barak River, Silchar, Assam.

Silurus attu Schneider, in Bloch & Schneider, 1801: 378 (type locality: Malabar).

Diagnosis. Body elongated and compressed, its depth at anal fin origin 14.4% SL. Head large, its length 21.1% SL; snout pointed; mouth wide, its gape extending beyond vertically through to posterior margin of orbit, eye in front of vertical level of corner of mouth, its diameter 14.2% HL; strong conical teeth on jaws, palatine and vomer; barbels two pairs: maxillary pair long, extending beyond anal origin, mandibular pair shorter; gill rakers 5 + 20, short and pointed; caudal peduncle height twice of its length. Dorsal fin small with one simple and four branched rays, its origin between opposite pectoral and pelvic fins; pectoral fin with one simple and 12 branched rays, its spine smooth and weak; pelvic fin with I,8 rays, reaching anal-fin origin; anal fin long with 95 rays; caudal fin with 17 principal rays, forked, upper lobe longer. Maximum size ~2000 mm SL.

Distribution. India: Chindwin and Brahmaputra drainages. Mekong basin in Laos, Thailand, Cambodia, Vietnam. Chao Phraya basin. Sumatra, Java, Malay Peninsula.

Family Chacidae
Angler or frogmouth fishes

Head broad and depressed; mouth terminal, wide; eyes dorsal, small; barbels short, three to four pairs; dorsal fin with one spine and four soft rays; pectoral spine serrated internally; adipose dorsal confluent with caudal fin; gill rakers absent (Fig. 4.380).

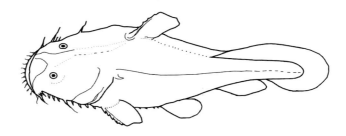

FIGURE 4.380

Outline diagram of Chacidae.

Genus *Chaca* Gray, 1831

Chaca Gray, 1831: 9 (type species: *Chaca hamiltonii* Gray). Gender: feminine.

Chaca: Roberts, 1982: 896 (Review).

Diagnosis. Body short, depressed ahead of anal fin, strongly compressed tapering behind; abdomen flat; head large, strongly depressed; eyes minute, superior, in anterior quarter of head; lips thick, fleshy; lower jaw prominent; teeth small, villiform on jaws, palate edentate; barbels three pairs, one of maxillary, two of mandibulars; gill membranes free up to pectoral base and united with isthmus; rayed dorsal fin short with a strong spine four rays; adipose dorsal confluent with caudal fin; pectoral fin with a short strong spine and four to five rays; pelvic fins large with six rays, margins rounded; anal short with 8−10 rays; caudal fin rounded with a long upper procurrent dorsal and a shorter ventral part; lateral line complete, marked by a prominent papillated and tuberculated ridge; numerous cutaneous flaps or cirri on dorsolateral surface of head, sometimes also on lower lip and body; skin of lips, pelvic and median fins, and dorsolateral surfaces of head and body with numerous fine fin granulations; air bladder large, concave anteriorly, lying across the bodies of the anterior vertebrae, not enclosed in bone.

Chaca burmenseis Brown & Ferraris, 1988

Chaca burmensis Brown & Ferraris, 1988: 3, fig. 1 (type locality: Sittang River, Myanmar).

Diagnosis. A species of *Chaca* most readily distinguished from its congeners by a greater number of serrations on the anterior edge of pectoral spines, by the absence of papillae surrounding the eye, and a temporal fossa which is bordered dorsally by pterotic and epiotic only; orbitosphenoid paired; vomer absent as independent element. Maximum size ∼150 mm SL.

Distribution. Irrawaddy and Sittang Rivers, Myanmar.

Chaca chaca (Hamilton, 1822)

(Fig. 4.381)

Platystacus chaca Hamilton, 1822: 140, 374, pl. 28, fig. 43 (type locality: rivers and ponds of north eastern parts of Bengal).

Diagnosis. A species of *Chaca* with a stumpy body and a flat, broad head; a small dorsal fin with two spines and four soft rays; anal fin long with one spine and 8−10 branched rays; body short, depressed in front of anal fin; posterior part of body strongly compressed, abdomen flat, head large and greatly depressed; mouth wide, lips thick and fleshy; lower jaw prominent; eyes minute;

FIGURE 4.381

Chaca chaca, 93.7 mm SL, (A) lateral view; (B) dorsal view; Jiri River, Manipur.

lateral line complete marked by a prominent papillated and tuberculated ridge; rayed dorsal fin short with a strong spine and three branched rays; adipose dorsal confluent with caudal fin pectoral with one spine and five branched rays, the spine with numerous serrae along its anterior edge. Maximum size ~150 mm SL.

Distribution. Barak and Brahmaputra basins in Assam and Manipur, India: Bangladesh.

Family Clariidae
Air breathing catfishes

Body smooth. Dorsal fin without spine, its base long, confluent or separate with caudal fin. Anal fin long, caudal fin rounded. Gill opening wide. Barbels four pairs, Labyrinthine organ arising from gill arches for air breathing (Fig. 4.382).

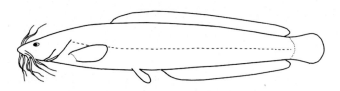

FIGURE 4.382

Outline diagram of Clariidae.

Key to genera:

1a. Rayed dorsal fin long with 70 or more rays; accessory respiratory organ as dendritic organ on the second and fourth branchial arches .... *Clarias*

1b. Rayed dorsal fin short with six to seven rays; accessory respiratory organ as air sacs in opercular chamber extended backward into caudal region .... *Heteropneustes*

Genus *Clarias* Scopoli, 1777

Clarias Scopoli, 1777: 455 (type species: *Silurus anguillaris* Linnaeus, 1758). Gender: masculine.

Diagnosis. Body elongated, compressed; head depressed, dorsally covered with osseous plates; mouth terminal, wide and transverse; eyes small, dorsolateral; teeth villiform, in bands on jaws and palate; barbels four pairs; accessory respiratory organ dendritic, present on second and fourth branchial arches; rayed dorsal without a spine, long, extending from behind occiput to caudal-fin base; adipose dorsal absent; pectoral fin with an internally strongly serrated spine; anal fin long, extending from middle of body to caudal-fin base; lateral line complete.

Clarias gariepinus (Burchell, 1822)
(Fig. 4.383)

FIGURE 4.383

Clarias gariepinus, 184.6 mm SL, Singjamei market, Imphal, Manipur.

Silurus (*Heterobranchus*) *gariepinus* Burchell, 1822: 425 (type locality: Vaal River, at Smidtsdrift, above confluence with Riet River, Cape Province, South Africa).

Diagnosis. Anterior edge of pectoral spine serrated; head length 2.9−3.8 times in SL; two color forms: a uniform dark grayish-greenish black form and a marmorated one; size up to 1500 mm SL. Maximum size ∼1500 mm SL.

Distribution. Widely distributed in South-East Asia. Introduced illegally in India.

Clarias magur (Hamilton, 1822)
(Fig. 4.384)

Macropteronotus magur Hamilton, 1822: 145, 374 (type locality: Ganges River, India).

Diagnosis. Snout broadly rounded, anterior margin of pectoral spine smooth and without bumps distally, and with conical and pointed serrae proximally; posterior margin with broad bumps distally and a few serrae upto the middle of the spine; dorsal fin with no spine, commencing from near occiput and extending to but not continuous with caudal fin and bearing 70−76 rays; head moderately depressed, covered with osseous cask covering a diverticula of the gill cavity; jaws subequal; upper jaw longer; lips fleshly, papillated; mouth

FIGURE 4.384

Clarias magur, 166.0 mm SL, wetlands near Loktak Lake, Manipur.

terminal, teeth villiform in broad crescentric; eyes small, dorso-lateral with free orbital margins; gill membranes deeply notched, partly united with each, free from isthmus; an accessory respiratory dendritic branchial organ attached to second to fourth branchial arches present; lateral line complete; barbels four pairs: one pair each of maxillary and nasal; mandibular two pairs. Maximum size ~250 mm SL.

Distribution. India, Bangladesh, Myanmar?

Genus *Heteropneustes* Müller, 1840

Heteropneustes Müller, 1840: 115 (type species: *Silurus fossilis* Bloch). Gender: masculine.

Diagnosis. Body elongated and compressed; abdomen rounded; head greatly depressed, sides and top covered with bony plates; mouth transverse, narrow and terminal; eyes small; rayed dorsal fin short, its origin slightly ahead of pelvic fin origin; barbels four pairs: each of maxillary, nasal and two mandibulars; adipose dorsal fin absent; pectoral spine serrated internally, epithelial gland secretion.

Heteropneustes fossilis (Bloch, 1794)
(Fig. 4.385)

FIGURE 4.385

Heteropneustes fossilis, 223.0 mm SL, wetland near Loktak Lake, Manipur.

Silurus fossilis Bloch, 1794: 46 (type locality: Tranquebar).

Diagnosis. Body elongated and compressed; dorsal profile arched; head moderate sized, greatly depressed, covered with thin skin; snout flat; jaws subequal; lips fleshy and papillated; mouth terminal, transverse, narrow; teeth villiform in broad bands on jaws and in two oval patches on palate; eyes small, laterally placed; gill membranes separated by a deep notch, not united with isthmus; gill chambers with accessory air sacs extended backward into caudal region; OP not reaching base of dorsal fin; anal fin separated from the caudal by a deep notch; uniformly yellow, occasionally with two longitudinal bands on body; line complete; barbels in four pairs: one maxillary, one nasal, and two mandibulars; dorsal fin short with six to seven branched rays, no spine; pectoral with a strong spine serrated posteriorly and seven branched rays; anal long with 60–79 rays, contiguous with caudal fin; caudal fin rounded. Maximum size ~250 mm SL.

Distribution. Pakistan, India, Andaman Is., Nepal, Bangladesh, Myanmar.

Family Ariidae
Sea catfishes

Catfishes with no nasal barbel, anal fin with 14–26 rays, nostrils close together, caudal fin deeply forked, not pointed, dorsal fin with a pungent spine; external posterior branch of lateral ethmoid columnar; a bony blade anteriorly connecting the nasal tubules; lateral ethmoid and frontal bones connected mesially and laterally delimiting a fontanel; presence of three infraorbitals; lachrymal well developed; space between transcapular process and otic capsule small; otic capsules enlarged; wing process of parasphenoid present; subvertebral process well developed; anterior portion of second basibranchial expanded and very conspicuous; third pharyngobranchial boomerang shaped; anterior portion of proximal cartilage of fourth ceratobranchial narrow about one-half as wide as posterior portion; dorsal processes of pharyngeal tooth plates long; anterior and posterior nostrils close together (Fig. 4.386).

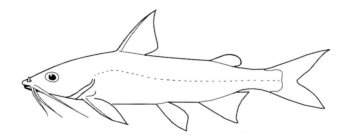

FIGURE 4.386

Outline diagram of Ariidae.

Genus *Cochlefelis* Whitley, 1941

Cochlefelis Whitley, 1941: 8 (type species: *Arius spatula* Ramsay and Ogilby, 1886). Gender: feminine.

Diagnosis. Body elongated, head depressed, occipital spine saddle shaped, granulated, extended close to nuchal plate; maxillary and mandibular barbels filamentous, fairly elongated, no nasal barbel; mouth wide, lips thick, upper lip produced posteriorly in a lobe; anterior and posterior nostrils close together; gill openings wide, gill membranes untied across isthmus; dorsal spine strong, serrated, adipose fin large, pectoral spine weakly serrated; caudal fin forked.

Cochlefelis burmanicus (Day, 1870)
(Fig. 4.387)

FIGURE 4.387

Cochlefelis burmanicus, 128.0 mm SL, Chindwin River at Kalemyo, Myanmar.

Arius burmanicus Day, 1870a: 618 (type locality: Irrawaddy River, Bassein District, Salween, Tenasserim Provinces, Myanmar).

Diagnosis. Body elongated and compressed, head depressed snout spatulate; upper jaw longer than the lower; nasal barbel absent; maxillary barbel reaching middle of pectoral fin; posterior nostril with a flap on the anterior margin; cranial fontanel not reaching base of OP; OP keeled, spotted and corrugated irregularly and almost reaching the fused anterior and medial nuchal plate; dorsal fin with one simple and seven branched rays, both dorsal and pectoral spines strong and serrated on both the edges; adipose base short, origin posterior to anal fin origin; anal fin with four to six simple rays and 16 branched rays; caudal fin forked, lateral line bifurcated at caudal region, one to upper and the other to the lower lobe of the fin (Fig. 4.388). Maximum size ∼350 mm SL.

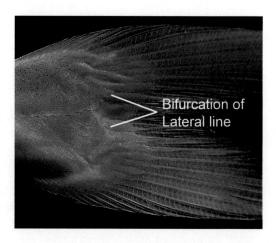

FIGURE 4.388

Side view of caudal fin of *Cochlefelis burmanicus* showing bifurcation of lateral line.

Distribution. Manipur River, Chin Hills, Myanmar, Chindwin, Irrawaddy and Salween drainages, Myanmar.

Family Ailiidae
Asian schilbeids

Dorsal fin absent; adipose fin small, anal fin very long 58−90 rays, eyes small, ventrolateral; body elongated, compressed; abdomen rounded; teeth present on premaxillaries, mandible and vomer; nostrils widely separated; gill openings wide, extends upto lateral line, membranes free from each other and also from isthmus; dorsal fin short with a spine or absent; adipose fin present; anal-fin base elongated; barbels four pairs; skin smooth; anal fin long separate from caudal.

Key to genera:

1a. Snout pointed, dorsal profile steeply rising to dorsal fin origin .... *Pachypterus*
1b. Snout blunt, dorsal profile moderately rising up to dorsal fin origin 2

2a. Dorsal fin absent .... *Ailia*
2b. Dorsal fin present 3

3a. Mouth cleft horizontal, not extending beyond anterior margin of eye .... *Clupisoma*
3b. Mouth cleft oblique, reaching or slightly extending beyond
    anterior margin of eye 4

4a. Teeth band on palate very closely set, bands on jaws laterally
    produced .... *Eutropiichthys*
4b. Teeth bands on palate in four oval patches forming semilunar band on
    palate .... *Proeutropiichthys*

Genus *Ailia* Gray, 1830

Ailia Gray, 1830: pl. 85 [type species: *Malapterus (Ailia) bengalensis* Gray, 1830]. Gender: feminine.

Diagnosis. Body compressed; head short, greatly compressed; mouth subterminal, crescentic moderately wide; eyes small, lateral; barbels four pairs; rayed dorsal fin absent; adipose fin small, pectoral fin with a spine and 13−16 soft rays; pelvic fin rudimentary, anal long with more than 40 rays; caudal fin forked; lateral line complete.

Ailia coila (Hamilton, 1822)
(Fig. 4.390)

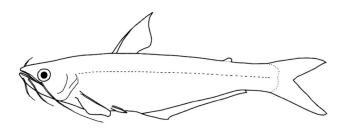

FIGURE 4.389

Outline diagram of Ailiidae.

FIGURE 4.390

Ailia coila, 113.6 mm SL, Brahmaputra River, Guwahati, Assam.

Malapterurus coila Hamilton, 1822: 158, 375 (type locality: Bengal).

Diagnosis. Body compressed and elongated; upper jaw slightly longer; maxillary barbel reaching anterior 1/3 of anal fin; dorsal fin absent; adipose fin small; pelvic fin rudimentary; pectoral fin with one simple and 14−16 branched rays; pelvic fin with one simple and five branched rays and anal with 72−75 rays; body silvery tinged with golden yellow; caudal fin edge dark brown. Maximum size ~250 mm SL.

Distribution. Ganga, Brahmaputra, Mahanadi and Surma-Meghna drainages in India.

Genus *Clupisoma* Swainson, 1838

Clupisoma Swainson, 1838: 347, 351, 354 (type species: *Silurus garua* Hamilton, 1822). Gender: neuter.

Diagnosis. Body compressed, elongated; abdomen edge partly keeled; mouth subterminal, transverse, its cleft not extending to front edge of eye; teeth villiform in bands on jaws and palate; barbels four pairs; rayed dorsal fin with a spine and six to nine soft rays, inserted above pectoral fin; pectoral fin with a spine serrated internally; a small adipose fin may or may not be present; caudal fin forked; lateral complete.

Clupisoma garua (Hamilton, 1822)
(Fig. 4.391)

FIGURE 4.391

Clupisoma garua, 221.0 mm SL, Brahmaputra River, Guwahati, Assam.

Silurus garua Hamilton, 1822: 156, 375. (type locality: Rivers of Gangetic provinces).

Diagnosis. Abdominal edge keeled between pelvic fin and anus; pectoral fin not reaching pelvic fin, bearing one spine and 11 branched rays, the spine serrated posteriorly; maxillary barbel extending beyond pectoral fin just reaching pelvic fin; anal fin with 29−36 branched rays; body color greenish gray above, silvery below. Maximum size ~600 mm SL.

Distribution. India: large rivers, upto the Mahanadi in the south; Bangladesh; Pakistan.

Clupisoma montanum **Hora, 1937**
(Fig. 4.392)

FIGURE 4.392

Clupisoma montanum, 197.0 mm SL, Teesta River, Rangeet, Sikkim.

Clupisoma montana Hora, 1937b: 673 (type locality: Teesta River below Darjeeling, India).

Diagnosis. Abdominal edge rounded; maxillary barbel short, extending to pectoral fin origin; pectoral fin extending to pelvic fin origin; anal fin with 39—43 rays. Body color olivaceous in the head region and dorsally upto one-third on sides, silvery on the sides. Maximum size ∼250 mm SL.

Distribution. Ganga-Brahmaputra basin in North and North East India.

Clupisoma prateri **Hora, 1937**
Clupisoma prateri Hora, 1937b: 671, figs. 2b, 3b, 6 (type locality: Burma).
Clupisoma prateri: Ferraris, 2004: 6, fig. 2 (redescription).

Diagnosis. A species of *Clupisoma* with pectoral spine extending to the base of pelvic fin, abdomen keeled from the level of the pectoral fin origin to the anus; pectoral fin with at least 12—13 segmented rays; anal fin with 37—41 branched rays; first gill arch with at least 20 rakers; body silvery, except for a dorsal greenish band that extends only slightly ventral of the mid-dorsal line of the body. Maximum size ∼240 mm SL.

Distribution. Yu River, Indo-Myanmar border, lower and middle reaches of the Irrawaddy River, lower reaches of the Salween River and the Bago and Sittang rivers, Myanmar.

Genus *Eutropiichthys* Bleeker, 1862

Eutropiichthys Bleeker, 1862a: 398 (type species: *Pimelodus vacha* Hamilton, 1822). Gender: masculine.

Diagnosis. Body elongated and compressed; abdomen rounded; an elongated mouth that extends posteriorly at least to the vertical through the anterior margin of the orbit, an elongated accessory premaxillary tooth patch, the palatal teeth arranged in a broadly parabolic patch that is continuous across the midline and consists of a central more or less transverse vomerine tooth plate and the elongated lateral accessory tooth plates that extend posteriorly past the limit of the accessory premaxillary tooth plates; barbels four pairs, one each of nasal, maxillary and two of mandibular; rayed dorsal fin with a spine and seven rays, inserted above pectoral fin; adipose dorsal short; pectoral fin with an internally serrated spine; anal long with more than 35 rays; caudal fin deeply forked; lateral line complete.

Eutropiichthys burmannicus **Day, 1877**
(Fig. 4.393)

FIGURE 4.393

Eutropiichthys burmanicus, 216.2 mm SL, Chindwin River, Kalemeu, Myanmar.

Eutropiichthys burmannicus Day, 1877: 490 (type locality: Myanmar).

Diagnosis. Gill rakers on the first arch 22−28; the number of branched pectoral-fin rays 15−17, rarely 15, the accessory premaxillary tooth patch extending posteriorly nearly to the terminus of the gape, the fleshy flap along the anterior margin of the posterior naris barely extending to the posterior margin of the posterior naris, the number of branched anal-fin rays 46−54, form of the snout of adults distinctly pointed in lateral view and acutely angular at dorsal view. Maximum size ∼340 mm SL.

Distribution. Confluence of Manipur River with Chindwin River; Irrawaddy and Salween River drainages, Myanmar and western Thailand.

Eutropiichthys cetosus Ng, Lalranliana, Lalronunga & Lalnuntluanga 2014

Eutropiichthys cetosus Ng, Lalranliana, Lalronunga & Lalnuntluanga, 2014a: 6074, Images 1 and 2 (type locality: Kaladan River, Kawlchaw village, Lawngtlai District, Mizoram, India).

Diagnosis. Gill rakers 25−35 on the first gill arch; shape of the snout moderately rounded in lateral view, slightly trilobed in dorsal view; head depth 68.7%−77.1% HL; branched pectoral-fin rays 13−15; a slenderer body depth at dorsal-fin origin 19.2%−23.5% SL; depth of body at anal-fin origin 17.5%−23.5% SL; total vertebrae 49−52; fleshy narial flap not extending medially much past medial margin of naris. Maximum size ∼128 mm SL.

Distribution. Kaladan River drainage, Mizoram.

Eutropiichthys murius (Hamilton, 1822)
(Fig. 4.394)

FIGURE 4.394

Eutropiichthys murius, 128.6 mm SL, Brahmaputra River, Goalpara, Assam.

Pimelodus murius Hamilton, 1822: 195, 378 (type locality: Mahananda River, West Bengal, India).

Diagnosis. Cleft of mouth extending to anterior rim of orbit; vomero-palatine band of teeth narrower than maxillary band; nasal barbel half of head length, palatine teeth band narrower than premaxillary teeth band; pectoral fin with 11−12 rays; dorsal-fin base 37.5%−45.0% HL; dorsal, pectoral, and caudal fins with dusky tips. Maximum size ∼240 mm SL.

Distribution. Brahmaputra and Ganga basin in India, Nepal, and Bangladesh.

Eutropichthys vacha **(Hamilton, 1822)**
(Fig. 4.395)

FIGURE 4.395

Eutropiichthys vacha, 236.4 mm SL, Barak River, Silchar, Assam.

Pimelodus vacha Hamilton, 1822: 196, 378 (type locality: Larger freshwater rivers, Gangetic provinces, India).

Diagnosis. Cleft of mouth reaching up to posterior edge of orbit; vomero-palatine band of teeth wider than maxillary band; nasal barbel as long as head, palatine teeth band wider than premaxillary teeth band; pectoral fin with 15 rays; dorsal-fin base 30.1%−30.2% HL. Maximum size ∼340 mm SL.

Distribution. Ganga, Brahmaputra and Surma-Meghna drainages, India, Nepal, and Bangladesh.

Family Horobagridae
Imperial or sun catfishes

Body compressed, moderately elongated; abdomen rounded; head depressed anteriorly; mouth subterminal, transverse, wide; villiform teeth in jaws and palate; barbels in four pairs: one pair each of nasal and maxillary and two of mandibulars; dorsal fin with a spine and five to seven branched rays, its origin above the middle of pectoral fin; pectoral fin with a spine serrated internally and eight to nine branched rays; adipose fin low; anal fin with 23−29; caudal fin forked or deeply emarginated; lateral line complete (Fig. 4.396).

Genus *Pachypterus* Swainson, 1838

Pachypterus Swainson, 1838: 346 (type species: *Silurus atherinoides* Bloch, 1794). Gender: masculine.

Diagnosis. Body compressed, elongated; abdomen rounded up to anal fin origin then keeled behind; head small, pointed and depressed; snout pointed; eyes large; mouth wide, ventral; teeth in jaws and palate villiform, vomero-palatine teeth in the form of two patches; teeth on upper jaw exposed; barbels in four pairs: nasal, maxillary, four pairs: nasal, maxillary, outer and inner mandibulars; lateral line complete; dorsal and pectoral fins with a spine each, finely serrated posteriorly, dorsal with six to seven branched rays and pectoral with seven; adipose dorsal short; anal fin long with 24−26; caudal fin deeply forked.

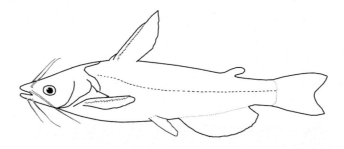

FIGURE 4.396

Outline diagram of Horobagridae.

Pachypterus atherinoides (Bloch, 1794)
(Fig. 4.397)

FIGURE 4.397

Pachypterus atherinoides, 105.0 mm SL, Brahmaputra River, Dibrugarh, Assam.

Silurus atherinoides Bloch, 1794: 48, pl. 371, fig. 1 (type locality: Tranquebar, India).

Diagnosis. Abdomen keeled; maxillary barbel reaching pelvic-fin, mandibular barbels slightly longer than head; dorsal and pectoral fin spines serrated posteriorly; lateral line complete; pectoral fin extending beyond the pelvic fin and almost reaching anal fin origin; pectoral fin with i,5 rays; pelvic fin with one simple and four branched rays and anal fin with two and 34–39 rays; two black blotches, one behind the operculum and another at caudal-fin base; body with black spots on the dorsum and others arranged to form three stripes on the sides. Maximum size ~145 mm SL.

Distribution. Ganga and Brahmaputra drainages in India, Nepal, and Bangladesh.

Genus *Proeutropiichthys* Hora, 1937

Proeutropiichthys Hora, 1937c: 353 (type species: *Eutropius macropthalmos* Blyth, 1860). Gender: masculine.

Diagnosis. The genus has characters as of other scielbeid genera; characteristic in having four patches of teeth on the palate; nostrils widely separated; posterior one slit-like with a flap; dorsal fin with a spine and a short base, adipose fin present; anal fin long, not confluent with caudal fin; four pairs of barbels: nasal, maxillary and outer and inner mandibulars.

Proeutropiichthys macropthalmos (Blyth, 1860)
(Fig. 4.398)

FIGURE 4.398

Proeutropiichthys macrophthalmos, 234.5 mm SL, confluence of Myitha and Chindwin rivers, Kalewa, Myanmar.

Eutropius macropthalmos Blyth, 1860: 156 (type locality: Tenasserim, Myanmar).

Diagnosis. Body compressed, rounded from behind isthmus to pelvic fin origin and then keeled as in other schilbeids; eyes very large, its diameter almost half or slightly more than the height of head; maxillary barbel reaching to the vent, mandibular barbels to the origin of pectoral fin; dorsal spine weak, pectoral weaker, both with minute antrose serrations posteriorly; branched dorsal fin rays 7½; anal fin rays 47−54; adipose dorsal fin small; body color bright silvery, opercles shot with golden tinge. Maximum size ∼240 mm SL.

Distribution. Yu River, Tamu, Myanmar; Irrawaddy and Sittang river drainages, Myanmar.

Family Pangasiidae
Shark catfishes

Catfishes usually growing to large size; body compressed; teeth present on premaxillaries, mandible and vomer; nostrils widely separated; barbels two pairs, maxillary and a single pair of mandibulars; nasal barbel absent; anal fin long with 26−46 rays, but free from caudal; lateral ethmoid facet for articulation of palatines rod like; ednate; ectopterygoid ednate; endopterygoid present; metapterygoid sutured and articulate with hyomandibula; posttemporals present, sphenotics alone provide articular facet ventrally for hyomandibula (Fig. 4.399).

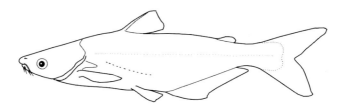

FIGURE 4.399

Outline diagram of Pangasiidae.

Genus *Pangasianodon* Chevey, 1931

Pangasianodon Chevey, 1931: 538 (type species: *Pangasianodon gigas* Chevey, 1931: 538, by monotype). Gender: masculine.

Diagnosis. Posterior nostril located near anterior nostril; barbels small, or absent; pelvic fin rays 8–9; vomerine with teeth, palatine teeth absent; fins black or dark gray; dorsal fin branched rays six; pelvic-fin rays six; large adults, uniformly gray.

Pangasianodon hypophthalmus (Sauvage, 1878)
(Fig. 4.400)

FIGURE 4.400

Pangasianodon hypophthalmus, 29.5 mm SL, fish farms in Imphal valley, Manipur.

Helicophagus hypophthalmus Sauvage, 1878: 235 (type locality: Laos; lectotype designated by Kottelat, 1984: 812).

Diagnosis. Body iridescent; fins dark gray or black; six branched dorsal-fin rays; gill rakers well developed; young with two black stripes, one along the lateral line and another below the lateral line; a black stripe along lateral line and a second long black stripe below lateral line; middle of anal fin with a dark stripe; also dark stripe in each caudal lobe; adults (about 600 mm SL) lose black stripes and become uniform gray coloration. Maximum size ~1300 mm SL

Distribution. Native of southeast Asia, now widely introduced for aquaculture.

Family Bagridae
Bagrid catfishes

Last undivided dorsal fin ray spiny, branched rays 6–7; pectoral spine serrated internally; adipose fin present; caudal fin forked and separated from anal fin; anterior and posterior nostrils widely separated; body naked; four pairs of barbels; nuchal plates 1, 2, and 3 present (except *Olyra*); occipital spine prominent, in contact with nuchal plate 1, may not be visible externally (Fig. 4.401).

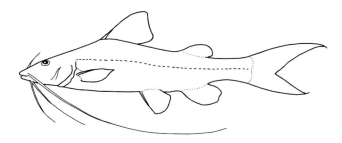

FIGURE 4.401

Outline diagram of Bagridae.

Key to genera:

1a. Body anguilliform, supraneural (interneural) bone, nuchal plate and
   dorsal spinelet absent, simple dorsal fin ray not ossified,
   dorsal-fin origin between 12th and 13th vertebrae .... *Olyra*
1b. Body not anguilliform, interneural shield present (Fig. 4.402);
   spinelet and second dorsal-fin spine ossified; dorsal-fin origin just
   behind complex vertebrae (Fig. 4.403) 2

2a. Body substrate, cubito-humeral process elongated (Fig. 4.404);
   mental barbel in a single pair .... *Rita*
2b. Body compressed, cubito-humeral process not elongated,
   mental barbel more than one pair 3

3a. Dorsal surface of first nuchal plate rugose and enlarged to form
   "interneural shield" and located between OP
   and basal bone of dorsal fin (Fig. 1) .... *Sperata*
3b. Dorsal surface of interneural smooth and slender 4

4a. Maxillary barbel short, not longer than head 5
4b. Maxillary barbel longer than head 6

5a. Adult maximum body size usually reaching upto 50 mm (TL);
   body translucent .... *Rama*
5b. Adult size slightly greater than 500 mm (TL), body opaque .... *Batasio*

6a. Head strongly depressed; interorbital region slightly convex to flat; OP slender; first
   nuchal plate reduced, mostly covered with skin, anterior tip of supraneural
   truncated .... *Hemibagrus*
6b. Head deep, interorbital region convex, broad OP, stout interneural shield .... *Mystus*

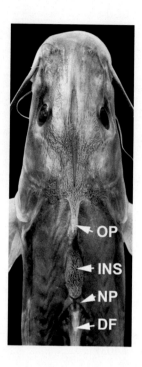

FIGURE 4.402

Dorsal view of anterior part of body of *Sperata* showing occipital process (OP), interneural shield (INS), nuchal plate (NP) and dorsal fin (DF).

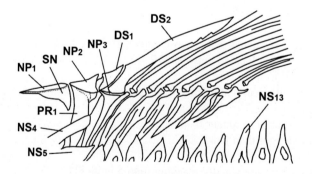

FIGURE 4.403

Lateral view of dorsal-fin skeleton of *Batasio affinis*: NP_{1-3}, nuchal plates 1–3; SN, supraneural; DS_1, dorsal spine 1 (dorsal fin spinelet); DS2, dorsal spine; PR_1, proximal radial 1; NS_4, NS_5, neural spine 4 and 5.

Genus *Batasio* Blyth, 1860

Batasio Blyth, 1860: 149 (type species: *Batasio buchanani* Blyth, 1860). Gender: masculine.

Diagnosis. Small bagrid with compressed head and body; head conical; mouth inferior; lips fleshy; villiform teeth in jaws and palate, teeth band on lower jaw continuous; barbels short and in

FIGURE 4.404

Side view of head of *Rita* showing cubito-humeral process.

four pairs: one each of nasal and maxillary and two mandibulars; all not extending beyond head; gill membranes free from isthmus; dorsal-fin with a spine and seven to eight branched rays, inserted above the middle of pectoral fin, its origin above middle of pectoral-fin; caudal fin deeply emarginated; lateral line complete; total vertebrae 35 or more; absence of a prominent anterolateral process of the pelvic girdle.

Batasio affinis **Blyth, 1860**
(Fig. 4.405)

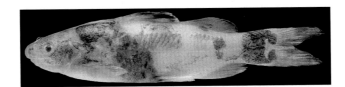

FIGURE 4.405

Batasio affinis, 79.1 mm SL, Khujairok stream, Chindwin drainage, Manipur.

Batasio affinis Blyth, 1860: 150 (type locality: Tenasserim, Myanmar).
Batasio affinis: Ng & Kottelat, 2008a: 290 (redescription).
Diagnosis. One round black spot near gill covers; three to four dark brown bars on a light brown body: one predorsal, obliquely upto pelvic-fin base, second and third bars fainter and a broad bar across the caudal peduncle extending to caudal-fin base, not reaching to the ventral margin of the body; dorsal-spine length 12.4%–14.8% SL, adipose-fin base length 25.5%–29.1% SL, dorsal-fin not reaching anterior origin of adipose-fin when adpressed, post-adipose distance 14.3%–16.9% SL.
Distribution. Chindwin drainage, Manipur; Irrawaddy, Sittang and Ataran River drainages, Myanmar.

Batasio batasio **(Hamilton, 1822)**
(Fig. 4.406)

FIGURE 4.406

Batasio batasio, 57.0 mm SL, Karnaphuli River, Mizoram, India.

Pimelodus batasio Hamilton, 1822: 179, 377 (type locality: Tista River, North Bengal, India).

Diagnosis. Body with two dark brown stripes: one above the lateral line extending from below the origin of dorsal fin upto one-third of the adipose fin; another on the lateral line extending upto caudal-fin base, with a dark brown oval spot below the dorsal fin; snout length 43.9%−46.2% HL. Maximum size ∼95 mm SL.

Distribution. Teesta, Karnafuli, Ganga and Brahmaputra drainages in India, Bangladesh and Nepal.

Batasio convexirostrum **Darshan, Anganthoibi &Vishwanath, 2011**
(Fig. 4.407)

FIGURE 4.407

Batasio convexirostrum, 78 mm SL, holotype, 88.1 mm SL, Mat River, Kaladan drainage, Mizoram.

Batasio convexirostrum Darshan, Anganthoibi &Vishwanath, 2011a: 53 (type locality: Mat River near May bridge, a tributary of the Kaladan River, Lunglei District, Mizoram, India).

Diagnosis. Convex snout region; head and body coloration consisting of only an oblique dark-brown vertical predorsal bar against a lighter-brown body; dorsal-fin dark gray at base and distal one-third hyaline in between; pectoral-fin rays 9−10; dorsal-fin to adipose-fin distance 1.7%−4.1% SL, snout length 39.2%−45.5% HL; total vertebrae 39−40. Maximum size ∼80 mm SL.

Distribution. Kaladan drainage, Mizoram, India.

Batasio fasciolatus **Ng, 2006**
(Fig. 4.408)

FIGURE 4.408

Batasio fasciolatus, 83.0 mm SL, Panye River, Brahmaputra drainage, Arunachal Pradesh.

Batasio fasciolatus Ng, 2006c: 107 (type locality: Market at Malbazar, West Bengal, India).

Diagnosis. Adult coloration of five to six dark brown oblique bars on a light-brown background (one passing through eye, second predorsal, third between dorsal and adipose-fin bases, fourth below anterior half of adipose-fin base, fifth below posterior quarter of adipose-fin base and last at caudal-fin base; dorsal-fin spine 13.6%−16.8% SL, pectoral spine 1.7%−14.3% SL, adipose-fin base 16.9%−22.2% SL; dorsal fin when adpressed reaching the anterior margin of adipose fin. Maximum size ~ 85 mm SL.

Distribution. Teesta River drainage, North Bengal, India.

Batasio macronotus **Ng & Edds, 2004**

Batasio macronotus Ng & Edds, 2004: 296 (type locality: Saptari/Sunsari, Kosi River drainage, Nepal).

Diagnosis. Body and head dark yellow with two horizontal stripes: mid-lateral stripe extending from humeral region to caudal-fin base, anterior end with a slight ovoid expansion; another fainter stripe on dorsal third of body, extending from supraoccipital to adipose fin origin; dorsal spine length 14.0%−16.1% SL; adipose-fin-base length 28.3%−31.8% SL; body depth 19.5%−21.2% SL and caudal-fin length 9.3%−9.7% SL. Maximum size ~ 89 mm SL.

Distribution. Kosi River drainage, Eastern Nepal.

Batasio merianiensis **(Chaudhuri, 2013)**
(Fig. 4.409)

FIGURE 4.409

Batasio merianiensis, 65.1 mm SL, Dikrong River, Brahmaputra drainage, Arunachal Pradesh.

Macrones merianiensis Chaudhuri, 2013: 253 (type locality: pond at Mariani Junction, Assam).
Batasio merianensis: Ng, 2009a: 253 (redescription).

Diagnosis. A body coloration having four dark brown oblique bars: one on head passing through eye, second predorsal extending down upto just below lateral line, third below adipose-fin origin, and fourth at caudal-fin base; adipose-fin base length 16.9%−22.2% SL, dorsal fin when adpressed not reaching anterior origin of adipose fin, body depth at anus 15.2%−18.4% SL, caudal peduncle depth 9.7%−11.5% SL, eye diameter 18.3%−25.9% HL and absence of dark mid-dorsal stripe. Maximum size ∼66 mm SL.

Distribution. Brahmaputra drainage, northeastern India.

Batasio spilurus **Ng, 2006**
(Fig. 4.410)

FIGURE 4.410

Batasio spilurus, 54.8 mm SL, Riang River, Brahmaputra drainage, Arunachal Pradesh.

Photo courtesy: A. Darshan.

Batasio spilurus Ng, 2006c: 110 (Dibrugarh District, Assam, India).

Batasio spilurus: Darshan et al., 2017; 159 (record from Arunachal Pradesh, India).

Diagnosis. Snout tapering when viewed laterally, adipose-fin base length 10.0%−12.8% SL, caudal peduncle length 5.7%−6.6% SL, body depth at anus 11.9%−14.1% SL, head width 14.7%−16.5% SL, narrow mid-lateral blackish stripe, one spot each on the tympanum and caudal-fin base. Maximum size ∼60 mm SL.

Distribution. Brahmaputra River, Dibrugarh, Assam and Siang River, Arunachal Pradesh, India.

Batasio tengana **(Hamilton, 1822)**
(Fig. 4.411)

FIGURE 4.411

Batasio tengana, 59.2.1 mm SL, Dikrong River, Brahmaputra drainage, Arunachal Pradesh.

Pimelodus tengana Hamilton, 1822: 176 (Goalpara, Assam, India).

Batasio tengana: Ng, 2006c: 102 (redecription).

Diagnosis. Dorsal surface of occipital region and nuchal shield dark brown, a faint mid-dorsal stripe; adipose-fin base length 14.5%−17.5% SL; pectoral fin rays not reaching the pelvic-fin origin, caudal peduncle depth 6.7%−8.2% SL; head width 12.5%−14.5% SL; rounded, bulbous snout when viewed laterally. Maximum size ~85 mm SL.

Distribution. Ganga and Brahmaputra drainages.

Genus *Hemibagrus* Bleeker, 1862

Hemibagrus Bleeker, 1862a: 9 (type species *Bagrusnemurus* Valenciennes In Cuvier and Valenciennes, 1839). Gender: masculine.

Hemibagrus: Ng & Kottelat, 2013: 208 (revision).

Diagnosis. Moderate to large sized bagrids with strongly depressed head; interorbital region slightly convex to flat; short and slender interneural shield; highly depressed mesethmoid; prominent dorso-posterior laminar extension of the mesethmoid, first infraorbital with a posterolateral spine; enlarged premaxilla, metapterygoid with a long, free posterior margin; seven to eight soft dorsal-fin rays.

Hemibagrus menoda (Hamilton, 1822)
(Fig. 4.412)

FIGURE 4.412

Hemibagrus menoda, Hamilton's (1822) pl. I, fig. 72 (mislabeled as *Mugil corsula*) reproduced.

Pimelodus menoda Hamilton, 1822: 203, 379, pl. 1, fig. 72 [type locality: Kosi, Mahananda, and other rivers in north of Bihar and Bengal; now: Sharighat bazaar, Sylhet-Shillong highway, Surma (Meghna) drainage, Bangladesh, by neotype designation].

Diagnosis. Head length 32.7%−33.5% SL; head depth 14.2%−15.3% SL, eye diameter 11.9%−12.3% SL; less deeply notched cleithral process; black dots in vertical columns on the sides of the body; snout convex and broad; shallowly incised humeral process; a black blotch on the posterior extremity of adipose fin; greyish fins in life. Maximum size ~400 mm SL.

Distribution. Ganga and Brahmaputra and Mahanadi drainages, India; Bangladesh.

Hemibagrus microphthalmus (Day, 1877)
(Fig. 4.413)

FIGURE 4.413

*Hemibagrus microphthalmus,*156 mm SL, Imphal River, Sugnu, Chindwin drainage, Manipur.

Macrones microphthalmus Day, 1877: 446 (type locality: Irrawaddy valley, Myanmar).

Mystus microphthalmus: Vishwanath & Tombi Singh, 1986b: 197 (redescription and record from Manipur, India).

Diagnosis. Head highly depressed; fins gray in life; head length 29.4%−31.0% SL; head width 18.0%−19.7% SL, predorsal distance 40.6%−44.1% SL; dorsal spine smooth, poorly ossified, thinner than soft dorsal−fin rays; pectoral spine strong, serrated internally, teeth retrose; caudal peduncle length 16.4%−18.1% SL, depth 6.8%−7.7% SL; dorsal to adipose distance 8.6%−14.2% SL; eye diameter 9%−11% HL; maxillary barbel reaching to at least the middle of the adipose-fin base; upper lobe of caudal-fin produced into filament; grows to big size weighing more than 40 kg. Maximum size ∼1500 mm SL.

Distribution. Chindwin basin in Manipur, India; Myanmar; Thailand.

Hemibagrus peguensis (Boulenger, 1894)
(Fig. 4.414)

FIGURE 4.414

*Hemibagrus peguensis,*161.5 mm SL, Lokchao River, Chindwin Drainage, Manipur.

Macrones peguensis Boulenger, 1894: 196 (type locality: Sittang River near Toungoo, Myanmar).

Hemibagrus peguensis: Ng & Kottelat, 2013: 227 (revision).

Diagnosis. *Hemibagrus peguensis* differs from its congeners in having 9−12 vertical columns of black dots on each side of the body and the biggest one on each column lies on the lateral line; head length 29.0%−32.5% SL; snout margin gently curved, deeply incised humeral process and upper caudal lobe with filamentous extension of uppermost principle ray; skin smooth; lateral line complete; dorsal fin with one spinelet, one spine and seven branched rays, the spine smooth anteriorly and serrated posteriorly at distal half; adipose fin short; caudal fin deeply forked, first principle ray of upper lobe modified into a filamentous extension. Maximum size ∼286 mm SL.

Distribution. Lokchao River, Manipur, India; Irrawaddy, Sittang and Pegu drainages, Myanmar.

Genus *Mystus* Scopoli, 1777

Mystus Scopoli, 1777: 451 (type species: *Bagrus halepensis* Valenciennes in Cuvier and Valenciennes, 1840). Gender: masculine.

Diagnosis. Body moderately elongated; abdomen rounded; dorsal surface of head with median longitudinal groove; mouth subterminal; lips thin, jaws subequal, teeth villiform in bands on jaws and palate; barbels in four pairs: one each of nasal and maxillary, two of mandibulars, maxillary barbel extended beyond head; gill membranes free from isthmus and also from each other; dorsal-fin with a spine and seven branched rays, its origin above the middle of pectoral-fin ray; pectoral-fin spine serrated internally; adipose-fin present; caudal-fin forked, lobed unequal; lateral line complete.

Mystus bleekeri (Day, 1877)
(Fig. 4.415)

FIGURE 4.415

Mystus bleekeri, 98.8 mm SL, Ganga River at Patna, Bihar.

Macrones bleekeri Day, 1877: 451 [type locality: Jumna (Yamuna) River, India].

Diagnosis. Smooth dorsal spine; OP extending to basal bone of dorsal-fin; maxillary barbel reaching posteriorly to anal-fin. Two light brownish stripe bands above and below lateral line; a black spot at the shoulder just behind operculum; fins edged black. Maximum size ~ 100 mm SL.

Distribution. Ganga and Brahmaputra drainages, headwaters of Mahanadi, India; Bangladesh; Nepal.

Mystus carcio (Hamilton, 1822)
(Fig. 4.416)

FIGURE 4.416

Mystus carcio, 39.0 mm SL, wetland of Comilla District, Meghana basin, Bangladesh.

Pimelodus carcio Hamilton, 1822: 181 (type locality: ponds in northern parts of Bengal).

Pimelodus batasius Hamilton, 1822: Pl. 23, fig. 60 (figure and caption).

Mystus carcio: Darshan, Anganthoibi & Vishwanath, 2010: 2177 (redescription).

Diagnosis. Small adult size, maturing at 44.0 mm SL; adipose-fin base shorter than or equal to dorsal-fin base (8.5%−11.9% SL); pelvic fin reaching anal-fin origin; pectoral girdle with coracoid shield exposed ventro-laterally below pectoral fin; eyes rounded, large (25.6%−30.7% HL), dorsally orientated on head; vomerine tooth-patch interrupted at midline. Maximum size ∼50 mm SL.

Distribution. Ganga-Brahmaputra and Meghana drainages in West Bengal, Assam, India, Bangladesh.

Mystus cavasius (Hamilton, 1822)
(Fig. 4.417)

FIGURE 4.417

Mystus cavasius, 78.2 mm SL, Brahmaputra River at Guwahati, Assam.

Pimelodus cavasius Hamilton 1822: 203, 379, pl. 11, fig. 67 (type locality: larger freshwater of Gangetic provinces, India).

Mystus cavasius: Chakrabarty & Ng, 2005: 2 (redescription).

Diagnosis. Adipose-fin base long commencing immediately from the base of last dorsal-fin ray and extended up to the middle of the caudal peduncle, tall dorsal fin, height almost equal to lateral head length, its tip not falcate, a straight or gently concave dorso-posterior margin; dorsal spine short and feebly serrate, a long maxillary barbels reaching to caudal-fin base; an ovoid humeral black spot and another faint spot in front of the dorsal-spine base; body without distinct midlateral stripes, gill rakers on the first gill arch 13−22. Maximum size ∼400 mm SL.

Distribution. Ganges, Brahmaputra, Mahanadi, Subarnarekhar and Godavari river drainages, India, Nepal, and Bangladesh.

Mystus cineraceus Ng & Kottelat, 2009
(Fig. 4.418)

Mystus cineraceus Ng & Kottelat, 2009: 245, figs. 1 and 2A (type locality: Nant Yen Khan Chaung, affluent of Lake Indawgyi, a little south of Lonton Village, Kachin state, Myanmar).

Diagnosis. Adipose-fin base long, contacting base of the last dorsal-fin ray anteriorly; maxillary barbel reaching to middle of anal-fin base; body color pattern consisting of a uniform brownish-gray with a diffuse dark mid-lateral line and a diffuse dark tympanic region, no distinct black spots at the tympanic region and at the base of the caudal-fin; head length 24.1%−27.2% SL; rakers on the first gill arch 13−15. Maximum size ∼90 mm SL.

FIGURE 4.418

Mystus cineraceus, 86.4 mm SL, stream near Tamu, Yu River Drainage, Myanmar.

Distribution. Chindwin drainage, Manipur, India; middle reaches of the Irrawaddy River drainage in northern Myanmar.

Mystus dibrugarensis (Chaudhuri, 1913)
(Fig. 4.419)

FIGURE 4.419

Mystus dibrugarensis, 67.8 mm SL, Dikrong River, Brahmaputra drainage, Arunachal Pradesh.

Photo courtesy: A. Darshan.

Macrones montanus var. *dibrugarensis* Chaudhuri, 1913: 254, pl. 9, figs. 2, 2a, 2b (type locality: Dibrugarh, Assam).

Diagnosis. Black tympanic spot and a black mid-lateral stripe extending to caudal-fin base, ending near a black circular blotch, supraoccipital process raised, long and touching proximal radials; gill rakers 24–25; adipose fin originating distinctly behind the last dorsal-fin ray; posterior cranial fontanel not reaching the base of occipital process. Maximum size ~90 mm SL.

Distribution. Brahmaputra drainage in Assam and Arunachal Pradesh.

Mystus falcarius Chakrabarty & Ng, 2005
(Fig. 4.420)

Mystus falcarius Chakrabarty & Ng, 2005: 13, figs. 3c, 7 (type locality: Myitkyina market, Kachin State, Myanmar).

Diagnosis. Adipose fin long, its origin contacting base of last dorsal-fin ray; dorsal fin tall, its first and second rays elongated, dorso-posterior margin concave (sickle-shaped or falcate appearance), dorsal spine short and feebly serrate; maxillary barbels reaching to caudal-fin base; body

FIGURE 4.420

Mystus falcarius, 112.0 mm SL, stream at Tamu, Yu River drainage, Myanmar.

without distinct mid-lateral stripes, a crescentic or ovoid dark humeral mark and a black spot in front of the dorsal-spine base; 22−29 rakers on the first gill arch. Maximum size ~190 mm SL.

Distribution. Chindwin Drainage, Manipur, India; Irrawaddy and Salween river drainages in Myanmar, drainages in Tenasserim, southern Myanmar.

Mystus ngasep **Darshan, Vishwanath, Mahanta & Barat, 2011**
(Fig. 4.421)

FIGURE 4.421

Mystus ngasep, 103.0 mm SL, Nambul River, Imphal, Chindwin basin, Manipur.

Mystus ngasep Darshan, Vishwanath, Mahanta & Barat, 2011b: 2178, fig. 1, Image 1 (type locality: Nambul River at Bijoy Govinda-Polem Leikai Bridge, Chindwin-Irrawaddy Drainage, Manipur, India).

Diagnosis. Three brown stripes separated by pale narrow longitudinal lines on the sides of the body, a distinct dark tympanic spot; max barbel not reaching anal-fin; cranial fontanel reaching the base of the OP; slender cleithral process; adipose-fin base long anteriorly contacting the base of the last dorsal-fin ray; eye diameter 16.5%−19.8% SL; pectoral spine serrated posteriorly with 9−11 serrae; pectoral-fin with 9−10 and anal-fin with eight to nine branched rays; gill rakers on first arch 16−19. Maximum size ~109 mm SL.

Distribution. Chindwin river basin in Manipur, India.

Mystus prabini **Darshan, Abujam, Kumar, Parhi, Singh, Vishwanath, Das, Pandey, 2019**
(Fig. 4.422)

FIGURE 4.422

Mystus prabini, 82.2 mm SL, holotype, Sinkin River, tributary of Siang River, Brahmaputra drainage, Arunachal Pradesh.

Mystus prabini Darshan, Abujam, Kumar, Parhi, Singh, Vishwanath, Das, Pandey, 2019b: 513 (type locality: Sinkin River, tributary of Siang at Anpun village, Lower Dibang District, Arunachal Pradesh).

Diagnosis. Long adipose-fin base that reaches the base of the last dorsal-fin ray; narrow blackish mid-lateral stripe extending from the anterior region of the tympanic spot to the rounded spot at the caudal-fin base; posterior cranial fontanel reaching the base of the OP. Gill rakers $2-3+8-9=10-12$; vertebrae $39-40$. Maximum size ~ 96 mm SL.

Distribution. Sinkin River at Anpum village in Lower Dibang valley District, Arunachal Pradesh, and also from the Dibang River in Arunachal Pradesh, India.

Mystus pulcher (Chaudhuri, 1911)
(Fig. 4.423)

FIGURE 4.423

Mystus pulcher, 69.9 mm SL, Chatrickong River, Kamjong District, Chindwin Drainage, Manipur.

Macrones pulcher Chaudhuri, 1911: 20 (type locality: Bhamo, close to Yunnan border, upper Myanmar).

Diagnosis. Occipital process extending to basal bone of dorsal fin; median longitudinal groove on head not extending to base of occipital process; a prominent dark spot on the shoulder and another one at the base of the caudal fin; dorsal-fin with one finely serrated weak spine; adipose-fin long, its anterior origin contacting base of last ray of dorsal-fin; pectoral with one spine and eight branched rays, the spine strong, longer than dorsal spine, finely serrated anteriorly and with 10−12 denticulations on the posterior side. Maximum size ∼80 mm SL.

Distribution. Chatrickong River, Ukhrul District, Chindwin drainage, Manipur, India; Irrawaddy drainage, Myanmar.

Mystus rufescens (**Vinciguerra, 1890**)
(Fig. 4.424)

FIGURE 4.424

Mystus rufescens, 96.3 mm SL, Lokchao River, Yu River Drainage, Indo-Myanmar border, Manipur.

Macrones rufescens Vinciguerra, 1890: 226, pl. 3, fig. 2 (type locality: Meetan, Tenasserim Provinces, Myanmar).

Diagnosis. Body rufescent, greenish to iron-rust coloration on body, humeral and caudal fin blackish, a black spot at base of caudal-fin seen in fresh specimens, nuchal spot at dorsal-fin base extremely faint, a large triangular mid-caudal peduncular spot; maxillary barbel extending to anal origin; 13−20 gill rakers on first gill arch; 40−42 vertebrae. Maximum size ∼146 mm SL.

Distribution. Lokchao River, tributary of Yu River in Manipur, Chindwin drainage, India; Irrawaddy, Sittang and lower Salween basin, Myanmar.

Mystus tengara (**Hamilton, 1822**)
(Fig. 4.425)

FIGURE 4.425

Mystus tengara, 96.4 mm SL, Meghna drainage, Sylhet District, Bangladesh.

Pimelodus tengara Hamilton, 1822: 181 (type locality: parts of Bengal, India).

Diagnosis. Long posterior cranial fontanel, its posterior tip reaching middle of supra-occipital; comparatively long adipose fin; light brown on top turning dull yellow on sides and beneath; about five parallel stripes on body present; occasionally a dark shoulder spot may also be seen. Maximum size ∼150 mm SL.

Distribution. Ganga, Brahmaputra and Barak-Surma and Meghna drainages.

Genus *Olyra* McClelland, 1942

Olyra McClelland, 1942: 588 (type species: *Olyra longicaudatus* McClelland). Gender: feminine.
Olyra: Hora, 1936b: 202 (redescription of the genus).

Diagnosis. Loach-like fishes with long and slender body, anteriorly depressed and compressed caudal region; nostrils wide apart, anterior tubular, posterior oval with a rim anteriorly produced into a long barbel; mouth small anterior, jaws equal, lips thin, continuous, labial groove widely interrupted, both jaws with a number of open pores; four pairs of long and thin barbels: one pair each of nasal and maxillary, two mandibulars; gill openings wide but functional part restricted by skins developed along lower edges of gill openings acting as valves; dorsal-fin short with seven to eight rays, no spine, its origin opposite that of pelvic fin; adipose-fin low; anal fin moderate, 16−23 rays, lateral line complete; nuchal plate, supraneural bone and dorsal spinelet absent; dorsal origin shifted posteriorly, first proximal radial inserted between the 12th and 13th or 13th and 14th vertebrae; upper caudal fin lobe enlarged or elongated.

Olyra kempi Chaudhuri, 1912
(Fig. 4.426)

Olyra kempi Chaudhuri, 1912: 443, Pl. 41, figs. 4−4b, (Dishnor River, Mangaldai District, Assam−Bhutan border).

FIGURE 4.426

Olyra kempi: 82.0 mm SL, tributary of Siang River, Anpum Village, Lower Divang District, Arunachal Pradesh.

Diagnosis. Head depressed, snout spatulate, maxillary barbel extending beyond pectoral fin; adipose fin small, low, elongated, posterior extremity rounded, slightly elevated with a knob-like end and not confluent with the caudal fin; pectoral fin with a serrated spine on both the sides, anterior fourth with no serrations and is pointed; Anal fin with 17−18 rays; caudal fin deeply forked, upper lobe longer; conspicuous glandular opening along the lateral line and on the chest and head. Maximum size ∼80 mm SL.

Distribution. Brahmaputra drainage, northeastern India.

Olyra longicaudata McClelland, 1842
(Fig. 4.427)

FIGURE 4.427

Olyra longicaudata, 86.6 mm SL, stream at Chirang District, Brahmaputra Drainage, Assam.

Olyra longicaudatus McClelland, 1842: 588, pl. 21, fig. 1 (type locality: India: Khasya, Boutan, and Mishmee mountains).

Olyra kempi Chaudhuri, 1912: 443, Pl. 41, figs. 4−4b, (Dishnor River, Mangaldai District, Assam−Bhutan border).

Diagnosis. Anal fin with vii,14 rays; caudal fin with 5 + 7, pelvic fin extending vertically through to beyond half-length of distance between ventral and anal fin, body brownish with three longitudinal bands, two on side of lateral line; posterior end of adipose-fin anterior to vertical level of posterior end of anal-fin base; body brownish with three stripes, one each above and below lateral line, third one overlapping the lateral line. Maximum size ∼110 mm SL.

Distribution. Brahmaputra Drainage in Arunachal Pradesh, Assam, Meghalaya, Nagaland, northeastern India.

Olyra parviocula Kosygin, Shangningam & Gopi, 2018

Olyra parviocula Kosygin, Shangningam & Gopi, 2018: 590, fig. 1 (type locality: Kameng River at Bhalukpong, Brahmaputra River basin, West Kameng District, Arunachal Pradesh, India).

Diagnosis. Body depth at vent 6%−7% SL; head length 15%−17% SL; eye diameter 5%−8% HL; maxillary barbel almost reaching pelvic-fin base; branched dorsal fin ray six; adipose-fin short; postadipose distance 14%−17% SL; not confluent with caudal fin, its base length 9%−12% SL; anal-fin rays viii−xi, 8−10, caudal fin deeply forked, its upper lobe two times longer than lower; total vertebrae 53. Maximum size ∼120 mm SL.

Distribution. Kameng River, Brahmaputra drainage, Arunachal Pradesh, India.

Olyra praestigiosa Ng & Ferraris, 2016
(Fig. 4.428)

FIGURE 4.428

Olyra praestigiosa, 78.2 mm SL, Dikrong River, Arunachal Pradesh.

Photo courtesy: *A. Darshan.*

Olyra praestigiosa Ng & Ferraris, 2016: 381, fig. 1 (type locality: Chel River, north of Gorubathan, Kalimpong subdivision, Darjeeling District, West Bengal, India).

Diagnosis. Interorbital distance 30%−37% HL; body depth at anus 6%−9% SL; adipose-fin base length relatively shorter from congeners, its length 9%−16% SL; adipose-fin short and low, separate from upper principal caudal-fin rays; postadipose distance 15%−18% SL; caudal peduncle length 14%−19% SL; and caudal peduncle depth 6%−8% SL; anal-fin rays 17−22. Maximum size ∼80 mm SL.

Distribution. Brahmaputra River drainage in Bangladesh and northeastern India.

Olyra saginata Ng, Lalramliana & Lalthanzara, 2014
(Fig. 4.429)

FIGURE 4.429

Olyra saginata, 74.6 mm SL, Kaladan River at Kawlchaw, Mizoram.

Olyra saginata Ng, Lalramliana & Lalthanzara, 2014b: 266, figs. 1−3 (type locality: Palak River in the vicinity of Phurra Village, Saiha District, Mizoram, India).

Diagnosis. Body depth at anus 10.5%−12.0% SL; adipose-fin-base length 17.3%−22.4% SL, fin separate from upper principal caudal fin rays; postadipose distance 11.8%−14.3% SL; caudal peduncle length 14.8%−17.7% SL, its depth 7.8%−8.9% SL; lateral line pore 54−66. Maximum size ~86 mm SL.

Distribution. Palak, Sala and Tuisi rivers, all tributaries of the Kaladan River, Mizoram, northeastern India.

Genus *Rama* Bleeker, 1862

Rama Bleeker, 1862b: 8 (type species: *Pimelodus rama* Hamilton, 1822).
Chandramara Jayaram, 1972: 816 (type species: *Pimelodus chandramara* Hamilton, 1822).

Diagnosis. Small bagrid of maximum size of about 50 mm SL; body compressed; head small; mouth subterminal; jaws subequal, teeth on jaws and palate in broad villiform bands; barbels small and slender, four pairs: one each of nasal and maxillary, two mandibular, barbels not extending beyond head; dorsal-fin origin above distal tip of cleitheral process, a slender spine and seven branched rays; pectoral spine strong serrated internally; anal fin 13−17 rays; caudal-fin forked, lobes equal; lateral line complete or not reaching caudal-fin base; pores on side and ventral surface of head; body with numerous black dots.

Rama chandramara (Hamilton, 1822)
(Fig. 4.430)

FIGURE 4.430

Rama chandramara, (A) 65.0 mm SL, northern Bengal; (B) 84.2 mm SL, Brahmaputra basin, Goalpara, Assam.

Pimelodus chandramara Hamilton, 1822: 162, 375 (type locality: Atreyi = Atrai River, North Bengal, India).

Pimelodus rama Hamilton, 1822: 176, 377, pl. 3, fig. 55 (Brahmaputra River, India).

Diagnosis. Small fish, eyes large and slightly protuberant; head blunt, smooth, dorsal fin I,6 rays, pectoral spines long and slender with 11−13 antrorse serrae posteriorly; adipose fin short, well behind origin of last rayed dorsal fin and ending well in advance of the procurrent rays of caudal fin; caudal fin forked, lobes equal, tips of lobes may be with black coloration; diaphanous fish marked with clusters of black dots; olivaceous and golden stripe along the lateral line in some; a black shoulder spot, skin on the snout tip and occiput darker due to melanophore pigments. Maximum size ~70 mm SL.

Distribution. Ganga and Brahmaputra basins in India and Bangladesh.

Genus *Rita* Bleeker, 1853

Rita Bleeker, 1853: 60 (type species: *Pimelodus rita* Hamilton, 1822).

Rita: Misra, 1976: 112 (character of genus).

Diagnosis. Body moderate to elongated, subterate; head and snout depressed, covered with skin; skin smooth, occipital region osseous; eyes dorso-lateral, subcutaneous; barbels in three pairs, nasal barbel minute; teeth in premaxilla, vomer and dentary villiform or molariform; OP reaching nuchal plate; cubito-humeral process much elongated; dorsal fin with a spine serrated posteriorly; adipose-fin present; pectoral-fin horizontal with a spine serrated on both edges.

Rita rita (Hamilton, 1822)
(Fig. 4.431)

FIGURE 4.431

Rita rita, 154.0 mm SL, Barak River, Silchar, Assam.

Pimelodus rita Hamilton, 1822: 165, 376, pl. 24, fig. 53 (type locality: Bengal estuaries).

Diagnosis. Dorsal-fin spine as long as or longer than head length, spine extends to or beyond adipose-fin origin when adpressed; palatal teeth in two elliptical patches, not meeting at midline; teeth on posterior extent of lower jaw and palate molariform, much larger than anterior teeth. Maximum size ~160 mm SL.

Distribution. Ganges and Brahmaputra drainages.

Rita sacerdotum Anderson, 1879
(Fig. 4.432)

FIGURE 4.432

Rita sacerdotum, 166.0 mm SL, Yu River, Tamu, Myanmar.

Rita sacerdotum Anderson, 1879: 864, pl. 79, fig. 3 (type locality: Irrawaddy River at location of pagoda of Thingadow, third defile of Irrawaddy River, Myanmar).

Rita sacerdotum: Ferraris, 1999: 19, figs. 1a, 2−5 (redescription).

Diagnosis. Dorsal-fin spine shorter than head length; its spine not reaching adipose-fin origin when adpressed; eye diameter 10%−13% HL; premaxillary large, single, crescent-shaped with large bluntly conical teeth of uniform size; palatal teeth in a single crescent patch that extends across midline of palate; teeth on palate of uniform size. Maximum size ∼195 mm SL.

Distribution. Chindwin-Irrawaddy and Sittang river drainages, Myanmar.

Genus *Sperata* Holly, 1939

Sperata Holly, 1939: 143 (type species: *Bagrus lamarrii* Valenciennes, 1840). Gender: feminine.

Sperata: Ferraris and Runge, 1999: 400 (redescription of the genus and revision).

Diagnosis. Body elongated, snout elongated and depressed, spatulate or rounded at tip; mouth subterminal, moderately wide; jaws subequal, maxilla elongated; lips thin; villiform teeth on jaws and palate; barbels extend posteriorly to pelvic fins or beyond to anal fin; dorsal surface of first nuchal plate (dorsally exposed supra neural) rugose and enlarged to form "interneural shield" and is located between OP and basal bone of dorsal fin (second nuchal plate, Fig. 4.433); OP long and often dorsal spine weakly serrated posteriorly; adipose-fin base short, about as long as the rayed dorsal-fin base or slightly greater; posterior surface of the posttemporal with a concavity to accommodate an extension of the swim bladder; caudal fin deeply forked; brownish-gray on back, silvery on flanks and belly; a large round or ovoid blue-black spot is near the posterior margin of the adipose fin.

Sperata acicularis Ferraris & Runge, 1999

(Fig. 4.434)

Sperata acicularis Ferraris & Runge, 1999: 403, figs. 2a, 3, 4a (type locality: South Oak-ka-lar-pa market, Yangon, Myanmar).

Diagnosis. Snout truncate; supraoccipital spine long and slender, tapers posteriorly and pointed at tip; interneural shield slender shorter and narrow, equals or slightly broader than the supraoccipital spine (Fig. 4.462A); position of orbit located in the middle of head; interdorsal length shorter than the length of rayed dorsal-fin base; adipose fin-base length 1½ the length of rayed dorsal-fin base; gill rakers on the first branchial arch 14−19; precaudal vertebrae 17−18; preanal vertebrae 29−30; anal-fin rays 10 or more; total vertebrae 51−54. Maximum size ∼1800 mm SL.

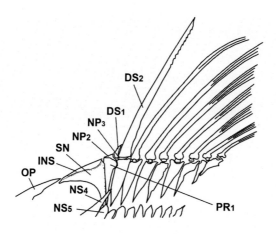

FIGURE 4.433

Lateral view of dorsal-fin skeleton of *Sperata aor*: OP, occipital process; INS, interneural shield; SN, supraneural; NP_{1-3}, Nuchal plate 1 to 3; DS_1, dorsal spine 1 (spinelet); DS_2, dorsal spine 2; PR_1, proximal radial.

FIGURE 4.434

Sperata acicularis, 192.0 mm SL, Lokchao River, tributary of Yu River, Manipur.

Distribution. Lokchao River, Chindwin drainage, Manipur, India; Irrawaddy drainage and Tenasserim River systems, Myanmar.

Sperata aor (**Hamilton, 1822**)
(Fig. 4.435)

FIGURE 4.435

Sperata aor, 403.0 mm SL, Barak River, Tamenglong, Manipur.

Pimelodus aor Hamilton, 1822: 205, 379 (type locality: rivers of Bengal and upper parts of Gangetic estuary).

Sperata aor: Ferraris & Runge, 1999: 410 (redescription).

Diagnosis. Snout tip broadly rounded; interneural shield large and ovoid, its length equals the length of supraoccipital spine (Fig. 4.462B); interdorsal length equals rayed dorsal-fin base length; gill rakers on first branchial arch 19−20; anal fin rays 8−10; maxillary barbels extending to base of anal fin; adipose-fin base length longer than the dorsal-fin base; pectoral fin rays 10−11; preanal vertebrae 28−30; precaudal vertebrae 16−19; total vertebrae 50−52. Maximum size ∼1500 mm SL.

Distribution. Barak−Surma−Meghna, Brahmaputra, Ganga River drainage, rivers of southern India up to the Krishna river system; Pakistan, Nepal, Bangladesh.

Sperata aorella **(Blyth, 1858)**
(Fig. 4.436)

FIGURE 4.436

Sperata aorella, 278.0 mm SL, Wetland of Comilla District, Meghna basin, Bangladesh.

Bagrus aorellus Blyth, 1858: 283 (type locality: Calcutta fish market, India).
Sperata aorella: Ferraris & Runge, 1999: 413 (revision of genus *Sperata*).

Diagnosis. Supraoccipital spine long and slender, constricted at base, broader posteriorly; the spine longer but not wider than the interneural shield; interneural shield short, ovoid (Fig. 4.462C); adipose-fin base length equals dorsal-fin base length; interdorsal length less than dorsal-fin base length; posterior margin of orbit at about the middle of the head length; snout tip truncated; barbels extend beyond adipose fin; pectoral-fin rays 9−10; branched anal-fin rays 9; gill rakers on first branchial arch 20−22; preanal vertebrae 28; precaudal vertebrae 16−19; total vertebrae fewer than 50. Maximum size ∼1200 mm SL.

Distribution. Ganga River Delta, India and Bangladesh.

Sperata lamarrii **(Valenciennes, 1840)**
(Fig. 4.437)

FIGURE 4.437

Sperata lamarrii, 315.0 mm SL, Brahmaputra River, Guwahati.

Bagrus lamarrii Valenciennes in Cuvier & Valenciennes, 1840: 407, Pl. 415 (type locality: Ganges River, India).
Sperata seenghala: Ferraris & Runge, 1999 (in part): 416 (redescription).
Sperata lamarrii: Kumar et al., 2000 (validation and redescription) (Fig. 4.438).

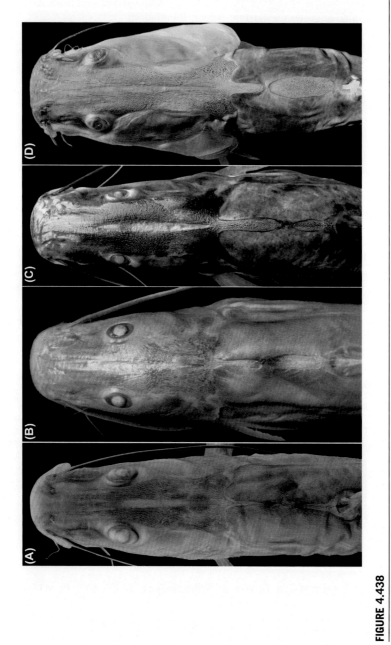

FIGURE 4.438

Dorsal views of heads of *Sperata*: (A) *Sperata acicularis*, (B) *S. aor*, (C) *S. aorella*, and (D) *S. lamarrii*.

Diagnosis. Snout spatulate; eyes positioned distinctly in the anterior half of head; supra-occipital spine shorter than interneural shield, the spine short with a blunt tip; shield rugose and ovoid (Fig. 4.462D); adipose fin short, its base equals or slightly longer than dorsal-fin base; pectoral-fin rays 8–9; branched anal-fin rays 9; gill rakers on first branchial arch 13–15; precaudal vertebrae 21–23; preanal vertebrae 29–32; total vertebrae 50–53. Maximum size ~1350 mm SL.

Distribution. Ganga, Brahmaputra, Indus rivers and river systems in southern India.

Family Salmonidae

Trouts

Body compressed, covered with minute scales; head with no scales; barbels absent; eyes large, superior; maxilla large, extending to front border of orbit; mouth cleft wide; vomerine and palatine teeth forming a continuous horse-shoe shaped band; accessory pelvic process present; adipose dorsal fin present; swim bladder with pneumatic duct (Fig. 4.439).

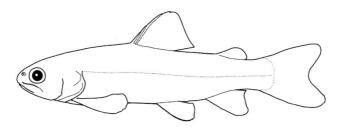

FIGURE 4.439

Outline diagram of Salmonidae.

Genus *Oncorhynchus* Suckley, 1861

Oncorhynchus Suckley, 1861: 313 (type species: *Salmo scouleri* Richardson, 1836). Gender: masculine.

Diagnosis. Body elongated, compressed; abdomen rounded; mouth wide, maxillary long, extending beyond eyes; eyes superior; lips thin, jaws equal; rayed dorsal fin with 12–15 rays; adipose fin low; anal fin long with 12–15 rays; lateral line straight.

Oncorhynchus mykiss (Walbaum, 1792)

(Fig. 4.440)

Salmo mykiss Walbaum, 1792: 59 (type locality: Kamchatka, Russia).

Diagnosis. Body elongated, compressed moderately; dorsal fin with three to four spines and 10½–12½ branched rays; anal fin with three to four spines and 8½ branched rays; wide pinkish stripe on mid-dorsal line; 10–12 faint brown oval blotches over-layered with randomly distributed darker spots on the sides; on the side scales in mid-lateral row 115–130; vertebrae 60–66. Maximum size ~1100 mm SL.

Distribution. Pacific basin of North America, Europe, Kamchatka to Amur drainage. Introduced widely in other countries including India.

FIGURE 4.440

Oncorhynchus mykiss, 105.0 mm SL, Brahmaputra basin, Arunachal Pradesh.

Family Mugilidae
Mullets

Spiny and soft dorsal fins present, widely separated; pelvic fin subabdominal; lateral line absent or faint; ctenoid scales on body; mouth moderate, teeth small or absent; gill rakers long; intestine exceedingly long (Fig. 4.441).

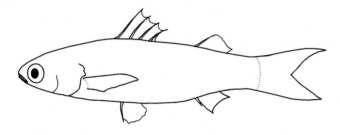

FIGURE 4.441

Outline diagram of Mugilidae.

Genus *Rhinomugil* Gill, 1863

Rhinomugil Gill, 1863: 169 (type species: *Mugil corsula* Hamilton, 1822). Gender: masculine.

Diagnosis. Body stout, elongated, compressed; mouth ventral, protrusible; eyes prominent, dorsal and in anterior half of head; opercular spine absent; teeth indistinct on jaws; two dorsal fins; fins well separated; scales cycloid in young, ctenoid in adult.

Rhinomugil corsula (Hamilton, 1822)
(Fig. 4.442)

Mugil corsula Hamilton, 1822: 221, 381 (type locality: most rivers of the Gangetic provinces, introduced in some ponds).

Diagnosis. First dorsal fin origin nearer to caudal-fin base than to tip of snout; caudal fin slightly emarginated; scales in lateral series 48–52. Maximum size ~70 mm SL.

Distribution. India, Bangladesh, Nepal.

FIGURE 4.442

Rhinomugil corsula, 67.3 mm SL, Brahmaputra River, Guwahati, Assam.

Genus *Sicamugil* Fowler, 1939

Sicamugil Fowler, 1939: 9 (type species: *Mugil hamiltoni* Day, 1870a). Gender: masculine.

Diagnosis. Body moderately compressed; head compressed, covered with scales; mouth protrussible; lower jaw with symphyseal knob; operculum with a strong spine; teeth absent on jaws and palatine; first dorsal fin with four spines, its origin opposite pectoral-fin origin; second dorsal fin with eight rays; anal fin with three spines and eight to nine rays; caudal fin forked; scales ctenoid; lateral line absent.

***Sicamugil cascasia* (Hamilton, 1822)**
(Fig. 4.443)

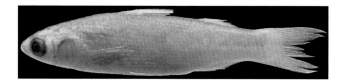

FIGURE 4.443

Sicamugil cascasia, 53.5 mm SL, Jiri River, Manipur—Assam border.

Mugil cascasia Hamilton, 1822: 217, 380 (type locality: Rivers of northern Bengal).

Diagnosis. Two dorsal fins widely separated, interdorsal space 15.5%−15.8% SL, first dorsal spiny and second dorsal soft; dorsal fin origin conspicuously nearer tip of snout than to caudal-fin base; anal fin origin opposite to second dorsal fin origin; lateral line absent. Maximum size ∼95 mm SL.

Distribution. Ganga and Brahmaputra river drainages, India; Bangladesh and Pakistan.

Family Belonidae
Needle fishes

Upper and lower jaws produced into a beak; jaws provided with numerous needle-like teeth; mouth opening large; scales small; dorsal fin origin above anal fin origin; caudal fin truncated or rounded (Fig. 4.444).

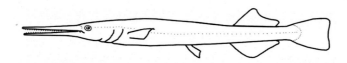

FIGURE 4.444

Outline diagram of Belonidae.

Genus *Xenentodon* Regan, 1911

Xenentodon Regan, 1911a: 332 (type species: *Belone cancila* Hamilton, 1822). Gender: masculine.

Diagnosis. Body elongated, subcylindrical, compressed; head and snout sharply pointed; jaws produced in the form of a beak, large canine teeth alternating with much more numerous conical teeth; a deep longitudinal groove on the upper surface of head; gill raker absent; dorsal fin without a spine, inserted above anal fin; caudal fin truncated.

Xenentodon cancila (Hamilton, 1822)
(Fig. 4.445)

FIGURE 4.445

Xenentodon cancila, 142.0 mm SL, Irang River, Noney, Barak drainage, Manipur.

Esox cancila Hamilton, 1822: 213, 380 (type locality: ponds and smaller rivers of Gangetic provinces).

Diagnosis. Upper and lower jaws extended into long beaks armed with sharp teeth to their tip; a silvery lateral band extending on flank of the body. Maximum size ∼350 mm SL.

Distribution. Widely distributed in India, Pakistan, Bangladesh, Sri Lanka, Myanmar, and Thailand.

Family Aplocheilidae
Rivulines

Body fusiform and compressed; upper jaw protrussible, bordered by only premaxilla; fins with no spines; dorsal fin origin far behind; scales large, cycloid; lateral line on head (Fig. 4.446).

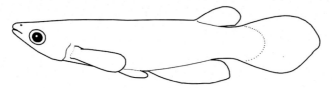

FIGURE 4.446

Outline diagram of Aplocheilidae.

Genus *Aplocheilus* McClelland, 1839

Aplocheilus McClelland, 1839: 301 (type species: *Aplocheilus chrysostigmus* McClelland, 1839). Gender: masculine.

Diagnosis. Body elongated, fusiform and compressed; head conical; mouth terminal, cleft not extending to front border of orbit; upper jaw protractile; barbels absent; dorsal fin origin above the last part of anal fin; caudal fin rounded; scales cycloid, moderate size; lateral line absent.

Aplocheilus panchax (Hamilton, 1822)
(Fig. 4.447)

FIGURE 4.447

Aplocheilus panchax, 34.0 mm SL, Loktak Lake, Manipur.

Esox panchax Hamilton, 1822: 211, 380 (type locality: ditches and ponds of Bengal).

Diagnosis. Eye diameter equal to interorbital space; pelvic fin without a prolonged ray; dorsal fin with four branched rays and a large black ocellus at its base; lateral line absent; cephalic sensory pores in the head region; dorsum of head with an oval shining spot (Fig. 4.448). Maximum size ~85 mm SL.

FIGURE 4.448

Aplocheilus panchax, dorsal views showing shining silvery spots on the occiput.

Distribution. India: throughout; Bangladesh, Malay Archipelago, Myanmar, Pakistan, and Thailand.

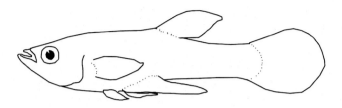

FIGURE 4.449

Outline diagram of Poeciliidae.

Family Poeciliidae
Livebearers

Body cylindrical, head and body covered with scales; premaxilla slightly protractile; Pectoral fin usually placed high on the side of the body; pelvic fin placed anteriorly; pleural ribs present on first several hemal arches; ventral hypohyal forms a bony cap over anterior facet of the anterior ceratohyal; gonopodium usually present; includes small species, maximum being 20 cm.

Genus *Poecilia* Bloch & Schneider, 1801

Poecilia Bloch & Schneider, 1801: 452 (type species: *Poecilia vivipara* Bloch and Schneider, 1801). Gender: feminine.

Diagnosis. Movable spatuliform teeth; male with elongated anterior anal-fin rays to form gonopodium, primarily the third, fourth, and fifth rays; length of anal-fin base as long as gonopodium in males; fish exhibiting internal fertilization; eggs with large yolks, have live birth (ovoviviparous) exoccipital condyles absent.

Poecilia reticulata Peters, 1859
(Fig. 4.450)

FIGURE 4.450

Poecilia reticulata: (A) male, 27.0 mm SL; (B) female, 30 mm SL.

Poecilia reticulata Peters, 1859: 412 (type locality: Guaire River, Caracas, Venezuela).

Diagnosis. Exhibit sexual dimorphism; females have gray body color; males exhibit spots, or stripes that can be any of a wide variety of colors, usually bright; males usually smaller. Maximum size: males ~35 mm; females ~60 mm.

Distribution. Native of northeastern South America; introduced in India.

Family Syngnathidae
Pipe fishes and sea horses

Body elongated, covered with a series of bony rings; opercular bones reduced to a single plate; gill openings small and rounded, gills in the form of small rounded tufts; pectoral fin may be absent; tail may be prehensile; lateral line absent; male with egg pouch on tail or abdomen (Fig. 4.451).

FIGURE 4.451

Outline diagram of Syngnathidae.

Genus *Microphis* (Kaup, 1853)

Doryichthys Kaup, 1853: 234, pl. 1 (type species: *Syngnathus deocata* Hamilton, 1822). Gender: masculine.

Diagnosis. Body elongated, covered with many serrated shields; abdomen rounded; trunk heptagonal, tail tetragonal; mouth terminal, narrow, jaws produced into a beak; barbels absent; anal fin minute with two to three rays; pectoral and caudal fins present; brood pouch from one to three rings; lateral line complete.

Microphis deocata **(Hamilton, 1822)**
(Fig. 4.452)

FIGURE 4.452

Microphis deocata, 167.6 mm SL, stream at Goalpara, Brahmaputra drainage, Assam.

Syngnathus deocata Hamilton, 1822: 14, 363 (Tista, Kuwarlayi and other rivers of Purania or Mithila in northern Bengal).

Diagnosis. Tail rings 29−33; lateral snout ridge poorly defined; adult female with distinct Y-shaped markings on lower half of body; males with indication of stripe on head. Maximum size ~170 mm SL.

Distribution. Assam, North Bengal, Bihar in India and Bangladesh.

Family Synbranchidae

Swamp eels

Paired fins absent; dorsal and anal fins reduced to rayless ridge; caudal fin small or absent; scales minute, embedded in skin or absent; eyes small; anterior and posterior nostrils widely separated; the only teleost with "amphistylic" jaw suspension; palatoquadrate articulating in two places; gill membranes united; branchiostegal rays four to six; swim bladder absent; ribs absent; no spines in anal fin (Fig. 4.453).

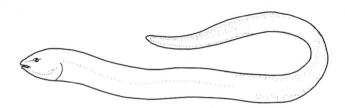

FIGURE 4.453

Outline diagram of Synbranchidae.

Genus *Monopterus* Lacepède, 1800

Monopterus LaCepède, 1800: 138 (type species: *Monopterus javanensis* LaCepède, 1800). Gender: masculine.

Diagnosis. Body elongated, eel like; head conspicuous; snout rounded; mouth wide, terminal, cleft reaching anterior margin of orbit; eyes superior, reduced, covered with skin; villiform teeth on jaws and palate; gill membranes triangular or crescentic, internally attached to isthmus; caudal fin absent; scales absent or minute, embedded in skin; lateral line present.

Monopterus cuchia (Hamilton, 1822)
(Fig. 4.454)

FIGURE 4.454

Monopterus cuchia, 670.0 mm SL, swamps in Agartala, Tripura.

Unibranchiapertura cuchia Hamilton, 1822: 16, 363 (type locality: south-eastern Bengal, India).

Diagnosis. Gill opening triangular; scales absent in skin; palatine teeth biserial; suprabranchial pouch absent; body elongated, eel like; head conspicuous; mouth wide; eyes superior; scales absent or minute, embedded in skin; lateral line present. Maximum size ~680 mm SL.

Distribution. North and northeastern India, Bangladesh, Pakistan, Nepal.

Monopterus ichthyophoides Britz, Lalremsanga, Lalrotluanga & Lalramliana, 2011

Monopterus ichthyophoides Britz, Lalremsanga, Lalrotluanga & Lalramliana, 2011: 52 (type locality: Sawleng River, a tributary of Tuirial River in the vicinity of Sawleng, Barak River drainage, Mizoram, India).

Diagnosis. Branchiostegal rays only two; gill filaments absent on all gill arches; scales present and restricted to posterior parts of the body extending far beyond the vent; total vertebrae 79 + 82 + 34−37 = 114−117. Maximum size ~187 mm SL.

Distribution. Sawleng River, a small stream with muddy bank and bottom in the vicinity of Sawleng, and a public well with a muddy bottom at Luangmual, Aizawl, Mizoram, India.

Monopterus rongsaw Britz, Sykes, Gower & Kamei, 2018

Monopterus rongsaw Britz, Sykes, Gower & Kamei, 2018: 3 (type locality: Nongriat Village, Sohra, Khasi Hills, Meghalaya, India).

Diagnosis. Eyes tiny, covered by skin, visible as tiny black spots under skin in life, barely visible externally in preserved specimens; skin pigmentation absent; total vertebrae 92 + 69 = 161; the shoulder girdle articulating with the skull, with the posttemporal contacting the epiotic and supracleithrum. Maximum size ~150 mm SL.

Distribution. Khasi Hills, Meghalaya, India.

Monopterus javanensis Lacepède, 1800
(Fig. 4.455)

FIGURE 4.455

Monopterus javanensis, 725.0 mm SL, swamps around Loktak Lake, Manipur.

Monopterus javanensis Lacepède, 1800: 139 (type locality: Indonesia: Sunda Strait, near the coasts of Java).

Diagnosis. Gill opening crescentic; minute scales embedded in skin; palatine teeth uniserial; synbranchial pouch for accessory respiration; body elongated eel like; head conspicuous; snout rounded. Maximum size ~900 mm SL.

Distribution. Swamps of the valley and eastern parts of Manipur (Chindwin basin), India; Myanmar; East Indies; Indo-Malayan archipelago; China; Japan.

Family Chaudhuridae

Earthworm eels

Small apodes with a fan shaped and free caudal fin provided with well-developed rays and supported by a pair of hypural bones; with pectoral fins; with minute scales; vent situated far backward from head; teeth arranged in bands on the jaws only; lateral nostrils; gill openings separate and the integument covering supported by few branchiostegal rays; pharyngeal slit wide; four fully developed gill-bearing branchial arches; heart close to the branchial arches; ethmoid and vomer distinct and the former separating the maxillaries in front with well-developed zygapophyses on the vertebrae. Chaudhuriidae is unique in having a strongly developed true tail (Fig. 4.456).

FIGURE 4.456

Outline diagram of Chaudhuridae.

Genus *Garo* Yazdani & Talwar, 1981

Garo Yazdani and Talwar, 1981:287 (type species: *Pillaia khajuriai* Talwar et al., 1977). Gender: masculine.

Diagnosis. Body elongated, eel-like and naked; head conical; mouth wide, terminal and with an indistinct fleshy rostral appendage; gill openings wide, separate, free from isthmus; eyes lateral; dorsal and anal fins confluent with the long caudal fin; spines absent in dorsal and anal fins; dorsal fin with 40−44 soft rays, its origin in the middle between tip of snout and caudal-fin base; anal fin with 37−38 soft rays; pectoral fin fairly large, with 19−20 rays; caudal fin with 12 unbranched rays; total vertebrae 65.

Garo khajuriai (Talwar, Yazdani & Kundu, 1977)

Pillaia khajuriai Talwar, Yazdani & Kundu, 1977: 53, fig. 1 (type locality: paddy field at Rongrengiri, Garo Hills District, Meghalaya, India).

Diagnosis. Body eel-like, naked, subcylindrical, tail compressed and tapers posteriorly; with a laterally compressed tail which tapers considerably. Head moderately elongated; mouth wide, gape extending to the level of anterior margin of eye; lips thick anteriorly, flap-like laterally, with weakly developed fleshy rostral appendages; anterior nostrils tabular; vent opposite slightly behind dorsal fin origin; lateral line not distinct; dorsal and anal fins well developed, confluent with the caudal fin; pectoral fin placed in the lower half of the body, their bases are partially concealed under the opercula. Maximum size ∼68 mm SL.

Distribution. Meghalaya and Brahmaputra valley, northeastern India.

Genus *Pillaia* Yazdani, 1972

Pillaia Yazdani, 1972: 134 (type species: *Pillaia indica* Yazdani, 1972). Gender: feminine.

Diagnosis. Small eel like fish with long depressed head, without spines before dorsal or anal or anywhere else on the body, with both dorsal and anal united with the caudal of 8−10 unbranched rays; without scales, lateral line clearly discernible on the head, and less distinct on the body; branchiostegals 6, a rather indistinct fleshy rostral appendage bearing anterior tubular nostrils, eyes fairly prominent, place dorsally, gill openings wide, extending dorsally to the level of pectoral origin, gill membranes free from each other and from isthmus, mouth wide and horizontal, teeth on jaws arranged in narrow bands, small, sharply pointed and curved inward pectoral fin small; ventral fin absent.

Pillaia indica Yazdani, 1972
(Fig. 4.457)

FIGURE 4.457

Pillaia indica, 115.0 mm SL, stream in Ri Bhoi District, Brahmaputra drainage, Meghalaya.

Pillaia indica Yazdani, 1972: 134, fig. 1 (Sumer stream, about 22 km north of Shillong, Khasi and Jaintia Hills, Meghalaya, India).

Diagnosis. Dorsal fin with 34−36 rays, anal with 34−36, pectoral with 7−9 and caudal with 8−10; body color in spirit variable, upper part of the body in light to dark purplish brown, lower part yellowish or very light brown, fins dirty white, series of dark color spots or dark lines on either side of the body; Maximum size - 120 mm SL. mm SL.

Distribution. West Bengal, Meghalaya, Assam, and Arunachal Pradesh, India.

Family Mastacembelidae
Spiny eels

Body elongated, eel like, compressed, covered with minute scales; snout elongated, supported by cartilaginous rod; soft dorsal fin preceded by series of isolated spines; anal fin with two to three spines; fleshy rostral appendage present; basisphenoid absent (Fig. 4.458).

FIGURE 4.458

Outline diagram of Mastacembelidae.

Key to genera:

1a. Rostral tooth plate absent (Fig. 4.459A), dorsal fin spine 33or more .... *Mastacembelus*
1b. Rostral tooth plate present (Fig. 4.459B) dorsal fin spine 32 or less .... *Macrognathus*

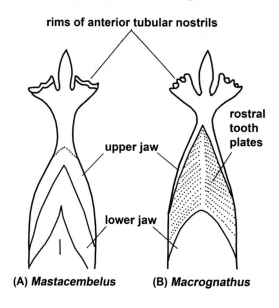

FIGURE 4.459

Figures showing: (A) simple rim of tubular nostril and plain rostrum in *Mastacembelus* and (B) anterior rim of tubular nostril with fingerlike tentacles and rostral tooth plates in *Macrograthus*.

Genus *Macrognathus* Lacepède, 1800

Macrognathus Lacepède, 1800: 283 (type species: *Ophidium aculeatum* Bloch, 1786). Gender: masculine.
Macrognathus: Sufi, 1956: 99 (revision).
Macrognathus: Roberts, 1986b: 97 (revision).

Diagnosis. Body elongated eel like; snout long, fleshy; upper jaw may be with paired series of tooth plates; preorbital and preoperculum smooth; gill rakers absent; depressible dorsal fin spines; anal fin with three spines; caudal fin rounded, separated from dorsal and anal fins; lateral line present; dorsal fin spines 32 or less, rim of anterior nostril with six finger like projections.

Macrognathus aral (Bloch & Schneider, 1801)
(Fig. 4.460)

FIGURE 4.460

Macrognathus aral, 42.0 mm SL, Jiri River, Manipur.

Rhynchobdella aral Bloch &and Schneider, 1801: 479 (type locality: Tranquebar, India).

Diagnosis. Rostral tooth plates 14−28; snout long; anal fin with three spines; caudal fin rounded; body elongated eel like; lateral line present; depressible dorsal fin spines; preorbital and preoperculum smooth; pectoral-fin rays 19−24; body usually with two or more broad pale longitudinal stripes extending its entire length, never with oblique bars; no black ocellus at base of caudal fin; four ocelli on base of dorsal fin. Maximum size ∼59 mm SL.

Distribution. India, Pakistan, Sri Lanka, Bangladesh, Nepal, and Myanmar.

Macrognathus lineatomaculatus **Britz, 2010**

Macrognathus lineatomaculatus Britz, 2010b: 304, fig. 7 (type locality: Jorau River, a tributary of the Sankosh at Laskapara, Barobisha town outskirts, Jalpaiguri District, West Bengal, India).

Diagnosis. Eye large; dorsal fin with 19−22 spines and 50−57 soft rays; anal fin with three spines: and 47−54 soft rays; 15−17 rostral tooth plates; black blotches along dorsal fin. Maximum size ∼184 mm SL.

Distribution. Asia: Jorai River in India and Rapti River basin in Nepal.

Macrognathus morehensis **Arunkumar & Tombi Singh, 2000**
(Fig. 4.461)

FIGURE 4.461

Macrognathus morehensis, 29.5 mm SL, Lokchao River, Manipur.

Macrognathus morehensis Arunkumar & Tombi Singh, 2000b: 119 (type locality: Maklang River, Manipur).

Diagnosis. Rostral tooth plates 8−11; a black ocellus at base of caudal fin; anal fin with three spines; 11−16 dorsal fin spines; 20−25 black broad transverse bars on the body; six black oval spots at the base of anal fin rays. Maximum size ∼155 mm SL.

Distribution. Lokchao River at Moreh, India, Chindwin-Irrawaddy Drainage, Myanmar.

Macrognathus pancalus **Hamilton, 1822**
(Fig. 4.462)

FIGURE 4.462

Macrognathus pancalus, 26.4 mm SL, Jiri River, Manipur.

Macrognathus pancalus Hamilton, 1822: 30, 364. (type locality: Ganges river drainage, India).

Macrognathus siangensis Arunkumar, 2016: 2004 (Siang River, Pasighat, East Siang District, Brahmaputra drainage, Arunachal Pradesh).

Diagnosis. Snout entirely scaly both above and on sides; behind this top of head naked as far as hind of preoperculum; two to five spines on the operculum, one strong preorbital spine piercing the skin; gape of mouth not extending to below nostrils; rostral tooth plates absent; spinous dorsal 26–31, originating above middle of pectoral fin; last spine small and not hidden beneath skin; soft dorsal 30–42; anal spines three, close together, last spine small and not hidden beneath the skin, soft anal originating in advance of soft dorsal; vent nearer to base of caudal than to snout; caudal fin distinctly separated from the dorsal and anal; body with round white spots arranged longitudinally, sometimes confined to posterior half of body; stripes join together to form network in some specimens; soft dorsal, anal and pectoral yellow with numerous black dots. Maximum size ~170 mm SL.

Distribution. India, Bangladesh, Pakistan.

Genus *Mastacembelus* Scopoli, 1777

Mastacembelus Scopoli, 1777: 458 (type species: *Psidium mastacembelus* Banks & Solander in Russell, 1794). Gender: masculine.

Diagnosis. Body eel like and pointed; snout long, conical without transversely striated tooth plates on the under surface; mouth inferior, cleft narrow; eyes small, superior; jaws and palate with minute teeth; dorsal fin origin above middle of pectoral fin, 32–40 spines and 67–90 soft rays; anal fin with three spines and 56–90 soft rays; caudal fin rounded, dorsal and anal fins may or may not be confluent with caudal; scales present; air bladder elongated; simple rim of tubular nostril and plain rostrum.

Mastacembelus armatus (Lacepède, 1800)
(Fig. 4.463)

FIGURE 4.463

Mastacembelus armatus, 36.0 mm SL, Jiri River, Manipur.

Macrognathus armatus Lacepède 1800: 283, 286 (locality unknown).

Mastacembelus armatus: Sufi, 1956: 134 (revision).

Diagnosis. Anal and dorsal fins confluent with caudal fin; gape of mouth extending to below posterior nostrils; dorsal fin origin opposite middle or posterior third of pectoral fin; dorsal spines 34–40, soft dorsal rays 64–92; scales between eyes and posterior nostrils; top of snout internasal space, interorbital space and top of head as far as hind edge of peoperculum naked; two to five spines on preoperculum, one strong preorbital spine usually piercing the skin; body usually with zig-zag lines, sometimes connecting to form a network. Maximum size ~900 mm SL.

Distribution. India: throughout; Pakistan; Sri Lanka; Nepal; Myanmar; Thailand; Malaya; and China.

Mastacembelus tinwini Britz, 2007
(Fig. 4.464)

FIGURE 4.464

Mastacembelus tinwini, 34.0 mm SL, Laniye River, tributary of Tizu, Jessami, Manipur.

Mastacembelus tinwini Britz, 2007: 258 (type locality: Thaton market, Mon State, Myanmar).

Diagnosis. Body with five regular and parallel black stripes, usually forming as series of interrupted lines or broken up into individual blotches; soft dorsal, anal, and caudal fins with white margins; head pointed with median fleshy tentacle projecting from upper jaw. Maximum size ∼300 mm SL.

Distribution. Chindwin Drainage, Manipur−Myanmar border, Sittang and Salween basins, Myanmar.

Family Ambassidae
Asiatic glassfishes

Body strongly compressed, semitransparent; first dorsal fin with seven to eight spines and second with a spine and 9−10 soft rays; preopercular, suborbital and lachrymal bones serrated; lateral line complete or in two rows, interrupted in between and extends to caudal fin (Fig. 4.465).

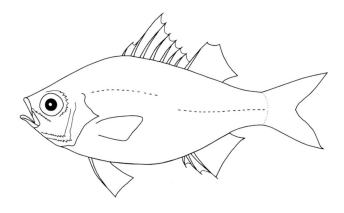

FIGURE 4.465

Outline diagram of Ambassidae.

Genus *Chanda* Hamilton, 1822

Chanda Hamilton, 1822: 103, 370 (type species: *Chanda nama* Hamilton, 1822). Gender: feminine.

Diagnosis. Body ovate, highly compressed, nearly transparent; suborbital ridge smooth with an indistinct spine; lower jaw strongly projecting; jaws, palate, and tongue with villiform teeth; lower limb of preopercle with double serrated edge; opercle without a prominent spine; a forwardly directed recumbent spine in front of dorsal fin; lateral line complete.

Chanda nama **Hamilton, 1822**
(Fig. 4.466)

FIGURE 4.466

Chanda nama, 34.2 mm SL, Barak River, Manipur.

Chanda nama Hamilton, 1822: 109, 371 (type locality: ponds throughout Bengal).

Diagnosis. Minute scales often irregularly arranged, 110−114 in longitudinal series; mouth particularly large, with very prominent lower jaw; supraorbital ridge smooth; mouth large, with a conspicuous lower jaw; teeth villiform on jaws, with canines on either side of lower jaw, tongue edentate; body silvery-yellowish, sparsely scattered minute black dots on body; eyes black; fin yellowish; caudal fin dusky and yellowish, with a pale outer border. Maximum size ∼90 mm SL.

Distribution. Ganga and Brahmaputra drainages.

Genus *Parambassis* **Bleeker, 1874**

Parambassis Bleeker, 1874: 86, 102 (type species: *Ambassis apogonoides* Bleeker, 1851b). Gender: feminine.

Diagnosis. Body moderately elongated and compressed; mouth large, gape oblique, extending to anterior border of orbit; jaws straight or slightly upturned; supraorbital ridge with one or two spines posteriorly; preorbital serrated on both ridge and edge; suborbital serrated; tongue devoid of tooth.

Parambassis baculis **(Hamilton, 1822)**
(Fig. 4.467)

FIGURE 4.467

Parambassis baculis, 46.0 mm SL, Agartala fish market, Tripura.

Chanda baculis Hamilton, 1822: 112, 371 (type locality: north-eastern parts of Bengal).

Diagnosis. Head gently rounded, body compressed, jaws short and equal, and nape distinctly concave; teeth villiform in jaws, vomer and palate; gill rakers on first gill arch 20−21; lateral line complete, preorbital strongly serrated along its lower edge and having a sharp spine directed toward the orbit at its anterior angle followed by several more along its upper edge, second spine of the first dorsal fin is highest and equal to the length of the head behind the hind edge of the orbit, scales very small, lateral series 80−100; preorbital edge with three small upper and three large lower serrae, ridge with large exposed primary serra plus four large lower spines; no suborbital serrae; supraorbital with a short portion without serrae posteriorly; ridge of preopercle nonserrate; horizontal limb of preopercle with 13 large serrae; body brownish to yellowish, no markings except a large brownish coloration on the nape. Maximum size ∼50 mm SL.

Distribution. India, Nepal, Bangladesh.

Parambassis bistigmata Geetakumari, 2012
(Fig. 4.468)

FIGURE 4.468

Parambassis bistigmata, 39.3 mm SL, holotype, Ranga River, Brahmaputra Drainage, Arunachal Pradesh; (A) lateral view; (B) ventral view between pelvic and anal fins showing dark-brown stigma.

Parambassis bistigmata Geetakumari, 2012: 60 (type locality: Ranga River, Kimin station, Brahmaputra drainage, Arunachal Pradesh, India).

Diagnosis. Presence of two distinct black spots on either side of the anal-fin origin, connected by a narrow black curved line formed of black dots; preorbital edge well developed with prominent, acute spines, partially obscuring the maxilla; caudal peduncle with short black stripes on dorsal and ventral sides; presence of a faint vertically elongated humeral spot. Maximum size ∼40 mm SL.

Distribution. Ranga River, Arunachal Pradesh and Umtrao River, Meghalaya, both Brahmaputra drainage.

Parambassis lala (Hamilton, 1822)
(Fig. 4.469)

FIGURE 4.469

Parambassis lala, 36.0 mm SL, Agartala fish market, Tripura.

Chanda lala Hamilton, 1822: 114, 371, pl. 16, fig. 39 (type locality: fresh waters of Gangetic provinces).

Diagnosis. Two distinct black spots on either side of the anal-fin origin, connected by a narrow black curved line; a well-developed preorbital edge with prominent, acute spines, partially obscuring the maxilla; dorsal and ventral side of caudal peduncle each with a black longitudinal stripe; and the presence of a vague, faint, vertically elongated humeral spot;13−14 gill rakers, 65−67 scales in lateral series. Maximum size ∼38 mm SL.

Distribution. Ranga River, Arunachal Pradesh, and Umtrao River, Meghalaya, of the Brahmaputra drainage, India.

***Parambassis ranga* (Hamilton, 1822)**
(Fig. 4.470)

FIGURE 4.470

Parambassis ranga, 38.2 mm SL, Agartala fish market, Tripura.

Chanda ranga Hamilton, 1822: 113, 371 (type locality: Freshwaters of all parts of Gangetic provinces).

Diagnosis. Preopercular hind edge smooth, often with few serrations, lateral line scales 51−52; body stout, deep, and compressed; preopercular hind edge smooth, with one or two serrations at angle; mouth oblique; scales small; cheek with five to six transverse scale rows; silvery-yellowish body; transverse lateral light-brown thin bands on sides of body; a definite dusky spot on shoulder; fins hyaline; dorsal and caudal fins with blackish edges. Maximum size ∼75 mm SL.

Distribution. Ganga and Brahmaputra drainages.

Parambassis serrata Dishma & Vishwanath, 2015
(Fig. 4.471)

FIGURE 4.471

Parambassis serrata, 51.9 mm SL, holotype, Kolo River, Kaladan Drainage, Mizoram.

Parambassis serrata Dishma & Vishwanath, 2015: 584 (type locality: Kolo River, Lawngtlai District, Kaladan drainage, Mizoram, India).

Diagnosis. A faint vertically elongated brown humeral blotch; posterior margin of most spines of first dorsal fin proximally black in color; absence of predorsal scales; 51−56 scales in lateral series; 14−16 dorsal- and anal-fin branched rays; a black longitudinal stripe on ventral surface of caudal peduncle; 17−18 gill rakers on first gill arch; four to nine large serrae on preopercular ridge; and 11 pectoral-fin rays. Maximum size ∼60 mm SL.

Distribution. Kolo River, Kawlchaw, Lawngtlai District, Kaladan drainage, Mizoram, India.

Parambassis waikhomi Geetakumari & Basudha, 2012
(Fig. 4.472)

Parambassis waikhomi Geetakumari & Basudha, 2012: 3328 (Loktak Lake, Chindwin basin, Manipur State, India).

Diagnosis. 58−60 lateral line scales; 9−10 pectoral fin rays; 19−20 gill rakers; two predorsal bones; presence of a vertically elongated humeral spot; maxilla reaches to one-third of the orbit; 8.2−10.9 interorbital width; four preorbital ridge; 11 preorbital edge; six supraorbital ridge; 18 serrae at lower edge of preoperculum; 24 serrae at hind margin of preoperculum. Maximum size ∼40 mm SL.

Distribution. Loktak Lake, Chindwin basin, Manipur, India.

FIGURE 4.472

Parambassis waikhomi, 33.2 mm SL, Loktak Lake, Manipur.

Family Sciaenidae
Drums, croakers

Dorsal fin long, a notch separating spinous and soft parts. Lateral line scale extending to end of caudal fin. Caudal fin emarginated to rounded. Bony upper edge of opercle forked, bony flap present above gill opening. Large lateral line canals on head. Pores on snout and lower jaw. Vomer and palatine toothless. Swim bladder with many branches. Otolith exceptionally large (Fig. 4.473).

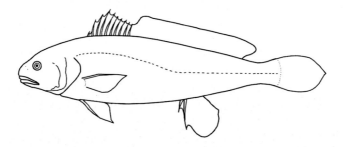

FIGURE 4.473

Outline diagram of Sciaenidae.

Genus *Johnius* Bloch, 1793

Johnius Bloch, 1793: 132 (type species: *Johnius carutta* Bloch, 1793). Gender: masculine.

Diagnosis. Body elongated; snout prominent, swollen superiorly; mouth inferior; upper jaw longer, lower jaw with villiform teeth, palate edentate; preopercle serrated at its angles, two weak opercular spines; two dorsal fins continuous; scales ctenoid on body, cycloid on opercle; swim bladder hammer headed.

Johnius coitor (Hamilton, 1822)
(Fig. 4.474)

FIGURE 4.474

Johnius coitor, 126.0 mm SL, Barak River, Silchar.

Bola coitor Hamilton, 1822: 75, 368, pl. 27, fig. 24 (type locality: Ganges River).

Diagnosis. Body elongated with prominent and projecting snout; mouth inferior lower gill rakers 11; ctenoid scales on body and head; cycloid on snout and below and immediately behind eyes and on anterior part of breast; ctenoid on top of head and body; teeth villiform; gill rakers 11 on lower arm of first arch; dorsal fin deeply notched, dorsal spines moderately weak; second anal spine strong; caudal fin acutely rhomboid; lateral line scales 51; body color silvery brown on dorsal and white on abdomen, spinous dorsal fin with a dusky edge; soft dorsal fin, anal and caudal fins with a dull gray border. Maximum size ~300 mm SL.

Distribution. East coast of India, Ganga and Brahmaputra drainages; Bangladesh, Myanmar, to the east coast of Australia.

Family Nandidae
Asian leaffishes

Mouth large, highly protractile; spiny and soft dorsal continuous; caudal fin rounded; lateral line in two rows, interrupted in between or absent; pelvic fin origin in anterior part of abdomen, usually with axillary scaly process; lachrymal and preopercular margin serrated, tongue (basihyal) toothed; body coloration as that dead leaves (Fig. 4.475).

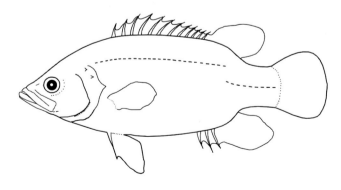

FIGURE 4.475

Outline diagram of Nandidae.

Genus *Nandus* Valenciennes, 1831

Nandus Valenciennes, in Cuvier & Valenciennes, 1831: 481 (type locality: *Nandus marmoratus* Valenciennes in Cuvier and Valenciennes, 1831). Gender: masculine.

Diagnosis. Body deep and compressed; snout pointed; eyes large; mouth terminal and protractile; its cleft wide, extending below posterior margin of eye; lower jaw longer; jaws with villiform teeth, teeth in palate and tongue; preopercle serrated; dorsal fin contiguous; posterior soft part elevated than anterior spines; pectoral spine weak; anal spine three; moderately strong; second spine largest; caudal fin rounded, scales ctenoid, lateral line interrupted.

Nandus andrewi Ng & Zaafar, 2008

Nandus andrewi Ng & Zaafar, 2008: 25, figs. 1, 2a, 3 (type locality: Ichamati River drainage, vicinity of Duttapulia, Nadia District, West Bengal, India).

Diagnosis. Uniform pale bluish-white body in life; slender body (24.3−29.1% SL); short pelvic fin (16.2%−18.9% SL); a dark spot on the caudal peduncle; lateral line scales 45−52, posterior edge of preopercle weakly serrated in smaller specimens 105 mm SL and below. Maximum size ∼125 mm SL.

Distribution. Ganga River drainage, West Bengal, India.

Nandus nandus (Hamilton, 1822)
(Fig. 4.476)

FIGURE 4.476

Nandus nandus, 105.0 mm SL, Barak River, Silchar, Assam.

Coius nandus Hamilton, 1822: 96, 370 (type locality: ponds of Gangetic provinces).

Diagnosis. Dorsal fin with 14 spines and 11 soft rays; operculum triangular, with a single prominent spine; preopercle strongly serrate in a continuous band near its angle; lateral line incomplete; mouth very large and highly protrusible; lower jaw projecting, maxilla reaching beyond hind edge of orbit; body color grayish brown; vertically three broad patchy blotches, a dusky blotch on caudal-fin base; some narrow dark bands radiate from eye; fins grayish; yellowish narrow bands of spots across soft portions of dorsal, anal, and caudal fins. Maximum size ∼150 mm SL.

Distribution. Ganga, Brahmaputra, Barak−Surma−Meghna drainage; Nepal and Pakistan.

Family Badidae
Chameleon fishes

Mouth relatively small and only slightly protractile; subocular shelf absent; spiny and soft dorsa fins continuous, spines 6 or 7, and soft rays 6−10; anal fin with three spines and six to eight soft rays; lateral-line scales 23−33; body with bright colorations that can change color very rapidly (Fig. 4.477).

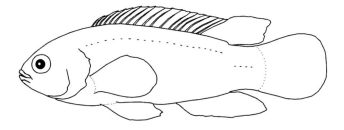

FIGURE 4.477

Outline diagram of Badidae.

Genus *Badis* Bleeker, 1853

Badis Bleeker, 1853: 106 (type species: *Labrus badis* Hamilton, 1822). Gender: masculine.
Badis: Kullander & Britz, 2002: 303 (redescription).

Diagnosis. Body compressed, moderately elongated; mouth terminal, upturned, slightly protractile; maxilla not extending behind anterior margin of orbit; postero-dorsal corner of opercle with sharp spine; soft and spinous dorsal contagious; lateral line pores on tip of horizontal tubes; hypobranchial three toothed; males with short dorsal fin lappets and short pelvic fin, the later not reaching beyond first anal spine; caudal fin rounded; middle of dorsal fin with a black stripe; dark bars on trunk modified into two narrow vertical lines in adults; dark pigment on caudal-fin base differentiated into three vertically aligned blotches.

Badis assamensis **Ahl, 1937**
(Fig. 4.478)

FIGURE 4.478

Badis assamensis, 44.7 mm SL, Brahmaputra River, Dibrugarh, Assam.

Badis badis assamensis Ahl, 1937: 118 (type locality: Brahmaputra River drainage, Assam, India).

Diagnosis. A dark blotch postero-dorsally on the opercle and two rows of irregular blackish blotches along the side; eyes large; mouth small, slightly upturned; lips thin; lower jaw longer than upper, teeth villiform on jaws; teeth on vomer, palatine, parasphenoid; tooth plate count 7−9; dorsal-fin origin opposite to pectoral fin, nearer the tip of snout than the base of caudal fin; caudal fin rounded; scales moderate, ctenoid, lateral line interrupted, lateral line row of scales 28−29, lateral transverse scales 12−13, predorsal scales 15, circumpeduncular scales 25; total vertebrae 29−30, caudal fin rays 14. Maximum size ∼68 mm SL.

Distribution. Brahmaputra Drainage, Assam, Meghalaya.

Badis badis (Hamilton, 1822)
(Fig. 4.479)

FIGURE 4.479

Badis badis, 36.7 mm SL, Hasila Beel, Brahmaputra drainage, Goalpara District, Assam.

Labrus badis Hamilton, 1822: 70, 368 (type locality: Ganges provinces).

Badis badis: Kullander et al., 2019: 313 (revision).

Diagnosis. A dark blotch on the exposed part of the cleithrum, absence of a dark blotch on the opercle, and sides of body with irregular narrow dark vertical bars formed by dark spots on scale bases; interorbital space 6.5%−8.3% SL; scales in a lateral row 25−27; dorsal-fin spines 16−17, young specimens and females with the prominent pattern of narrow bars, large males may be black, but with white scale centers arranged in rows or irregularly; anterior part of dorsal fin with no dark blotch, may have a black stripe along the middle of the dorsal fin. Maximum size ∼65 mm SL.

Distribution. Ganga, Brahmaputra, Padma, Barak−Surma−Meghna drainage.

Badis blosyrus Kullander & Britz, 2002
(Fig. 4.480)

Badis blosyrus Kullander and Britz, 2002: 339 (type locality: Janali River, Raimana, Brahmaputra River drainage, Kokrajhar District, Assam, India.)

Diagnosis. Body moderately elongated, relatively low and slightly compressed, abdomen rounded, snout moderately pointed; eyes large; mouth small, slightly upturned, lower jaw longer than upper, lips thin, teeth villiform on jaws, tooth plate count 10−13; dorsal spines slender; dorsal

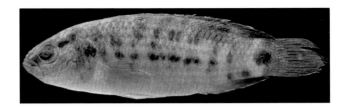

FIGURE 4.480

Badis blosyrus, 36.8 mm SL, Teju River, Teju District, Arunachal Pradesh.

fin origin opposite to pectoral fin, nearer the tip of snout than the base of caudal fin; caudal fin rounded; scales moderate, ctenoid; lateral line interrupted, lateral line row of scales 27—28, predorsal scales 13, lateral transverse scales 13, vertebral count 28, pectoral fin rays 13, caudal fin rays 15, circumpeduncular scales 24, four to five rows of cheek scales, branchiostegal rays six, anal fin with three spiny and seven soft rays; a round or blackish brown spot postero-dorsally on opercle, brownish bar across caudal peduncle, and sides with nine bars. Maximum size ∼55 mm SL.

Distribution. Brahmaputra Drainage, Assam and Arunachal Pradesh, India.

Badis chittagongis Kullander & Britz, 2002
(Fig. 4.481)

FIGURE 4.481

Badis chittagongis, 38.5 mm SL, Kaladan River, Kawlchaw, Mizoram.

Badis chittagongis Kullander & Britz, 2002: 314 (type locality: stream in Lama town, Matamohuri River drainage, Bandarban Hill tracts, Chittagong Division, Bangladesh).

Badis chittagongis: Kullander et al., 2019: 315 (revision).

Diagnosis. A conspicuous dark brown or black blotch on superficial part of cleithrum above pectoral-fin base; no blotch on dorsolateral aspect of caudal peduncle; interorbital width 5.5%—6.7% SL; lateral line row of scales 27—28; 13 pectoral fin rays; 20 circumpeduncular scales; 28 vertebrae; a series of prominent dark blotches along the middle of dorsal fin presence of a distal extrascapular. Maximum size ∼40 mm SL.

Distribution. Karnafuli drainage, Mizoram; hill streams near Chittagong, small streams on Maheshkhali Island and in small coastal streams draining to the Bay of Bengal from near Cox's Bazar south to Teknaf Game Reserve, Bangladesh.

Badis dibruensis Geetakumari & Vishwanath, 2010
(Fig. 4.482)

FIGURE 4.482

Badis dibruensis, (A) 39.3 mm SL, holotype, male; (B) 37.3 mm SL, paratype, female, both from Dibru River, Dibrugarh, Assam.

Badis dibruensis Geetakumari & Vishwanath, 2010: 645 (type locality: Dibru River, Dibru, Brahmaputra drainage, Assam, India).

Diagnosis. Two predorsal bones; small oval black blotch in the mid-base of caudal fin; interorbital width 9.9−15.0 upper jaw length 6.1%−6.9% SL and lower jaw length 7.1%−8.3% SL; eye diameter 7.6%−9.4% SL; vertebrae 27; absence of dark black or brown vertical bars on sides; preorbital stripe dark gray, continued across chin; postorbital stripe blackish, formed by a single blotch close to orbit; dark pigment also on one scale posterior to that blotch; no supraorbital stripe; a conspicuous black blotch covering the superficial part of the cleithrum above pectoral-fin base. Maximum size ~40 mm SL.

Distribution. Presently known from Dibru River at Dibrugarh, Assam, Brahmaputra drainage.

Badis ferrarisi **Kullander & Britz, 2002**
(Fig. 4.483)

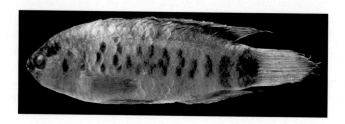

FIGURE 4.483

Badis ferrarisi, 32.0 mm SL, Yu River, Tamu, Myanmar.

Badis ferrarisi Kullander & Britz, 2002: 332 (type locality: Chindwin River Drainage, Kalaymyo market).

Diagnosis. A distinct blotch on a superficial portion of cleithrum; a moderately elongated body, head short, snout rounded, predorsal contour curved, mouth large, slightly projecting, short opercular spine, vomer, parasphenoid and palatine are toothed dentary pore two, supraorbital pore two, predorsal scales three to four anterior to coronalis pore, lateral line interrupted forming upper and lower lateral lines; transverse scales row 1½ above, seven below upper lateral line; circumpeduncular scales rows seven above, eight below lateral line totaling 17; sides with 11 narrow dark brown to blackish vertical bars, blotch on opercle dorsally, dark brown blotch on cleithrum, dorsal and anal fins dark. Maximum size ∼40 mm SL.

Distribution. Chindwin basin in Manipur, India and Myanmar.

Badis kanabos **Kullander & Britz, 2002**
(Fig. 4.484)

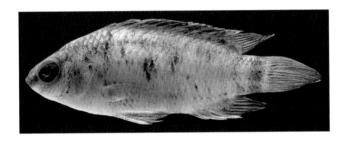

FIGURE 4.484

Badis kanabos, 48.7 mm SL, Barak River, Manipur.

Badis kanabos Kullander & Britz, 2002. 316 (type locality: Raimana in western Assam, India).

Diagnosis. Distinct narrow bars on side, a conspicuous black blotch between third and fourth dorsal spine, presence of distal extrascapular, circumpeduncular scale rows 7; body elongated, moderately compressed laterally; lower jaw only slightly projecting before upper jaw; dorsal fin with 16 spines and nine soft rays; pectoral fin with 12 rays; anal with seven rays; lateral row scales 26, predorsal scales four to five anterior to coronalis pore; pale brownish to yellowish with brown to black markings, preorbital stripe dark gray, continued across chin, postorbital stripe indistinct, dark blotches on opercle, male dark with dark fins, female yellowish ground color and soft fins hyaline and caudal-fin base with a small rounded brownish blotch at the middle. Maximum size ∼50 mm SL.

Distribution. Barak−Surma−Meghna and Brahmaputra drainages, Northeastern India.

Badis rhabdotus **Kullander, Norén, Rahman & Mollah, 2019**
Badis rhabdotus Kullander, Norén, Rahman & Mollah, 2019: 322 (type locality: Piyain River at Jaflong, Meghna River drainage, Sylhet District, Bangladesh).

Diagnosis. Presence of dark on the exposed cleithrum; absence of dark blotch dorsolaterally on the caudal peduncle; absence of short black bars along the middle of the side, absence of a dark blotch on the gill cover; absence of a black botch anteriorly on the dorsal fin; narrow curved bars on the sides of the body; vertebrae 28–29; lateral row of scales 27–29; dorsal fin rays 16–17, 7½–8½; interorbital space 4.8%–5.5% SL; body depth 25.7%–32.0% SL. Maximum size ~50 mm SL.

Distribution. Piyain River at Jaflong and headwaters of Karnafuli River and Meghna River basin including the lower Barak River in Bangladesh; tributaries of the Meghna in the Meghalaya, India.

Badis singenensis **Geetakumari & Kadu, 2011**
(Fig. 4.485)

FIGURE 4.485

Badis singenensis, 38.2 mm SL, Singen River, Arunachal Pradesh.

Badis singenensis Geetakumari & Kadu, 2011: 2086 (type locality: Singen River, Brahmaputra drainage, Saku-Kadu Village, East Siang District, Arunachal Pradesh, India).

Badis triocellus Khariam and Sen, 2011: 65 (Subansiri River, Lower Subansiri District, stream in Lohit District, Arunachal Pradesh; Dipali River, Dhemaji District, Assam India; all from Brahmaputra drainage).

Diagnosis. A black blotch posterodorsally on the opercle, at the base of opercle spine, round and usually covering portion of several scales; three distinct dark blotches on the body: two at dorsal-fin base: first: behind the third spine, second, behind the fifth and sixth soft dorsal fin ray, and the third at the base of the anal fin behind the fifth soft anal fin ray; scales in lateral row 25–26; interorbital width 9.2%–13.3% SL. Maximum size ~40 mm SL.

Distribution. Brahmaputra drainage in East Siang District, Lower Subansiri District, and Lohit District, Arunachal Pradesh, Dipali River, Dhemaji District, Assam, India.

Badis tuivaiei **Vishwanath & Shanta, 2004 Vishwanath and Shanta, 2004**
(Fig. 4.486)

FIGURE 4.486

Badis tuvaiei, 53.5 mm SL, 40.8 mm SL, Tuivai River, Manipur.

Badis tuivaiei Vishwanath & Shanta, 2004b: 1619 (type locality: Tuivai River, Barak−Surma−Meghna drainage, Manipur, India).

Diagnosis. A conspicuous black blotch covering the superficial part of the cleithrum above pectoral-fin base, a black blotch between third and fourth dorsal spine, a mid-basal rounded black spot on caudal fin; 20 circumpeduncular scales; total vertebra 30−31; tooth plate six to eight; jaws almost equal anteriorly, lower jaw slightly projecting, maxilla reaching to one-third of orbit; opercular spine slender with a simple sharp tip; palatine, vomer and parasphenoid toothed. Maximum size ∼60 mm SL.

Distribution. India; Tuivai and Irang River, tributaries of the Barak, Manipur.

Genus *Dario* Kullander & Britz, 2002

Dario Kullander & Britz, 2002: 354 (type species: *Labrus dario* Hamilton, 1822). Gender: masculine.

Diagnosis. Absence of a postcranial lateral line; dorsal fin with 13−15 spines; truncated caudal fin; absence of lateral line pores from the dentary, infraorbital series and anguloarticular; absence of teeth from the palatine, basihyal and basibranchial 3; and fewer scales 24 vs 26 or more) along the middle of the side.

Dario dario (Hamilton, 1822)

Labrus dario Hamilton, 1822: 72 (type locality: ponds and rivers of northern parts of Bengal and Behar).

Diagnosis. Absence of palatine dentition, 8½ scales in transverse row, 13−14 dorsal-fin spines, presence of pre-, post-, and supraorbital stripes; absence of anguloarticular lateralis canal; prominent bars crossing sides in male, absence of infraorbital bones, absence of black spot anteriorly in dorsal fin; anal fin rays modally six. Maximum size ∼350 mm SL.

Distribution. Jarnali River at Raimana, western Assam, Brahmaputra river drainage, India−Bhutan border.

Dario kajal Britz & Kullander, 2013

Dario kajal Britz & Kullander, 2013: 332 (type locality: Seinphoh stream at Umolong, Jaintia Hills, Meghalaya, India).

Dario kajal: Kullander et al., 2019: 310 (revision).

Diagnosis. A small species, not exceeding 20 mm SL, with a combination of characters: presence of a postorbital stripe that continues behind eye in line with preorbital stripe; presence of a series of double bars restricted to the upper half of the body in males; absence of a caudal-peduncle blotch and absence of a horizontal suborbital stripe; dorsal fin with 15 simple and 5½ branched rays; dorsal-fin lappets in males extending beyond the spine tip; vertebrae number 24−26; scales in lateral row 24; transverse scale count eight; absence of palatine teeth: presence in males of a black spot anteriorly in the dorsal fin; absence of an angulo-articular lateral-line canal; in life, males are pale red or rosy, with irregularly red vertical stripes on the dorsal sides; in preservative: pale beige, abdominal sides pale maroon; short dorsal bars are formed from dark scale bases, and a dark and a dark brown short postorbital stripe runs posterior from the orbit; females are pale gray in life and in preservative. Vertebrae 13 + 12 = 25; scales in lateral row 24. Maximum size ∼250 mm SL.

Distribution. Barak−Surma−Meghna drainages in Jaintia hills, Meghalaya, India and Sylhet, Bangladesh.

Family Cichlidae

Cichlids

Body deep and compressed; nostril single on each side. Inferior pharyngeal bones triangular with median longitudinal suture; dorsal fin spinous and soft part; lateral line interrupted; no subopercular shelf; caudal fin truncated or rounded (Fig. 4.487).

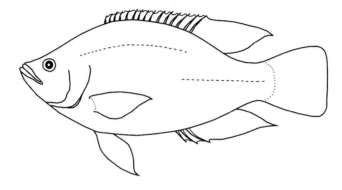

FIGURE 4.487

Outline diagram of Cichlidae.

Genus *Oreochromis* Günther, 1889

Oreochromis Günther, 1889: 70 (type species: *Oreochromis hunteri* Günther, 1889). Gender: masculine.

Diagnosis. Body deep, moderately elongated and compressed; mouth terminal, large, cleft extending to below anterior border of orbit; no subopercular shelf inferior pharyngeal bones triangular with median longitudinal suture; jaws with three to five rows of teeth; spinous part of dorsal fin longer than soft part which may be produced into a filament; lateral line interrupted; caudal fin truncated or rounded.

Oreochromis mossambica (Peters, 1852)
(Fig. 4.488)

FIGURE 4.488

Oreochromis mossambica, 112.5 mm SL, Loktak Lake, Imphal Market, Manipur.

Chromis (Tilapia) mossambicus Peters, 1852: 681 (type locality: Mozambique).

Diagnosis. Body elongated, fairly deep, and compressed; upper profile of body more convex than lower; mouth large; maxillary ending between nostril and eye in females and immature males, but below anterior edge of eye in breeding males; teeth three to five series on jaws; longest soft dorsal ray extending to above proximal part of caudal fin in females and immature males, but breeding males to half or three-quarter length of caudal fin; caudal fin truncated, often with rounded corners; scales cycloid, 30−32 in lateral line series; in life, females and nonbreeding males watery-gray to yellowish, with three or four dark blotches often apparent along flanks; body of males in breeding season deep black; lower part of head chalky or pale grayish-white; upper lip bluish; dorsal fin black with a red margin; pectoral fins translucent red; caudal fin with a broad red margin. Maximum size ∼350 mm SL.

Distribution. East Africa; an introduced species in India, Pakistan, Sri Lanka, etc.

Oreochromis niloticus **(Linnaeus, 1758)**
(Fig. 4.489)

FIGURE 4.489

Oreochromis niloticus, 96.5 mm SL, Fish farm, Manipur Valley, Manipur.

Perca nilotica Linnaeus, 1758: 290 (type locality: Nile River).

Diagnosis. Body deep, small head; jaws not greatly enlarged; spiny dorsal with 15−18 spines, soft with 11−13 rays; anal fin with three spines and 9−11 soft rays; body color brownish or graying overall, characteristic in having regular broad bars and caudal fin with narrower bars; bars wider in smaller fishes; males become reddish in color during breeding season, especially the fins; Maximum length 60 cm, males reaching larger sizes. Maximum size ∼450 mm SL.

Distribution. Native of coastal rivers in Israel, Nile basin, widely introduced for aquaculture.

Family Gobiidae
Gobies

Body oblong; pelvic fins united to form an adhesive disk; spinous and soft dorsal separate; scales cycloid or ctenoid; lateral line absent on body; caudal fin rounded (Fig. 4.490).

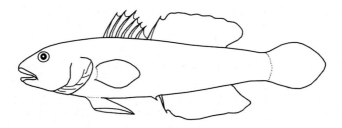

FIGURE 4.490

Outline diagram of Gobiidae.

Genus *Awaous* Valenciennes, 1837

Awaous Valenciennes in Cuvier and Valenciennes, 1837: 97 (type locality: *Gobius ocellaris* Broussonet, 1782). Gender: masculine.

Diagnosis. Gobiid genera with a bilobed tongue supported by a Y-shaped glossohyal bone, its diverged anterior limbs greatly expanded; a supra-occipital spine, 25−26 vertebrae; first and second anal pterygiophores origin anterior to the first hemal spine; mouth subterminal with very prominent upper lip; body with small ctenoid scales, interdorsal region and belly cycloid. Longitudinal scale series 62−63, and transverse, 17−18; scale series from origin of first dorsal fin to upper pectoral fin origin 14−15; predorsal scales 26−27; pelvic fins completely united with a pair of spine, pelvic frenum folded forward and a fleshy lobe present around each spine; pectoral fin with 16 branched rays, anal fins with one simple and 11 branched rays and caudal fin rounded with 15 branched rays. The genus is also characteristic in having paired anterior interorbital cephalic canal and transversely arranged suborbital sensory papillae; three median interorbital sensory pores between anterior and posterior cephalic canals; sizes of the pores are intermediate between cephalic canal and sensory papillae.

Awaous grammepomus (Bleeker, 1849)
(Fig. 4.491)

Gobius grammepomus Bleeker, 1849: 34 [type locality: Purworejo (Purworejo), Bogowonto River, Java, Indonesia].

Diagnosis. Body cylindrical anteriorly and compressed posteriorly; eyes large, dorsolateral; snout pointed, mouth subterminal, upper lip very prominent; Both jaws with three to four rows of conical teeth, and outer rows enlarged; tongue margin bilobed; anterior nostril in short tube and posterior nostril rounded with very shallow rim; cheek fleshy; gill opening restricted, extending ventrally near vertical midline of opercle; vertebrae 25−26; first dorsal fin rays six, spinous; second dorsal fin rays i, 12−13; anal fin rays i, 11; pectoral fin rays 16; origin of anal fin inserted below origin of second branched ray of second dorsal fin; the rear tips of both second dorsal fin and anal fin rays when depressed do not reach the procurrent rays of caudal fin; pectoral fin large and oblong; pelvic fin disk rounded spinous rays with pointed membranous lobe; caudal fin rounded; body with small ctenoid scales, anterior predorsal region and belly cycloid; longitudinal scale series 62−63; transverse scale series 17−18; scales between pectoral and first dorsal fin origin 14−15. Head and prepelvic region naked. Interdorsal region and belly with cycloid scales. Maximum size ∼150 mm SL.

FIGURE 4.491

Awaous grammepomus, 116.5 mm SL, Kaladan River Kolchow, Mizoram: (A) lateral view; (B) ventral view showing united pelvic fins.

Distribution. Coastal areas of Sri Lanka to New Guinea, freshwater visiting, reaches inland waters of India and adjoining countries.

Genus *Glossogobius* Gill, 1859

Glossogobius Gill, 1859: 46 (type species: *Gobius platycephalus* Richardson). Gender: masculine.

Diagnosis. Body elongated, rounded anteriorly and compressed posteriorly; head depressed; cheek and operculum naked; mouth small, oblique, not extending to anterior margin of orbit; eyes large, superior, in the middle of head; lips thick, jaws with villiform teeth in several rows; tongue bilobate; gill openings extend in advance of orbit below; pelvic fins oblong united midventrally (Fig. 4.492B); scales ctenoid on body, cycloid on head.

Glossogobius giuris (Hamilton, 1822)
(Fig. 4.492)

FIGURE 4.492

Glossogobius giuris, 96.7 mm SL, Loktak Lake, Manipur.

Gobius giuris Hamilton, 1822: 51 (type locality: Gangetic provinces).

Diagnosis. Body translucent yellowish gray with four to six blotches along lateral line; eye placed on top of head; dorsal fin with nine rays; pelvic fins united at base to form disk; scales ctenoid on body, cycloid on head; mouth small, oblique, not extending to anterior margin of orbit; dorsal fins separated: anterior dorsal with vi rays, its base length 15.7%−18.9% SL, posterior one with I, 9 rays, its base length 17.5%−21.0% SL; caudal fin rounded. In life, yellowish brown with fine dark blotches on sides, side of head with irregular dark spots. Maximum size ∼300 mm SL.

Distribution. Freshwater visiting species, widely distributed in the fresh and brackish waters from East Africa to Indo-Aurstralian archipelago.

Family Anabantidae
Climbing perches

Body oblong, compressed. Dorsal-fin base longer than anal-fin base; operculum, suboperculum and interopercular bordered by long, radiating spines. Jaws, prevomer and parasphenoid with fixed conical teeth; lateral line interrupted below dorsal fin (Fig. 4.493).

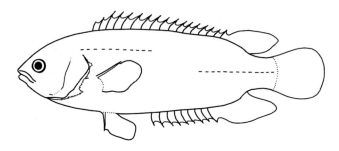

FIGURE 4.493

Outline diagram of Anabantidae.

Genus *Anabas* Cloquet, 1816

Anabas Cloquet, 1816: 35 (type species: *Perca scandens* Daldoff, 1797). Gender: masculine.

Diagnosis. Body compressed, oblong, abdomen rounded; mouth small, terminal and oblique; upper portion of the first epibranchial arch accommodate a labyrinthine suprabranchial organ for accessory aerial respiration; operculum, suboperculum and interoperculum bordered by long radiating spines; jaws, prevomer and parasphenoid with fixed conical teeth dorsal-fin base longer than anal-fin base; dorsal fin origin above pectoral fin and with 17−18 spines and 8−10 rays; anal fin with 8−10 spines and 9−11 rays; caudal fin rounded; scales ctenoid; lateral line in two rows, the anterior row upper and the posterior row, the lower.

Anabas testudineus (Bloch, 1792)
(Fig. 4.494)

FIGURE 4.494

Anabas testudineus, 42.0 mm SL, wetland of Manipur valley, Chindwin basin.

Anthias testudineus Bloch, 1792: 121 (type locality: Java).

Diagnosis. Body depth 28.6%−33.0% SL; snout length 5.1%−8.1% SL; head length 35.2%−37.8% SL; mouth terminal, fairly large; teeth villiform on jaws; scales large, lateral line incomplete, 21−29 in lateral line series; dorsal fin origin opposite pectoral fin origin, with 16−17 spinous rays and eight to nine soft rays; anal fin with 10−11 spinous and 9−10 soft rays, caudal fin with 15−16 rays. Maximum size ∼170 mm SL.

Distribution. India; Pakistan; Sri Lanka; Bangladesh; Myanmar; Malay Archipelago; Singapore; Philippines.

Family Osphronemidae
Gouramies

Body short and compressed; head and body covered with ctenoid scales; a supra-branchial cavity present; first ray of pelvic fin modified into filiform ray; dorsal-fin base shorter than anal-fin base; prevomer and palatine toothless; upper jaw toothless; lateral line vestigial or absent; caudal fin, emarginated, truncated, or rounded (Fig. 4.495).

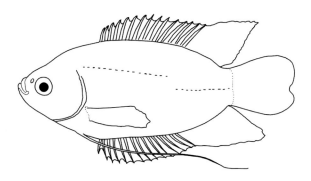

FIGURE 4.495

Outline diagram of Osphronemidae.

Genus *Trichogaster* Bloch & Schneider, 1801

Trichogaster Bloch & Schneider, 1801: 164 (type species: *Trichogaster fasciata* Bloch and Schneider, 1801). Gender: feminine.

Diagnosis. Bracnhiostegals five; branchial arches with toothed tubercles; opercle entire, preopercle usually serrated; jaws little protractile; vomer and palatine edentulous; lateral line interrupted if present. Body oval shaped and strongly compressed, mouth small, slightly protrussible.

Trichogaster fasciata Bloch & Schneider, 1801
(Fig. 4.496)

FIGURE 4.496

Trichogaster fasciata, 65.0 mm SL, wetlands near Agartala, Tripura, Meghna basin.

Trichogaster fasciatus Bloch & Schneider, 1801: 164 (type locality: Tranquebar, India).

Diagnosis. Body oval shaped and strongly compressed, its depth 37.0%−44.4% SL; mouth small, slightly protrusible; serrations on preorbital ranges from 5−13, number varies from left to right. lips not papillated but highly protrusible; lateral line interrupted, first one 15 scales from behind opercle, second one after two nonperforated scales of the first one; scales between mid-dorsal line to lateral line 5−6; predorsal scales 7−8; dorsal and anal fins long based, the soft portion in some is rounded, in others pointed; dorsal fin with xv−xvii, 9−14 rays; pectoral fin with i−ii, seven or nine rays, pelvic fin with a modified filamentous ray; anal fin with xv−xvii, 14−19 rays, contiguous with caudal fin, caudal fin truncated; body with 11−13 dark blue bars descending obliquely downward and backward; anal fin with a red margin, dorsal and caudal spotted with orange, a green spot on distal part of opercle. Maximum size ∼90 mm SL.

Distribution. West Bengal, northeastern India; Bangladesh; Nepal; Pakistan.

Trichogaster labiosa Day, 1877
(Fig. 4.497)

Trichogaster labiosus Day, 1877: 374 (type locality: Rangoon, Myanmar).

Diagnosis. Lips thick and papillated with villi-like projections; tip of soft dorsal and anal fins produced reaching both dorsal and ventral base of caudal fin, respectively; lateral line interrupted, first behind opercle to 12th−14th scale, second after one to four nonperforated scales; scales from mid-dorsal line to lateral line 5 or 6, predorsal scales 8−9; dorsal fin origin at vertical level of anal origin, bearing xv−xviii, 9−11 rays, its third to fifth rays elongated extending beyond vertical of tip of soft anal fin; pectoral fin with iv, 7−9 rays; pelvic fin modified into a single filamentous ray, extending upto posterior extremity of

FIGURE 4.497

Trichogaster labiosa, 52.0 mm SL, ponds at Mayang Imphal, Chindwin basin, Manipur.

caudal fin; anal fin with xv−xviii, 15−18 rays, contiguous with caudal fin; caudal fin slightly emarginated or wedge shaped, bearing 15 rays; body with 8−10 oblique orange brown bars; greenish, lighter ventrally; fins dark, outer edge of anal fin yellowish red. Maximum size ∼80 mm SL.

Distribution. Chindwin basin, Manipur; Southern Myanmar.

Trichogaster lalius **(Hamilton, 1822)**
(Fig. 4.498)

FIGURE 4.498

Trichogaster lalius, (A) male, 42.0 mm SL; (B) female, 38.5.0 mm SL; both from wetlands near Agartala, Meghna basin.

Trichopodus lalius Hamilton, 1822: 120, 372 (type locality: Gangetic provinces, India).

Diagnosis. Body oval, strongly compressed, mouth small, strongly protrussible, lips normal, preorbital denticulate, lower limb of opercle serrated, subopercle entire; head blunt and unarmed; eyes moderate size; lateral line absent; scales in longitudinal series 27–28; caudal fin fan-shaped; pectoral fins low and each with nine rays; ventral fin extending to end of anal fin with 19 spines and 11 rays; caudal fin with 16 rays; body with oblique scarlet and light blue, half of each scale being of either color; dorsal and caudal fins barred in scarlet dots; anal with a dark band along its base, and a red outer edge. Maximum size ~80 mm SL.

Distribution. Ganga, Brahmaputra, Surma-Meghna basins, India and Bangladesh.

***Trichogaster sota* (Hamilton, 1822)**
(Fig. 4.499)

FIGURE 4.499

Trichogaster sota, 24.3 mm SL, Jiri River, Barak drainage, Manipur.

Trichopodus sota Hamilton, 1822: 120, 372 (type locality: Gangetic provinces, India).

Diagnosis. Size small, body oblong and compressed, mouth small, upturned and highly protrusible; a broad black stripe on side; pelvic fin filamentous, extending up to posterior extremity of anal fin; caudal fin slightly emarginated; predorsal scales 7–8; body color dull greenish, lighter along abdomen from the eye on the side toward lower half of the base of the caudal fin, a black stripe consisting of black dots and shining with golden gloss; caudal fin with a black spot at its base. Maximum size ~50 mm SL.

Distribution. Ganga and Brahmaputra drainages.

Family Channidae
Snakeheads

Body elongated, cylindrical anteriorly and slightly compressed posteriorly; body covered with scales, head with plate like scales; lower jaw longer; teeth on jaws, vomer and palate; lateral line complete; dorsal and anal fins long; no fin spines; caudal fin rounded; suprabranchial organ for air breathing (Fig. 4.500).

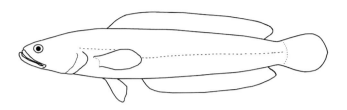

FIGURE 4.500

Outline diagram of Channidae.

Genus *Channa* Scopoli, 1777

Channa Scopoli, 1777: 459 (type species: *Channa orientalis* Bloch in Bloch and Schneider, 1801). Gender: feminine.

Diagnosis. Body elongated, cylindrical anteriorly; head depressed, large, covered with plate like scales; mouth opening wide; lower jaw longer; teeth on jaws and palate; suprabranchial chamber with bony lamellae as accessory respiratory organ; dorsal and anal fins long, without spines, both free from caudal fin; cadual fin rounded.

Channa amari Dey, Roy Choudhury, Nur, Sarkar, Kosygin & Barat, 2019
(Fig. 4.501)

FIGURE 4.501

Channa amari, 111.5 mm SL, stream in Bhalka forest, Brahmaputra basin, West Bengal.

Photo courtesy: A. Dey.

Channa amari Dey, Roy Choudhury, Nur, Sarkar, Kosygin & Barat, 2019: 230, figs. 1 and 2 (type locality: Bhalka forest, Alipurduar District, West Bengal, India).

Diagnosis. Pelvic fin absent; dorsal fin with 37, pectoral fin with 12−13 and anal fin with 23−25 rays; lateral line scales 44−45, vertebrae 40−41, transverse scale rows; body color creamy with brown to brownish-red marmorations; dorsal, pectoral, and caudal fin edges orange; anal fin edge dark blue and whitish tinge in the external extremity, dorsal and caudal fins with bluish stripes; pectoral fin with four to five blue black bars, outermost bar in the form of elongated spots on the rays, not on the membranes in between; caudal fin with four to five round orange markings. Maximum size ∼ 120 mm SL.

Distribution. Small streams or water bodies in Dooars area of Alipurduar District, Brahmaputra basin, West Bengal, India.

Channa amphibeus (McClelland, 1845)
(Fig. 4.502)

FIGURE 4.502

Channa amphibeus, 184.6 mm SL, northern Bengal, India.

Ophiocephalus amphibeus McClelland, 1845: 275 (type locality: vicinity of Chail River, one of the tributaries of the Teesta at the foot of the Bouton mountains).

Diagnosis. A species of *Channa* with lateral line scales 81, predorsal scales 17, circumpeduncular scales 31, cephalic sensory pores single; two large scales on each side of lower jaw; gular part of head without patch of scales; mouth large, maxilla extending far beyond posterior margin of eye. Maximum size ~200 mm SL.

Distribution. Chel River basin, Brahmaputra River drainage, northeastern India and Bhutan.

Channa andrao Britz, 2013

Channa andrao Britz, 2013: 288 (type locality: Lefraguri swamp, Jalpaiguri District, West Bengal, India).

Diagnosis. Absence of pelvic fins; total vertebrae 41−43; lateral-line scales 42−43; scales in lateral row 3½/5½; preanal scales 21−22; pectoral fin rays 14−15; principal caudal fin rays 12−13; dorsal fin length 64.6%−68.8% SL; anal-fin base 40.8%−44.5% SL; pectoral-fin base with two to three bars; absence of whitish cream (orange in life) blotches on the caudal fin; mouth brooder reproductive behavior. Maximum size ~250 mm SL.

Distribution. Lefraguri swamp, Jalpaiguri District, West Bengal, India.

Channa aurantimaculata Musikasinthorn, 2000
(Fig. 4.503)

FIGURE 4.503

Channa aurantimaculata, 125.0 mm SL, Dibrugarh, Assam.

Channa aurantimaculata, Musikasinthorn, 2000: 27 (type locality: streams near Dibrugarh town, Dibrugarh, Assam).

Channa pomanensis Gurumayum &Tamang, 2016: 177, figs. 1−5 (description from Poma River at Poma, about 10 km west to capital town, Itanagar, Papum Pare District, Arunachal Pradesh).

Diagnosis. A species of *Channa* with lateral line scales 51−54, dorsal fin rays 45−47, anal fin rays 28−30; two large cycloid scales on each side of the lower jaw undersurface; patch of scale on gular part of head absent; upper half of body dark brown to black with seven to eight large irregular orange blotches; pectoral fins with a black blotch at base and five vertical broad vivid black bars. Maximum size ∼400 mm SL.

Distribution. Brahmaputra drainage in Arunachal Pradesh and Dibrugarh District, Assam.

*Channa aurantipectoralis***Lalhlimpuia, Lalrounga & Lalramliana, 2016**
(Fig. 4.504)

FIGURE 4.504

Channa aurantipectoralis, 89.3 mm SL, Keisalam River, Karnafuli drainage, Mizoram.

Photo courtesy: Lalramliana.

Channa aurantipectoralis Lalhlimpuia, Lalrounga & Lalramliana,2016: 344 (type locality: Keisalam River, tributary of Karnafuli River, in the vicinity of Phuldungsei, Mamit District, Mizoram, India).

Diagnosis. The species is characteristic in having uniformly bright colored pectoral fins with no spots, stripes, or bars, a dark brown V-shaped mark on the head, its apex over occiput and the two bases behind the eyes, absence of scales on the gular region, presence of a large scale on each side of the ventral surface of the lower jaw, 51−64 lateral line scales, 34−37 dorsal fin rays, 23−25 anal fin rays, 13−14 pectoral fin rays, 5½−6½/1/7½−8½ transverse scale rows. Maximum size ∼170 mm SL.

Distribution. Seling and Keisalam Rivers, small tributaries of Karnafuli River of Dampa Tiger Reserve, Karnafuli drainage, Mizoram, India.

Channa aurolineata **(Day, 1870)**
(Fig. 4.505)
Ophiocephalus aurolineatus Day, 1870b: 99 (type locality: Moulmein, Myanmar).
Channa aurolineata: Adamson & Britz, 2018: 545 (treated valid).

Diagnosis. Head depressed; lateral line complete with 65−71 scales; dorsal-fin rays 55−58; black ocellus in the upper half of caudal fin, indistinct in juveniles, becoming distinct in adults; juveniles with an orange stripe extending from the snout to the upper half of caudal fin through the eye and the side of the body above the lateral line; eye diameter much larger in juveniles; color of the scales forming the dark lateral blotches in larger specimens (about 200 mm SL and more) with conspicuous white posterior margins. Maximum size ∼800 mm SL.

Distribution. Yu River drainage, Indo-Myanmar border, Manipur, India; Chindwin and Irrawaddy drainages, Myanmar and Salween drainage, Myanmar and Thailand.

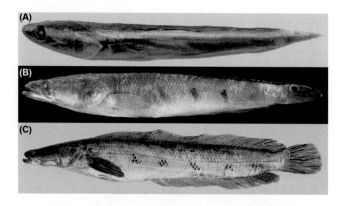

FIGURE 4.505

Channa aurolineata, (A) 70 mm SL, (B) 185 mm SL, both from Lokchao River, Moreh, Yu River drainage, Indo-Myanmar border, Manipur; (C) 800 mm SL, Irrawaddy River, Myanmar.

Photo courtesy: (C) M. Kottelat.

Channa barca (Hamilton, 1822)
(Fig. 4.506)

FIGURE 4.506

Channa barca, 295.0 mm SL, swamps at Morigaon, Assam.

Channa barca Hamilton, 1822: 67, 367 (type locality: Brahmaputra R. near Goyalpara, Assam, India).

Diagnosis. A species of *Channa* with lateral line 62−63, lives in vertical burrows; two large cycloid scales on each side of the lower jaw undersurface; the scales are roughened in lines, forming irregular arches and springing from the center of the base or middle of each scale; those on the anterior-posterior of the body are rather smaller than those in its posterior portion; dorsal and flanks covered with numerous black spots, pectorals reddish with numerous black spots. Maximum size ∼900 mm SL.

Distribution. Brahmaputra basin, northeast India, Bangladesh.

Channa bleheri Vierke, 1991
(Fig. 4.507)

FIGURE 4.507

Channa bleheri, 149.1 mm SL, Dikrong River, Doimukh, Arunachal Pradesh.

Channa bleheri Vierke, 1991: 22 (type locality: Assam).

Diagnosis. A species of *Channa* with a slightly broad head, pelvic fin absent; sides of lower jaw have one large cycloid scale, 4−11 red or orange markings on caudal fin, eyes moderate, and caudal fin rounded, mouth large, six to seven markings on the pectoral fin; branchial tooth plate count five; tooth plates are present only on the outer side of the first gill arch and absent on the inner side of the same arch; gray to brown on sides, pale yellow to white ventrally, pectoral fin with seven to eight bars of alternating black and white bars. Maximum size ∼170 mm SL.

Distribution. Brahmaputra Basin, Arunachal Pradesh and Assam.

Channa gachua (Hamilton, 1822)
(Fig. 4.508)

FIGURE 4.508

Channa gachua, 112.8 mm SL, Nambul River, Singda, Manipur.

Ophicephalus gachua Hamilton, 1822: 68, 367 (type locality: Ponds and ditches of Bengal).

Diagnosis. A species of *Channa* with dorsal fin rays 32−37; lateral line scales 39−48, pelvic fin shorter than half the pectoral fin length, pectoral fin rays 15−17, anal fin rays 21−27, and caudal fin rays 12, maxilla and premaxillary process extending to vertical level of the posterior end of the orbit, one or two large cycloid scale on each side of lower jaw undersurface, cephalic sensory pores single, total vertebral count 43, branchial tooth plate count nine; lateral line interrupted, running in two adjacent rows; anal and caudal fin margins white; often a large ocellus with a light edge on the last five dorsal rays in the young; body black getting lighter ventrally and abdomen creamish. Maximum size ∼200 mm SL.

Distribution. Widely distributed in South Asia: Afghanistan, Iran, Pakistan, India (West Bengal, Tamil Nadu, Andhra Pradesh, Maharashtra, Goalpara, Visakhapatnam), Nepal, Bangladesh, Sri Lanka, and Myanmar.

Channa lipor **Raveenraj, Uma, Moulitharan & Singh, 2019**

Channa lipor Raveenraj, Uma, Moulitharan and Singh, 2019: 62 (type locality: Umraling River, Umraling village, Ri-Bhoi District, Meghalaya, India).

Diagnosis. Small *Channa* species, adult size not exceeding more than 120.0 mm SL; unique coloration of orange to bronze-brown body with six brown oblique bars in the flank, dorsal fin orange with a series of irregular, dark-brown blotches in the membrane gradually become intense toward basal margin; possessing an orange to yellow pectoral fin with five faint semicircular brown bands alternating only on rays; 11−12 black to brown spots or blotches running parallel along the dorsal-fin membrane base; caudal fin with brown zigzag bars; 29−32 dorsal-fin rays; 20 anal-fin rays; lateral-line scales 35−40, scales extending from the shoulder girdle in a horizontal row dropping one scale row at scale 10 or 11; transverse scale row 3/1/6; absence of an ocellus in the posterior-most part of the dorsal fin in juveniles; presence of spots or blotches horizontally along the dorsal fin and zigzag bars on the dorsal-, anal-, and caudal-fin membranes; and fewer vertebrae 40. Maximum size ∼110 mm SL.

Distribution. Umraling River at Umraling village, Rhi Bhoi District, Meghalaya, India.

Channa marulius **(Hamilton, 1822)**
(Fig. 4.509)

FIGURE 4.509

Channa marulius, 176.0 mm SL, Barak River, Vanchengphai, Tamenglong District; Manipur.

Ophicephalus marulius Hamilton, 1822: 65, 367 (type locality: India).

Channa marulius: Adamson & Britz, 2018: 542 (diagnostic characters and distribution).

Diagnosis. A species of *Channa* with a large black ocellus on upper caudal-fin base, indistinct in juveniles, becoming conspicuous in adults; three white spots on body; presence of a sharp distinct pointed ridge of isthmus and anterior to it many cephalic sensory pores more than one; total vertebrae count 62; lateral line complete with 62−65 scales; dorsal-fin rays 52−56; juveniles with a pale white stripe on the upper half of the side of the body commencing from behind the eye to caudal-fin base; eye diameter larger in juveniles; adults with black blotched on the sides, the scales forming the blotches with white spots in the middle. Maximum size ∼1000 mm SL.

Distribution. Ganga, Brahmaputra, Krishna, Cauvery drainages and those of northeastern India and drainages in the eastern slip of the Rakhine Yoma, Myanmar.

Channa melanostigma **Geetakumari & Vishwanath, 2011**
(Fig. 4.510)

FIGURE 4.510

Channa melanostigma, 112.5 mm SL, paratype, Brahmaputra Drainage, Tezu, Arunachal Pradesh.

Channa melanostigma Geetakumari & Vishwanath, 2011: 231 (type locality: Lohit River, Tezu, Brahmaputra drainage, Assam, India).

Diagnosis. A species with a distinct 14−15 black zig-zag transverse bars at irregular intervals (when stretched); total vertebrae 50−51; tooth plate count seven; last dorsal fin inserted in between 41st and 43rd vertebrae; cephalic sensory pore single, without satellite openings. Maximum size ∼150 mm SL.

Distribution. Lohit River at Tezu, Lohit District, Brahmaputra Drainage, Arunachal Pradesh.

Channa pardalis **Knight, 2016**
(Fig. 4.511)

FIGURE 4.511

Channa pardalis, ∼ 100 mm SL, Stream in Nongstoin, West Khasi Hills, Meghalaya.

Photo courtesy: M. Beta.

Channa pardalis Knight, 2016: 8584 (type locality: streams in Nongstoin, West Khasi Hills, Meghalaya, India).

Diagnosis. A unique color pattern consisting of numerous large black spots on the postorbital region of the head, opercle and body; a broad white and black margin to the dorsal fin, anal fin and caudal fin; 36–37 dorsal fin rays; 24–25 anal fin rays; 44–45 pored scales on the body and two scales on the caudal-fin base; 4½ scales above lateral line and 6½ scales below lateral line; 45 vertebrae and the palatine with two rows of teeth: outer row with numerous minute teeth and inner row with short, stout inward curved teeth. Maximum size ~180 mm SL.

Distribution. Khasi Hills, Meghalaya, India.

Channa punctata (Bloch, 1793)
(Fig. 4.512)

FIGURE 4.512

Channa punctata, 106.0 mm SL, Nambul River, Singda, Manipur.

Ophicephalus punctatus Bloch, 1793: 139 (type locality: unknown).

Diagnosis. A species of *Channa* with scales on cheek 4–6, pelvic fin longer than half pectoral fin length, pectoral fin with no bars, lateral line scales 35–40, dorsal fin rays 28–32, lower jaw with three to six canines behind a single row of villiform teeth, anal fin rays 19–21, body with two rows of bars, maxilla and premaxillary processes extending to vertical level of beyond the middle of orbit, sides of lower jaw with one large cycloid scale, cephalic sensory pores single, total vertebral count 35, branchial tooth plate count eight. Maximum size ~170 mm SL.

Distribution. India; Afghanistan to Myanmar through Sri Lanka; Pakistan; Nepal; Bangladesh and Yunnan, China.

Channa stewartii (Playfair, 1867)
(Fig. 4.513)

FIGURE 4.513

Channa stewartii, 260.0 mm SL, Guijan; Assam.

Ophiocephalus stewartii Playfair, 1867: 14 (type locality: Cachar, Assam).

Diagnosis. A species of *Channa* with small black spots scattered on sides of body, dorsal fin rays 37–41, and anal fin rays 24–27, lateral line scales 45–53, eyes moderate, mouth large, pelvic fin about one-third as long as pectoral fin, four to five bars on the pectoral fin, head slightly broad with blunt snout, sides of lower jaw with two large cycloid scale, cephalic sensory pores single, 44 total vertebrae and seven branchial tooth plate. Maximum size ∼250 mm SL.

Distribution. Swamps of the Eastern Himalayan region.

Channa stiktos **Lalramliana, Knight, Lalhlimpuia & Singh, 2018**
(Fig. 4.514)

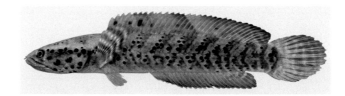

FIGURE 4.514

Channa stiktos, ∼90.0 mm SL, Tiau River, Mizoram.

Photo courtesy: Lalramliana.

Channa stiktos Lalramliana, Knight, Lalhlimpuia & Singh, 2018a: 167 (type locality: Tianu River in the vicinity of Zokhawthar Village, Kaladan River drainage, Mizoram, India).

Diagnosis. *Channa stiktos* is distinguished in having 34−36 dorsal fin rays; 47−49 pored scales on the lateral line; presence of pelvic fin; unique color pattern consisting of numerous large, well defined black spots on the head: both dorsal and ventral, and on the body: well distributed above and below the lateral line; absence of dark spot on the anal fin of juveniles; striped pectoral fins: four to seven alternating dark brown and white bars; absence of zig-zag dark brown-black cross bars on the caudal fin. Maximum size ∼190 mm SL.

Distribution. Kaladan River Drainage, Mizoram, India.

Channa striata **(Bloch, 1793)**
(Fig. 4.515)

FIGURE 4.515

Channa striata, 328 mm, Naharlagun Bazar, Papum Pare District, Arunachal Pradesh.

Ophicephalus striatus Bloch, 1793: 141(type locality: Tranquebar, India).

Diagnosis. A species of *Channa* with dorsal fin rays 42−45, anal fin rays 25−29, lateral line scales 55−65, mouth large, and lower jaw four to seven canines behind a single row of villiform teeth, dorsal and anal fins slightly darker in color than body maxilla and premaxillary process extending to vertical level of beyond posterior margin of orbit, presence of a sharp pointed ridge at the mid-ventral part of isthmus and anterior to it many longitudinal striae are present, cephalic single pores not single, total vertebral count 54, branchial tooth plate count 13. Maximum size ~1200 mm SL.

Distribution. India; Pakistan in the west through China, Thailand, Malaysia, and Indonesia.

Channa torsaensis Dey, Nur, Raychowdhury, Sarkar, Kosygin Singh & Barat, 2018
(Fig. 4.516)

FIGURE 4.516

Channa torsaensis, 211.00 mm SL, Torsa River basin, Alipurduar, West Bengal.

Photo courtesy: A. Dey.

Channa torsaensis Dey, Nur, Raychowdhury, Sarkar, Kosygin Singh & Barat. 2018: 498 (type locality: Barjhar forest, Alipurduar district, West Bengal, India).

Diagnosis. Body with five to six greyish blue bars from dorsum to the middle of body, bars oblique, descending anteriad; dark brown marmorations on the head and ventral half of body; all fins with characteristic broad dark blue border, whitish edges in dorsal, anal, and caudal fins. Pectoral fin with three to four dark brown bars, bars contiguous with the dark border distally; lateral line scales 46, dorsal-fin rays 36–38; anal fin rays 22–25. Maximum size ~250 mm SL.

Distribution. Tributaries of Torsa River, Dakshin Barajhar forest, Alipurduar District, West Bengal, India.

Family Tetraodontidae
Puffers, globefishes

Body short, rounded in cross section, inflatable; skin naked or with prickles; fine conical teeth in jaws; ribs and epineurals absent; pelvic fin absent; caudal fin emarginated to rounded; lateral line absent (Fig. 4.517).

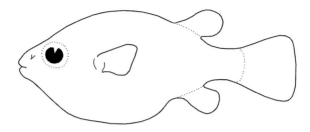

FIGURE 4.517

Outline diagram of Tetraodontidae.

Genus *Leiodon* Swainson, 1839

Leiodon Swainson, 1839: 194 (type species: *Tetraodon marmoratus* Swainson, 1839). Gender: masculine.

Diagnosis. Body short, head oval; snout blunt; mouth small, terminal and transverse; lips fleshy; eyes large, superior in the posterior half of head; dorsal fin short, without a spine, its origin slightly ahead of anal fin origin; pelvic fin absent; body with fine dermal spines; esophagus inflatable.

Leiodon cutcutia (Hamilton, 1822)
(Fig. 4.518)

FIGURE 4.518

Leiodon cutcutia, 72.0 mm SL, Yu River, Tamu, Myanmar.

Tetraodon cutcutia Hamilton, 1822: 362 (type locality: Ganga River and its branches).

Diagnosis. Body short, oval, tapering behind dorsal and anal fin origins; snout slightly protruding with terminal mouth; body naked; lateral line absent; gill opening as a slit anterior to pectoral fin origin; caudal fin rounded; dorsal part of body brownish, pale white ventrally; a black ocellus laterally midway between pectoral and dorsal fin origins. Maximum size ~120 mm SL.

Distribution. Widely distributed in India, Bangladesh, Sri Lanka, and Myanmar.

Miscellaneous Notes

Oromandibular structures of *Bangana*

Bangana bears a snout smooth that is blunt, pendulous, and inferior. The rostral fold is smooth and separated from the upper lip by a deep grove and is disconnected from lower lip around the corner of the mouth. The lateral lobe of upper lip is smooth or slightly papillose and is laterally connected with the lower lip; lower lip anteriorly separated from lower jaw by a transverse groove extending along length of entire lower jaw, with a free anterior margin containing numerous papillae on dorsal surface. Lower jaw is heavily cornified, with a sharp cutting edge. Postlabial groove is interrupted and present on the side of the lower jaw (Fig. 5.1).

Bangana dero

The rostral fold of *Bangana dero* is thick, smooth, and slightly fimbriae at the tip. It is separated from upper lip by a deep groove, disconnected from the lower lip around the corner of the mouth; upper lip is smooth with its lateral extremities highly papillose, connected with lower lip at the corner of the mouth; lower lip separated from lower jaw around the corner of the mouth; lower jaw cornified, with sharp cutting edge; lower lip equally thick with papillae arranged regularly; postlabial groove interrupted medially forming a deep groove on each side of the lower lip separating the lower lip from the mental margin laterally.

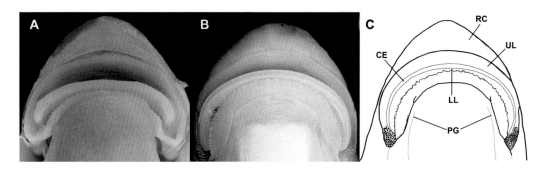

FIGURE 5.1

Oromandibular structure of *Bangana:* (A) *B. dero*; (B) *B. devdevi*; (C) Illustration indicating CE, cornified edge of lower lip; LL, lower lip; PG, postlabial groove; RC, rostral cap; UL, upper lip.

Freshwater Fishes of the Eastern Himalayas. DOI: https://doi.org/10.1016/B978-0-12-823391-7.00003-5
© 2021 Elsevier Inc. All rights reserved.

Bangana devdevi

The rostral fold of *Bangana devdevi* is thick, smooth, pendulous, and well developed, separated from upper lip by a deep groove, disconnected from lower lip around the corner of the mouth. The upper lip is smooth with its lateral extremities slightly papillose, connected with lower lip laterally at the corner of the mouth. The lower lip is separated from lower jaw anteriorly by a transverse groove extending along the length of the entire jaw; lower jaw cornified, with sharp cutting edge. The lower lip is thick or slightly thinner with medium-sized papillae arranged regularly. Postlabial groove is interrupted medially forming a deep groove on each side of the lower lip separating the lower lip from the mental margin laterally.

Labeo bata has a lip structure characteristic of genus *Labeo*. This is not considered under *Bangana*.

Barilius versus *Opsarius*

It is felt that there is no clear-cut differentiation between the genus *Barilius* and *Opsarius*. The characters are overlapping. The genus *Barilius* was established as a subdivision of *Barilius* under the division *Cyprinus* Hamilton (1822). Later, Bleeker (1863a) designated *Barilius barila* as the

Table 5.1 Extent of mouth cleft and lower jaw in relation to tip of upper jaw.

	Mouth cleft	Lower jaw tip vs upper jaw tip
B. barila	Extends up to one-fourth of the length of orbit	Lower jaw tip slightly extends beyond the tip of upper jaw
B. bendelisis	Extends up to one-third of the length of orbit	Lower jaw never extends beyond the tip of upper jaw
B. profundus	Extends up to one-third of the length of orbit	Lower jaw tip slightly extends beyond the tip of upper jaw
B. sachra	Nearly touches the anterior-most margin of the orbit	Lower jaw tip never extends beyond the tip of upper jaw
B. vagra	Extends beyond the middle of the orbit	Lower jaw tip slightly extends beyond the tip of upper jaw
O. barna	Extends slightly the anterior-most margin of the orbit	Lower jaw tip slightly extends beyond the tip of upper jaw
O. barnoides	Ends before the anterior-most margin of the orbit	Lower jaw tip slightly extends beyond the tip of upper jaw
O. dogarsinghi	Nearly touches the anterior-most margin of the orbit	Lower jaw tip never extends beyond the tip of upper jaw
O. lairokensis	Extends up to one-fourth of the length of the orbit	Lower jaw tip slightly extends beyond the tip of upper jaw
O. tileo	Extends up to one-fifth of the length of the orbit	Lower jaw tip extends beyond the tip of upper jaw

Table 5.2 Comparison of some selected osteological characters.

Characters	B. barila	B. bendelisis	B. profundus	B. sachra	B. vagra	O. barna	O. lairokensis	O. barnoides	O. dogarsinghi	O. tileo
Dorsal fin origin between vertebrae	17 – 18	15–16	14–15	15–16	17–18	13–14	16–17	13–14	17–18	14–15
Anal fin origin between vertebrae	23–24	25–26	18–19	21–22	22–23	20–21	23–24	21–22	21–22	21–22
Total vertebra	23 + 16 = 39	25 + 17 = 42	18 + 15 = 33	21 + 18 = 39	22 + 17 = 39	20 + 18 = 38	23 + 19 = 42	21 + 16 = 37	21 + 17 = 38	21 + 20 = 41

type species of the genus. Representatives of the genus *Barilius* are diagnosed by their relatively elongate, compressed body, rounded belly, presence of bars on the flank, 9−17 total anal-fin rays, and lateral line in the lower half of the body (Hamilton, 1822; Talwar and Jhingran, 1991a, 1991b; Rainboth, 1996a, 1996b). The genus can also be separated from the barred *Devario* by the absence of a "danionin notch" (Fig. 4.38) and lack of a pigmented, wide P-stripe (Fang, 2001). Males of many species are colorful and have tubercles on various parts of the body (Fowler, 1934b; Talwar and Jhingran, 1991a). Generally, *Barilius* species are found in small, clean, clear mountain streams, but some species live in large rivers as well (Smith, 1945; Talwar and Jhingran, 1991a, 1991b; Tejavej, 2010). Barbels may be present or absent. The length and thickness of the bars vary in different species. From these observations, it is clear that the genus *Barilius* displays variations in external morphology and anatomical features.

The genus *Opsarius* was erected by McClelland (1838) with the following diagnostic characters: body long and slender, considerably compressed; mouth large, that is, its cleft extends about the middle of the orbit; presence of a prominent symphyseal knob on the tip of lower jaw which fits into a notch present on upper jaw; ventral profile more prominent than dorsal profile; sides of the body marked with either bars or spots; dorsal fin short and placed posterior to middle of the body, slightly opposite to a long anal fin. However, McClelland (1838) did not designate a type species. Jordan (1919a, 1919b) designated *O. maculatus* McClelland as type species and Kottelat (2013) considers the species a synonym of *O. tileo* (Hamilton).

Howes (1980) opined that *Barilius* is paraphyletic and attempted the phylogenetic study and grouped the genus into two groups: *B. barila* group and *B. gatensis* group. His study (Howes, 1983) contradicted some of his groupings with his previous publication. He did not examine the types of *Barilius* and *Opsarius* and thus the classification is confusing.

Tang et al. (2010) studied the phylogenetic relationships of barilines in their collection based on molecular informations. However, the type species of *Barilius*, that is, *B. barila* and of *Opsarius*, that is, *Opsarius tileo* were not included in their study. Thus the phylogenetic problem of *Barilius* and *Opsarius* is incomplete and remains unresolved.

The present study of 10 barilines from northeast India including, *B. barila* and *O. tileo* in respect of the extent of mouth cleft and lower jaw in relation to upper jaw, danionin notch and mandibular symphyseal knob, insertion of dorsal and anal fins and vertebrae did not show any significant difference between the two nominal genera. The results are shown in the tables below (Tables 5.1 and 5.2).

Danionin notch and mandibular symphyseal knob

Insignificant group	*B. barila, O. dogarsinghi, B. profundus, B. sachra*
Intermediate group	*B. bendelisis, O. lairokensis*
Well-developed group	*O. barna, O. barnoides, O. tileo, B. vagra*

The differences are considered as interspecific variations. Thus species presently put under genus *Opsarius* by some authors have been presently recognized under *Barilius* Hamilton pending further detailed studies.

Devario horai

Devario horai Barman (1984) is probably a synonym of *Danio rerio*. The species is described to have 28–30 scales in lateral series, with no lateral line pores; no barbel, and no stripe on body. Darshan et al. (2019a) in their book on the fishes of Arunachal Pradesh did not represent the species. Our collections from the Namdapha River are represented by only *D. rerio*.

Devario aequipinnatus

The intensity of the color stripes on the body of the species is variable. Specimens from Tuivai River, Manipur (Barak drainage) has faint color stripes (Fig. 4.46C). McClelland's (1839) drawing of the species reproduced here in Fig. 4.46D also does not show the stripes.

Snout morphology of *Garra*

Nebeshwar and Vishwanath (2017) divided *Garra* into five species groups based on snout morphology, namely, (1) a smooth snout species group; (2) a transverse lobe species group; (3) rostral flap species group; (4) rostral lobe species group, and (5) proboscis species group. A modified division is given here with the species group with rostral flap and rostral lobe incorporated into the smooth snout species group. Thus the revised division is species group with (1) snout with smooth dorsal surface; (2) snout with transverse lobe, and (3) snout with proboscis (Fig. 5.2).

Snout with smooth dorsal surface group

Dorsal surfaces of snout in this group are usually rounded or flattened. In some species, there may be a transverse depression anterior to nostrils and a slightly humped dorsal surface. Snout may

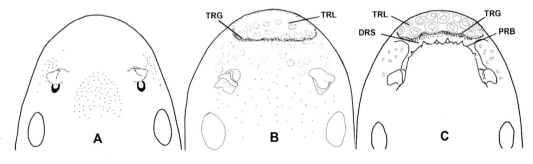

FIGURE 5.2

Diagram showing snout of *Garra*: (A) smooth surface, (B) with transverse lobe, and (C) with proboscis; *TRG*, Transverse groove; *TRL*, transverse lobe; *DRS*, depressed rostral surface; *PRB*, proboscis.

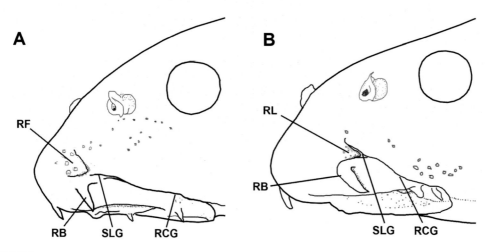

FIGURE 5.3

Figure showing: (A) Rostral flap and (B) Rostral lobe: *RF*, Rostral flap; *RB*, rostral barbel; *SLG*, sublachrymal groove; *RCG*, rostral cap groove; *RL*, rostral lobe

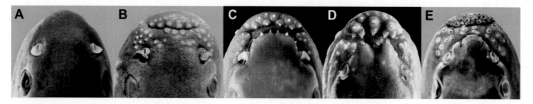

FIGURE 5.4

Snouts of Garra with: (A) smooth dorsal surface; (B) transverse lobe; (CE) proboscis: (C) unilobed quadrate, (D) bilobed, (E) trilobed.

have sensory pores or minute to medium-sized tubercles (Fig. 5.3). A sublachrymal groove originates from the base of the rostral barbel and connects to the groove of the rostral cap posteriorly. This group of *Garra* may have a rostral flap or a rostral lobe.

Rostral flap (*sensu* Zhang and Kottelat, 2006) is a triangular or broadly rounded paired fleshy structure at the tip of the snout. It is movable forward along a vertical axis and is located in a depression on the lateral side of the snout, close to the tip. The flap has a number of small conical tubercles along its lower edge (Fig. 5.4A).

The rostral lobe (*sensu* Zhang and Kottelat, 2006) is an elongated conical structure, with a conspicuous conical tubercle at its tip and a series of conical and starry tubercles along its lower edge This elongated triangular rostral lobe is narrowly pointed and not elevated from the surface of the snout. It is located at the end of the deep widened sublachrymal groove that passes along the base of the rostral barbel and surrounds it anteriorly. The posterior end of the rostral lobe is usually enclosed in the sublachrymal groove (Fig. 5.4B).

Snout with transverse lobe group

The transverse lobe is an elongated fleshy fold posteriorly demarcated transversely by a shallow or deep transverse groove over the anterodorsal surface just behind the tip of the snout. The lobe is covered with small- to large-sized tubercles. The surface of the snout behind the transverse groove may be smooth or tuberculate. The tuberculate snout surface behind the transverse groove may be swollen or slightly elevated upward and outlined from the lateral surface by a horizontal crease. The groove may extend laterally to make the transverse lobe a prominent knob-like protuberance, while in some, the lobe is indicated by a transversely elongated patch of small- to medium-sized tubercles over the anterodorsal surface of the snout just behind the snout tip.

Snout with proboscis group

The proboscis is a fleshy structure protruding anterodorsally over the dorsal surface of the snout in front of the nostrils. The degree of protrusion of the proboscis varies from a slight elevation from the snout to the level of the snout tip. The lower surface of the proboscis in highly protruded ones is not in contact with the depressed rostral surface. The depressed rostral surface is the surface of the snout between the transverse lobe and the proboscis. The surface of depressed rostral can be flat, bulgy or pleated, tuberculate and depressed from the level of the proboscis. The snout with proboscis group has a transverse lobe and a depressed rostral surface.

The shapes of the proboscis may be unilobed, bilobed, or trilobed. The shapes are variable among species and thus proboscis-shapes are good taxonomic characters. The anterior margin of the proboscis is usually delineated from the depressed rostral surface by a crease, shallow, or deep groove. Generally, the anterior margin of the proboscis is tuberculated with small- to large-sized tubercles. The tubercles may be present on the lateral margin, anterodorsal region, and ventral surface of the proboscis. The lateral surface of the snout, anteroventrally to the nostril on each side, is usually covered with a patch of small- to large-sized tubercles and elevated as a lobe. The sublachrymal groove is shallow, usually curves ventrally, and originates from the base of the rostral barbel, and posteriorly continuous or discontinuous to the rostral cap groove.

Garra nasuta

McClelland (1838) described *Platycara nasuta* (now *Garra*) from Kasya mountains, Assam (now Khasi Hills, Meghalaya). The type species does not exist. Neither McClelland's (1838) text nor the accompanying illustration provides sufficient information to diagnose the species. He, however, mentioned the presence of a pit between the nares.

While redescribing the species, Menon (1964) used specimens from Assam; Sittang drainage in southern Shan State, Myanmar and from parts of China. He described the species to have a snout with a prominent, trilobed proboscis, the lateral lobes being small and are in front of the nostrils and the tip of snout marked off into a transverse lobe; the free extremity of the proboscis, the transverse lobe and the lateral sides of the head in front of nostrils covered with several horny tubercles.

A stenotopic *Garra* species is not supposed to be widely distributed in different drainages of the northeastern India, Myanmar, and China. McClelland (1838) have not mentioned the trilobed proboscis in his description. Thus the identity of *G. nasuta* is not clear.

Poropuntius shanensis

Roberts (1998) reported the species to have 33 scales on the lateral series. He examined specimens from the Inle Lake, Myanmar that is a part of the Salween basin. Thus the specimens he examined may be some other species.

Semiplotus

Bănărescu and Herzig-Straschil (1995) considered *Semiplotus* to be different from *Cyprinion* Heckel in the absence of barbels, in having a higher number of branched dorsal-fin rays (20−25 vs 12−15) in the species of *C. macrostomus* group, (20−25 vs 9−11) in those of the *C. watsonimicrophthalmus* group (Fig. 4.161c) and in having five versus seven branched anal-fin rays (Fig. 4.161d1). My examination shows that all three species of *Semiplotus* in the Eastern Himalaya have a small pair of maxillary barbels. Genus *Cyprinion* is distributed from the western Syria and the south of the Arabian Peninsula to the western boundary of the Indus River in Punjab of Pakistan. Only three species of *Semiplotus* are known: *S. semiplotus* in the Ganga and Brahmaputra, *S. modestus* in the Kaladan, and *S. cirrhosus* in the Chindwin drainage. Our observation of the three *Semiplotus* species shows that there is a small pair of maxillary barbel and 6½−8½ branched anal-fin rays.

Some authors still use the species name as *Cyprinion semiplotum* (probably following Eschmeyer's Catalog of Fishes). The catalog was accessed on October 6, 2020, and it still maintains the name as valid. This note clarifies that the species name is *Semiplotus semiplotus*.

Tor yingjiangensis Chen and Yang, 2004

Chen and Yang (2004) diagnosed the species with a head length considerably longer than the depth of body. However, the drawing of the species in their Fig. 1.1A shows a head length that is similar to the body depth.

Type locality of *Lepidocephalichthys irrorata*

Type locality of *Lepidocephalichthys irrorata* is lakes and streams of Manipur valley, Chindwin basin (Hora, 1921a: 197). The type specimen, F 9904/1 he mentioned is from Manipur Valley but did not mention any paratype. He ended up the description of the species stating "*L. irrorata* is widely distributed in the lakes and streams of the Manipur Valley." Thus Havird and Page's (2010: 152) description of the species based on the specimens from the Brahmaputra basin and also referring to those specimens as paratypes are questionable.

Aborichthys cataracta and *A. verticauda*

Darshan et al. (2019a) treated *Aborichthys cataracta* Arunachalam et al (2014) and *A. verticauda* Arunachalam et al (2014) as junior synonyms of *A. boutanensis*. My observation of *Aborichthys* species from Arunachal Pradesh also has similar opinion. Thus I agree with Darshan et al.'s (2019a) decision. Moreover, the availability of the two names following the International Code of Zoological Nomenclature, art 8.5.9 is not clear.

Aborichthys kempi

Thoni and Hart (2015) treated *Aborichthys kempi* Chaudhuri as a synonym of *A. boutanensis*. *Aborichthys kempi* is characteristic in having no bars on the caudal peduncle, but *A. boutanensis* has bars on the caudal peduncle. Maurice Kottelat (pers. comm.) is of the similar opinion. Species of *Aborichthys* are stenotopic and have small distribution ranges; the occurrence of *A. kempi* in Bhutan is not likely. My examination of specimens (Fig. 4.224B) has the characteristic color pattern as figured by Chaudhuri (1913) which is reproduced here as Fig. 4.224C. Thus *A. kempi* is treated here as valid. The caudal fin coloration in the species of two different lengths is also seen. In smaller specimens (Fig. 4.224A) the distal margin of caudal fin is plain while the larger specimen (Fig. 4.224B) has a dark distal marginal band. Chaudhuri's (1913) figure reproduced as in Fig 4.224C is of specimen having more than 78 mm SL and has a similar caudal fin coloration pattern.

Schistura chindwinica

Tilak and Husain (1990), based on a single specimen, described *Schistura chindwinica* in which the type locality was stated as Manipur, but the exact location was not given. Probably they thought that all the rivers in Manipur belong to the Chindwin drainage and thus the name. The collector (MG Sharma, based on his collection, the species was described) had specimens having a similar description which was collected from a stream draining into the Irang River, a tributary of the Barak River at Lankha, Tamenglong District, Manipur. The species was represented in subsequent collections from the Barak drainage in Manipur. Thus species name associated with Chindwin is erroneous.

Paracanthocobitis marmorata

The coordinates of the type locality of the species as presented by Singer *et al* (2017) is very close to Kangchup Hills, Manipur, from where Hora (1921a) described *Schistura kangjupkhulensis*. The locality is drained by feeder streams of the Nambol River, Manipur, which is a part of the Chindwin drainage.

Blyth (1860) described *Cobitis zonalternans* (now *Acanthocobitis*) from Tenasserim province, Myanmar. Kottelat (1990) designated neotype of the species from the Tak Province, Thailand. Singer et al. (2017) restricts the distribution of the species in the Salween drainage of Myanmar and Thailand. *Nemacheilus zonalternans* as described by Hora (1921a) from Manipur valley, Chindwin drainage, most probably refers to *Acanthocobitis marmorata*. Hora's (1921a) drawing of *N. zonalternans* is reproduced here as Fig. 4.244C.

Rhyacoschistura manipurensis

Chaudhuri (1912) described the species to have a uniform steel gray on the upper two-thirds and dorsal and caudal fins with wavy bars. Hora and Mukerji (1935) found two forms: mottles or plain forms inhabiting ponds and tanks or other stationary or sluggish waters and banded forms inhabiting rapids with pebbly beds. However, they maintained that the bars on the dorsal and caudal fins were present in both.

Glyptothorax burmanicus and G. cavia

Premaxillae of some species of *Glyptothorax* (Tilak, 1963; de Pinna, 1996; Diogo et al, 2002); *Bagarius* (Gauba, 1962) and of *Glyptosternum* (Gauba, 1969) have proximal elements connected laterally to distal elements (Fig. 5.5). Fragmentation of distal element of *Euchiloglanis kishinouvei* into tripartite or multipartite structures have also been reported (de Pinna, 1996). In most of the species *Glyptothorax* of northeast India except *G. burmanica* and *G. cavia* (Vishwanath et al., 2010) premaxillae have a pair of medially located proximal elements and another pair of laterally located distal elements. Gauba (1966) recorded the premaxilla of *G. cavia* as being generally segmented or fused to form an enormously broad band that extends a considerable distance posteriorly across the palate. Vishwanath et al. (2010) found *G. burmanica* and *G. cavia* to have a pair of proximal elements, a pair of fragmented distal elements and in addition, a posterior element consisting of numerous tooth plates tightly attached by connective tissue (Fig. 5.5).

In most of the species of *Glyptothorax*, Weberian laminae extend along the margin of the fifth vertebrae (de Pinna, 1996). The lateral margins of the anterior and posterior portions of Weberian laminae are almost equal in many of the species examined, while the lateral margin of the posterior portion of the lamina in *G. burmanica* and *G. cavia* extend farther laterally, almost reaching the distal tip of the parapophysis of the fifth vertebrae (Vishwanath et al, 2010).

Rama and Chandramara

Hamilton's (1822: 162) *Pimelodus chandramara* from north Bengal measuring about 38 mm in length is probably a younger specimen of *Rama*. Our examination of specimens from the

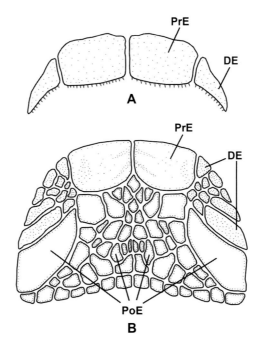

FIGURE 5.5

Premaxilla of (A) *Glyptothorax manipurensis* and (B) *G. burmanica*; *PrE*, Proximal element; *DE*, distal element; *PoE*, posterior element.

Brahmaputra basin in Goalpara, Assam (65.0 mm SL) has similar characters: prominent belly, and similar coloration. Larger specimens from the locality bear similar characters of Hamilton's (1822: 167) *Pimelodus rama* which is yellowish in color.

Mystus carcio

Mystus carcio is often confused with the south Indian species, *M. vittatus* (Bloch, 1794) which is distributed in southern India. *Mystus vittatus* was described from Tharangambadi, formerly known as Tranquebar, Tamil Nadu, in the vicinity of the mouth of the Kaveri River. Shaw and Shebbeare's (1937) *M. vittatus* from the streams in Terai and Duars, northern Bengal is probably *M. carcio* and Atur Rahman's (2005) *M. vittatus* from Bangladesh is probably *M. tengara*.

Olyra kempi

Chaudhuri (1912) described *Olyra kempi* from the Brahmaputra drainage in Mangaldai, Assam−Bhutan border, Darrang district, Assam. He described the species to have a long and low

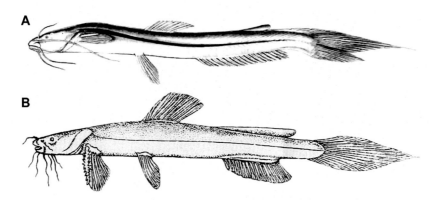

FIGURE 5.6

(A) *Olyra kempi*, Chaudhuri's (1912) Pl. 12, fig. 4 reproduced; (B) *Olyra longicaudata*, McClelland's (1842), Pl. 21, fig. 1 reproduced.

adipose fin which terminates in a raised knob-like end just above the termination of the anal fin. However, the drawing as in Pl. 41, Fig. 1.4 shows an adipose fin of medium length, not a short one, but terminating in a raised knob-like end just above the termination of the anal fin.

McClelland (1842) described *Olyra longicaudata* but did not mention its type locality. The next species he described in the same paper (pp. 588−589), *O. laticeps* was from the Kasyah Mountains (Khasi Hills). Both the species probably were from the same drainage. Most of the collections of *O. longicaudata* are from the Surma-Meghna basin in Meghalaya and Bangladesh. McClelland's (1842) drawing of *O. longicaudata* (Pl. 21, fig. 1) reproduced here as Fig. 5.6B shows an adipose fin ending well behind the termination of the base of the anal fin.

The specimen of *Olyra* collected from Sinkin River, a tributary of the Siang, Lower Divang Valley District, which is very near to the type locality of *O. kempi* agrees with characters in the drawing of Chaudhuri (1912) in all respects. The syntype of the species ZSIF 5387/1, 157 mm SL was examined and found to have a long adipose fin. However, it seems that the collections in the syntype involved more than one species and the syntype representing Chaudhuri's (1912) drawing is not traceable.

Although Ng et al. (2014a, 2014b) treated *O. kempi* a junior synonym of *O. longicaudata*, I feel that *O. kempi* should be treated valid until further studies (Fig. 5.6).

Channa aurolineata and *C. marulius*

Number of dorsal-fin rays and lateral line scales are the reliable characters for identifying *Channa aurolineata* and *C. marulius*. Adamson and Britz (2018) examined specimens of a wide range of sizes and found the scales of the larger juveniles and adults forming the black blotches on the sides of the body with conspicuous white posterior margins. They differentiated *C. marulius*, found in India and the western slope of the Rakhine Yoma of Myanmar in having white spots in the middle of the scales forming the blotches. Day (1870c) described *C. aurolineata* based on a single juvenile specimen measuring 79.0 mm SL from Moulmein, Myanmar. He described the species to have an orange stripe extending from the snout to the upper side of the caudal fin through the side above the lateral line (thus the name probably). My examination of specimens of *C. marulius* less than 150 mm SL from Barak River in Manipur and Assam and also from the Brahmaputra drainage in Arunachal Pradesh have pale stripes in the upper half of the side of the body. Blotches and ocelli are not clearly visible in the juveniles of both the species.

Bibliography

Abell R, Thieme ML, Revenga C, Bryer M, Kottelat M, Bogutskaya N, et al: Freshwater ecoregions of the world: a new map of biogeographic units for freshwater biodiversity conservation, *Bioscience* 58(5):403−444, 2008.

Abu Ahmed AT, Rahman MM, Mandal S: Biodiversity of hillstream fishes in Bangladesh, *Zootaxa* 3700 (2):283−f92, 2013.

Adamson EAS, Britz R: The snakehead fish *Channa aurolineata* is a valid species (Teleostei: Channidae) distinct from *Channa marulius, Zootaxa* 4514(4):5425−5552, 2018.

Ahl E: Neue Süßwasserfische aus dem Indischen und Malaiischen Gebiet, *Zoologischer Anz* 11(5/6):113−119, 1937.

Allen DJ, Molur S, Daniel BA: *The status and distribution of freshwater biodiversity in the Eastern Himalaya*, Cambridge, Gland, and Coimbatore, 2010, IUCN, Zoo Outreach Organisation.

Anderson J: Anatomical and zoological researches: comprising an account of the zoological results of the two expeditions to western Yunnan in 1868 and 1875; and a monograph of the two cetacean genera, Platanista and Orcella. Vol. 2. London, 1879, Bernard Quaritch. vol. 1: i−xxv + 1−984 + 1; vol. 2: Pls. 1−84. [Fishes; 1, pp. 861−69 and 2, Pl. 79].

Anganthoibi N, Vishwanath W: A new catfish, *Hara koladynensis* from northeastern India (Siluriformes: Erethistidae), *J Threat Taxa* 1(9):466−470, 2009.

Anganthoibi N, Vishwanath W: *Glyptothorax chimtuipuiensis*, a new species of catfish (Teleostei: Sisoridae) from the Koladyne basin, India, *Zootaxa* 2628:56−62, 2010a.

Anganthoibi N, Vishwanath W: *Pseudecheneis koladynae*, a new sisorid catfish from Mizoram, India (Teleostei: Siluriformes), *Ichthyol Explor Freshw* 21(3):199−204, 2010b.

Anganthoibi N, Vishwanath W: Two new species of *Glyptothorax* from the Koladye basin, Mizoram, India (Teleostei: Sisoridae), *Ichthyol Explor Freshw* 21(4):323−330, 2011.

Anganthoibi N, Vishwanath W: *Glyptothorax pantherinus*, a new species of catfish (Teleostei: Sisoridae) from the Noa Dehing River, Arunachal Pradesh, India, *Ichthyol Res* 60(2):172−177, 2013.

Annandale N: Fish and fisheries of the Inlé Lake, *Rec Indian Mus* 14:33−64, 1918. Pls. 1−7.

Arunachalam M, Muralidharan M, Sivakumar P: *Psilorhynchus amplicephalus*, a new species from Balishwar river of Assam, India, *Curr Sci* 92(10):1352−1354, 2007.

Arunachalam M, Raja M, Punniyam M, Mayden RL: New species of *Aborichthys* (Cypriniformes: Balitoridae) from Arunachal Pradesh, India, *Species* 7(18):33−47, 2014.

Arunkumar L: Fishes of the genus *Chela* Hamilton−Buchanan (Cyprinidae: Cultrinae) from Manipur, India, with description of a new species, *Aquacult* 1(2):121−124, 2000a.

Arunkumar L: *Laguvia manipurensis*, a new species of sisorid cat fish (Pisces: Sisoridae) from the Yu River system of Manipur, *Indian J Fish* 47(3):193−200, 2000b.

Arunkumar L: Loaches of the genus *Lepidocephalicthys* from Manipur, with description of a new species, *J Fish Biol* 57:093−104, 2000c.

Arunkumar L, Moyon WA: *Schizothorax chivae*, a new schizothoracid fish from Chindwin basin, Manipur, India (Teleostei: Cyprinidae), *Internatl J Fauna Biol Stud* 3(2):65−70, 2016. Part B.

Arunkumar L, Tombi Singh H: Fishes of the genus *Danio* (Hamilton-Buchanan) from Manipur, with description of a new species, *J Nat Conserv* 10(1):1−6, 1998.

Arunkumar L, Tombi Singh H: Bariliine fishes of Manipur, India, with description of a new species: *Barilius lairokensis, J Bombay Nat Hist Soc* 97(2):247−252, 2000a.

Arunkumar L, Tombi Singh H: Spiny eels of the genus *Macrognathus* Lacepède from Manipur, with description of a new species, *J Bombay Nat Hist Soc* 97(1):117−122, 2000b.

Arunkumar L, Tombi Singh H: Two new species of puntiid fish from the Yu River system of Manipur, *J Bombay Nat Hist Soc* 99(3):481−487, 2003.

Atur Rahman AK: *Freshwater fishes Bangladesh*, 2nd Edn, Dhaka, 2005, Zoological Survey of Bangladesh, I − xviii +1−394.

Bănărescu PM: Croizat's biogeographical principles, Pangaea and freshwater zoogeography, *Crisia* 15:1−12, 1985.

Bănărescu PM, Herzig-Straschil B: A revision of the species of the *Cyprinion macrostomus* − group (Pisces: Cyprinidae), *Ann Naturhist Mus Wien* 97B:411−420, 1995.

Bănărescu PM, Nalbant TT: Cobitidae (Pisces, Cypriniformes) collected by the German India expedition, *Mitteilungen aus dem Hamburgischen Zoologischen Mus und Inst* 65:327−351, 1968. Pls. 1−2.

Bănărescu PM, Nalbant TT: New data about the Malayan-Indochinese affinities of the aquatic fauna of the Western Ghat, South India, *Rev Roum Biol Biol Anim* 27:23−27, 1982.

Barman RP: A new species of the genus *Danio* Hamilton from India (Pisces: Cyprinidae), *Curr Sci* 52(4):177−178, 1983.

Barman RP: A new freshwater fish of the genus *Danio* Hamilton (Pisces: Cyprinidae) from Assam, India, with the key to the identification of the Indian species of the subgenus *Danio*, *Bull Zool Surv India* 6(1−3):163−165, 1984. Pl. 7.

Barman RP: A new cobitid fish of the genus *Aborichthys* Chaudhuri (Pisces: Cobitidae) from India, *J Bombay Nat Hist Soc* 81(3):680−683, 1985.

Barman RP: On a new cyprinid fish of the genus *Danio* Hamilton (Pisces: Cyprinidae) from Manipur, India, *J Bombay Nat Hist Soc* 84(1):172−174, 1987.

Bleeker P: Siluroideorum bataviensium conspectus diagnosticus. [Inside title: Overzigt der Siluroieden, welke te Batavia voorkomen.], *Verhandelingen van het Bataviaasch Genoot van Kunsten en Wetenschappen* 21(5):1−60, 1846.

Bleeker P: Bijdrage tot de kennis der Blennioïden en Gobioïden van den Soenda-Molukschen Archipel, met beschrijving van 42 nieuwe soorten, *Verhandelingen van het Bataviaasch Genoot van Kunsten en Wetenschappen* 22(6):1−40, 1849.

Bleeker P: Nieuwe bijdrage tot de kennis der ichthyologische fauna van Borneo met beschrijving van eenige nieuwe soorten van zoetwatervisschen, *Natuurkundig Tijdschr voor Nederlandsch Indië* 1(3):259−275, 1851a.

Bleeker P: Vierde bijdrage tot de kennis der ichthyologische fauna van Borneo, met beschrijving van eenige nieuwe soorten van zoetwatervisschen, *Natuurkundig Tijdschr voor Nederlandsch Indië* 2(2):193−208, 1851b.

Bleeker P: Nalezingen op de ichthyologische fauna van Bengalen en Hindostan, *Verhandelingen van het Bataviaasch Genoot van Kunsten en Wetenschappen* 25(8): 1−164 + 165−166, 1853. Pls. 1−6.

Bleeker P: De visschen van den Indischen Archipel beschreven en toegelicht. Deel I. Siluri, *Acta Societatis Regiae Scientiarum Indo−Neêrlandicae* 4(2): i−xii + 1−370, 1858a.

Bleeker P: Sur les genres de la famille des Cobitioïdes. Verslagen en Mededeelingen der Koninklijke Akademie van Wetenschappen, *Afd Natuurkunde* 15:32−44, 1858b.

Bleeker P: Enumeratio specierum piscium hucusque in Archipelago indico observatarum, adjectis habitationibus citationibusque, ubi descriptiones earum recentiores reperiuntur, nec non speciebus Musei Bleekeriani Bengalensibus, Japonicis, Capensibus Tasmanicisque, *Acta Societatis Regiae Scientiarum Indo−Neêrlandicae* 6(3): i−xxxvi + 1−276, 1859a.

Bleeker P: Negende bijdrage tot de kennis der vischfauna van Banka, *Natuurkundig Tijdschr voor Nederlandsch Indië* 18:359−378, 1859b.

Bleeker P: Conspectus systematis Cyprinorum, *Natuurkundig Tijdschr voor Nederlandsch Indië* 20(3):421−441, 1860.

Bleeker P: Notice sur les genres *Parasilurus*, *Eutropiichthys*, *Pseudeutropius*, et *Pseudopangasius*. Verslagen en Mededeelingen der Koninklijke Akademie van Wetenschappen, *Afd Natuurkunde* 14:390−399, 1862a.

Bleeker P: Atlas ichthyologique des Indes Orientales Néêrlandaises, publié sous les auspices du Gouvernement colonial néêrlandais. Tome II. Siluroïdes, Chacoïdes et Hétérobranchoïdes. F. Muller, Amsterdam. 1862b; 63: 1−112, Pls. 49−101.

Bleeker P: Notice sur les noms de quelques genres de la famille des Cyprinoïdes. Verslagen en Mededeelingen der Koninklijke Akademie van Wetenschappen, *Afd Natuurkunde* 15:261−264, 1863a.

Bleeker P: Sur les genres de la famille des Cobitioïdes. Verslagen en Mededeelingen der Koninklijke Akademie van Wetenschappen, *Afd Natuurkunde* 15:32−44, 1863b.

Bleeker P: Révision des espèces d'*Ambassis* et de *Parambassis* de l'Inde archipélagique, *Natuurkundige Verhandelingen van de Hollandsche Maatsch der Wetenschappen te Haarlem (Ser 3)* 2(2):83−106, 1874.

Bloch ME: Naturgeschichte der ausländischen Fische, *Berlin* 2:145−180, 1786. i−viii + 1−160, Pls.

Bloch ME: Naturgeschichte der ausländischen Fische, *Berlin* 6:289−323, 1792. i−xii + 1−126, Pls.

Bloch ME: Naturgeschichte der ausländischen Fische, *Berlin* 7:325−360, 1793. i−xiv + 1−144, Pls.

Bloch ME: Naturgeschichte der ausländischen Fische, *Berlin* 8:361−396, 1794. i−iv + 1−174, Pls.

Bloch ME: Naturgeschichte der ausländischen Fische, *Berlin* 9:397−429, 1795. i−ii + 1−192, Pls.

Bloch ME, Schneider JG: ME Blochii, Systema Ichthyologiae Iconibus cx Ilustratum. Post obitum auctoris opus inchoatum absolvit, correxit, interpolavit Jo. Gottlob Schneider, Saxo. Berolini, *Sumtibus Auctoris Impressum et Bibliopolio Sanderiano Commissum*:1−110, 1801. i−lx + 1−584, Pls.

Blyth E: Report of curator, zoological department, *J Asiatic Soc Bengal* 27(3):267−290, 1858.

Blyth E: Report on some fishes received chiefly from the Sitang River and its tributary streams, Tenasserim Provinces, *J Asiat Soc Bengal* 29(2):138−174, 1860.

Bohlen J, Šlechtová V: A new genus and two new species of loaches (Teleostei: Nemacheilidae) from Myanmar, *Ichthyol Explor Freshw* 22(1):21−30, 2011.

Bohlen J, Šlechtová V: *Schistura puncticeps*, a new species of loach from Myanmar (Cypriniformes: Nemacheilidae), *Ichthyol Explor Freshw* 24(1):85−92, 2013a.

Bohlen J, Šlechtová V: Two new species of *Schistura* from Myanmar (Teleostei: Nemacheilidae), *Ichthyol Explor Freshw* 24(1):21−30, 2013b.

Bohlen J, Šlechtová V: *Schistura shuensis*, a new species of loach from Myanmar (Teleostei: Nemacheilidae), *Ichthyol Explor Freshw* 24(3):217−223, 2014. (for 2013).

Boulenger GA: List of the fishes collected by Mr. E. W. Oates in the southern Shan States, and presented by him to the British Museum, *Ann Mag Nat Hist* 12(69):198−203, 1893.

Boulenger GA: Description of a new siluroid fish from Burma, *Ann Mag Nat Hist (Ser 6)* 14(81):196, 1894.

Briggs JC: Ostariophysan zoogeography: an alternative hypothesis, *Copeia* :111−118, 1989.

Britz R: Two new species of *Mastacembelus* from Myanmar (Teleostei: Synbranchiformes: Mastacembelidae), *Ichthyol Explor Freshw* 18(3):257−268, 2007.

Britz R: *Danionella priapus*, a new species of minature cyprinid fish from West Bengal, India (Teleostei: Cypriniformes: Cyprinidae), *Zootaxa* 2227:53−60, 2009.

Britz R: A new earthworm eel of the genus *Chaudhuria* from the Ayeyarwaddy River drainage, Myanmar (Teleostei: Synbranchiformes: Chaudhuriidae), *Zootaxa* 2571:62−68, 2010a.

Britz R: Species of the *Macrognathus aculeatus* group in Myanmar with remarks on *M. caudiocellatus* (Teleostei: Synbranchiformes: Mastacembelidae), *Ichthyol Explor Freshw* 20(4):295−308, 2010b.

Britz R: *Channa andrao*, a new species of dwarf snakehead from West Bengal, India (Teleostei: Channidae), *Zootaxa* 3731(2):287−294, 2013.

Britz R: Book Review: Fishes of the World. In Nelson JS, Grande TC, Wilson MVH, editors: *J Fish Biol*, 5th Edn, Hoboken, NJ, 2016, John Wiley & Sons, Inc., p 752. 2017; 90: 451−9.

Britz R, Kullander SO: *Dario kajal*, a new species of badid fish from Meghalaya, India (Teleostei: Badidae), *Zootaxa* 3731(3):331−337, 2013.

Britz R, Maclaine J: A review of the eel−loaches, genus *Pangio*, from Myanmar (Teleostei: Cypriniformes: Cobitidae), *Ichthyol Explor Freshw* 18(1):17−30, 2007.

Britz R, Lalremsanga HT, Lalrotluanga, Lalramliana: *Monopterus ichthyophoides*, a new species of scaled swamp eel (Teleostei: Synbranchiformes: Synbranchidae) from Mizoram, India, *Zootaxa* 2936:51−58, 2011.

Britz R, Sykes D, Gower DJ, Kamei RG: *Monopterus rongsaw*, a new species of hypogean swamp eel from the Khasi Hills in Northeast India (Teleostei: Synbranchiformes: Synbranchidae), *Ichthyol Explor Freshw* 28(4):315−326, 2018.

Brookfield ME: The evolution of the great river systems of southern Asia during the Cenozoic India−Asia collision: rivers draining southwards, *Geomorphology* 22:285−312, 1998.

Broussonet PMA: Ichtyologia, sistens piscium descriptiones et icones. Decas I. London. 1782.

Brown BA, Ferraris CJ Jr.: Comparative osteology of the Asian catfish family Chacidae, with the description of a new species from Burma, *Am Mus Novit* 2907:1−16, 1988.

Burchell WJ: Travels in the interior of southern Africa.2 vols. London. 1822; 1: i−xi + 1−582 + 1−4, 1 Map; 2: 1−648.

CEPF: Ecosystem Profile: Eastern Himalayas Region. Critical Ecosystem Partnership Fund. 2005. <http://www.cepf.net/Documents/final.ehimalayas.ep.pdf>.

Chakrabarty P, Ng HH: The identity of catfishes identified as *Mystus cavasius* (Hamilton, 1822) (Teleostei: Bagridae), with a description of a new species from Myanmar, *Zootaxa* 1093:1−24, 2005.

Chatterjee S, Scotese CR, Bajpai S: The restless Indian plate and its epic voyage from Gondwana to Asia: its tectonic, paleoclimatic and paleobiogeographic evolution, *Geol Soc Am, Spl pap* 529:1−149, 2017.

Chaudhuri BL: Description of a new species of *Nemachilus* from northern India, *Rec Indian Mus* 5(3):183−185, 1910.

Chaudhuri BL: Contributions to the fauna of Yunnan based on collections made by J. Coggin Brown, B.Sc., 1909−1910. Part II.−Fishes, *Rec Indian Mus* 6(1):13−24, 1911. Pl. 1.

Chaudhuri BL: Descriptions of some new species of freshwater fishes from north India, *Rec Indian Mus* 7(5):437−444. Pls. 38−41.

Chaudhuri BL: Zoological results of the Abor Expedition, 1911−12. XVIII. Fish, *Rec Indian Mus* 8(3):243−257. Pls. 7−9.

Chaudhuri BL: Report on a small collection of fish from Putao (Hkamti Long) on the northern frontier of Burma, *Rec Indian Mus* 16(4):271−287, 1919. Pl. 22.

Chen Z-M, Yang J-X: A new species of the genus *Tor* from Yunnan, China (Teleostei: Cyprinidae), *Env Biol Fishes* 70(2):185−191, 2004.

Chen X-Y, Yang J-X, Chen Y-R: A review of the cyprinoid fish genus *Barbodes* Bleeker, 1859, from Yunnan, China, with descriptions of two new species, *Zool Stud* 38(1):82−88, 1999.

Choudhury H, Dey A, Bharali RC, Sarma D, Vishwanath W: A new species of stone loach (Teleostei: Nemacheilidae: *Schistura*) from Arunachal Pradesh, India, *Zootaxa* 4551(1):40−52, 2019.

Clark MK, Schoenbohm LM, Royden LH, Whipple KX, Burchfiel BC, Zhang X, et al: Surface uplift, tectonicss, and erosion of eastern Tibet from largescale drainage patterns, *Tectonics* 23:TC106, 2004. Available from: https://doi.org/10.1029/2002TC001402.

Cloquet H: *Accounts: Anabas,* Anabas. *Archer,* Toxotes. Supplément 2 *Dictionnaire des sciences naturelles*, 35−36, Strasbourg (Levrault), Paris (Le Normant), 1816, AMA-ARGE, p 116.

Conway KW, Mayden RL: *Psilorhynchus breviminor*, a new species of psilorhynchid fish from Myanmar (Ostariophysi: Psilorhynchidae), *Ichthyol Explor Freshw* 19(2):111−120, 2008a.

Conway KW, Mayden RL: Description of two new species of *Psilorhynchus* (Ostariophysi: Psilorhynchidae) and redescription of *P. balitora*, *Ichthyol Explor Freshw* 19(3):215−232, 2008b.

Conway KW, Mayden RL: *Balitora eddsi*, a new species of hillstream loach (Ostariophysi: Balitoridae) from Nepal, *J Fish Biol* 76(6):1466−1473, 2010.

Conway KW, Dittmer DE, Jezisek LE, Ng HH: On *Psilorhynchus sucatio* and *P. nudithoracicus*, with the description of a new species of *Psilorhynchus* from northeastern India (Ostariophysi: Psilorhynchidae), *Zootaxa* 3686(2):201−243, 2013.

Croizat L: Manual of phytogeography, Junk, the Hague 1952.

Croizat L: Panbiogeography, 1, 2a, 2b. Caracas 1958.

Croizat L: Space, time, form; the biological synthesis, Caracas 1962.

Croizat L: Vicariance/vicariisam, panbiogeography, vicariane biogeography etc., a clarification, *Syst Zool* 31(3):291−304, 1982.

Croizat L, Nelson G, Rosen DE: Centres of origin and related concepts, *Syst Zool* 23:265−287, 1974.

Cuvier G: Le Règne Animal distribué d'après son organisation pour servir de base à l'histoire naturelle des animaux et d'introduction à l'anatomie comparée. Les reptiles, les poissons, les mollusques et les annélides. Edition 1. 1816; 2: i−xviii + 1−532, [Pls. 9−10, in v. 4].

Cuvier G, Valenciennes A: Histoire naturelle des poissons. Tome septième. Livre septième. Des Squamipennes. Livre huitième. Des poissons à pharyngiens labyrinthiformes. F. G. Levrault, Paris. 1831; 7: ivxxix + 1−531, Pls. 170−208.

Cuvier G, Valenciennes A: Hist nat poiss. Tome douzième. Suite du livre quatorzième. Gobioïdes. Livre quinzième. Acanthoptérygiens à pectorales pédiculées. 1837; 12: i-xxiv + 1−507 + 1 p., Pls. 344−68.

Cuvier G, Valenciennes A: Histoire naturelle des poissons. Tome quatorzième. Suite du livre seizième. Labroïdes. Livre dix-septième. Des Malacoptérygiens. Pitois-Levrault, Paris 1840; 14: i−xxii + 2 pp. + 1−464 + 4 pp., Pls. 389−420.

Cuvier G, Valenciennes A: Hist nat poiss, Cyprinoïdes 1844a; 17: i−xxiii + 1−497 + 2 pp., Pls. 487−519.

Cuvier G, Valenciennes A: Histoire naturelle des poissons. Tome dix-septième. Suite du livre dix-huitième, *Cyprinoïdes* :17, 1844b. I−xxiii + 1−497 + 2 pp., Pls. 487−519.

Daldorff DCD: Natural history of *Perca scandens, Trans Linn Soc Lond* 3(14):62−63, 1797.

Darlington PJ: *Zoogeography: the geographical distribution of animals*, New York, 1957, Wiley, p 675.

Darshan A, Anganthoibi N, Vishwanath W: Redescription of the striped catfush *Mystus carcio* (Hamilton) (siluriformes: Bagridae), *Zootaxa* 2475:48−54, 2010.

Darshan A, Anganthoibi N, Vishwanath W: *Batasio convexirostrum*, a new species of catfish (Teleostei: Bagridae) from Koladyne basin, India, *Zootaxa* 2901:52−58, 2011a.

Darshan A, Vishwanath W, Mahanta PC, Barat A: *Mystus ngasep*, a new catfish species (Teleostei: Bagridae) from the headwaters of Chindwin drainage in Manupur, India, *J Threatened Taxa* 3(11):2177−2183, 2011b.

Darshan A, Dutta R, Kachari A, Gogoi B, Aran K, Das DN: A new species of glyptosternine catfish (Siluriformes: Sisoridae) from Yomgo River, Arunachal Pradesh, India, *Aqua, Internatl J Ichthyol* 20(2):73−80, 2014.

Darshan A, Dutta R, Kachari A, Gogoi B, Das DN: *Glyptothorax mibangi*, a new species of catfish (Teleostei: Sisoridae) from the Tisa River, Arunachal Pradesh, northeast India, *Zootaxa* 3962(1):114−122, 2015.

Darshan A, Kachari A, Dutta R, Ganguly A, Das DN: *Amblyceps waikhomi*, a new species of catfish (Siluriformes: Amblycipitidae) from the Brahmaputra Drainage of Arunachal Pradesh, India, *PLoS One* 11(2):1−10, 2016. e0147282.

Darshan A, Dutta R, Abujam S, Das DN: First record of *Batasio spilurus* Ng from the Siang River of Arunachal Pradesh, Northeastern India (Teleostei: Bagridae), *J Fish Sciencescom* 11(2):21−27, 2017.

Darshan A, Abujam S, Das DN: *Biodiversity of fishes in Arunachal Himalaya*, 2019a, Academic Press-Elsevier, p 270.

Darshan A, Abujam S, Kumar R, Parhi J, Singh Y, Vishwanath W, et al: *Mystus prabini*, a new species of catfish (Siluriformes: Bagridae) from Arunachal Pradesh, north-eastern, India, *Zootaxa* 4648(3):511−522, 2019b.

Darshan A, Abujam S, Wangchu L, Kumar R, Das DN, Imotomba RK: A new species of glyptosternine catfish (Siluriformes: Sisoridae) from the Tawangchu River of Arunachal Pradesh, northeastern India, *Aqua, Internatl J Ichthyol* 25(1):17−24, 2019c.

Darshan A, Vishwanath W, Abujam S, Das DN: *Exostoma kottelati*, a new species of catfish (Teleostei: Sisoridae) from Arunachal Pradesh, India, *Zootaxa* 4585(2):369−377, 2019d.

Das DN, Darshan A: *Physoschistura harkishorei*, a new species of loach from Arunachal Pradesh, north−eastern India (Teleostei: Nemacheilidae), *Zootaxa* 4337(3):403−412, 2017.

Datta AK, Barman RP: On a new species of *Noemacheilus* (Pisces: Cobitidae) from Arunachal Pradesh, India, *Bull Zool Surv India* 6(1−3):275−277, 1984. Pl. 14.

Datta AK, Barman RP, Jayaram KC: On a new species of *Kryptopterus* (Pisces: Siluroidea, family: Siluridae) from Namdapha Wildlife Sanctuary, Arunachal Pradesh, India, *Bull Zool Surv India* 8(1−3):29−31, 1987.

Day F: On the freshwater fishes of Burma.−Part I, *Proc Zool Soc Lond* 3(3):614−623, 1870a. (Dec. 1869).

Day F: Remarks on some of the Fishes in the Calcutta Museum.−Part II, *Proc Zool Soc Lond* 3(7):548−560, 1870b.

Day F: On the freshwater fishes of Burmah −Part II, *Proc Zool Soc Lond* 1(3):99−101, 1870c.

Day F: Notes on the genus *Hara*, *J Asiatic Soc Bengal* 39(2):37−40, 1870d. Pl. 4.

Day F: Remarks on some of the Fishes in the Calcutta Museum −Part I, *Proc Zool Soc Lond* 3(7):611−614, 1870e.

Day F: Remarks on some of the Fishes in the Calcutta Museum −Part I, *Proc Zool Soc Lond* 3(7):511−527, 1870f.

Day F: Monograph of Indian Cyprinidae. Parts 1−3, *J Asiat Soc Bengal* 40: (pt 2, no. 1−4): 95−143, 277−336, 1871. Pls. 9, 21−23.

Day F: The fishes of India; being a natural history of the fishes known to inhabit the seas and fresh waters of India, Burma, and Ceylon. 1877; 3: 369−552, Pls. 79−138.

Day F: The fishes of India; being a natural history of the fishes known to inhabit the seas and fresh waters of India, Burma, and Ceylon. 1878; 4: i−xx + 553−78, Pls. 139−95.

Day F: The fishes of India; being a natural history of the fishes known to Inhabit the seas and fresh waters of India, Burma, and Ceylon. Suppl. 1888; 779−816.

de Pinna MCC: A phylogenetic analysis of the Asian catfish families Sisoridae, Akysidae, and Amblycipitidae, with a hypothesis on the relationships of the Neotropical Aspredinidae (Teleostei: Ostariophysi), *Fieldiana Zool (N Ser)* 84:1−83, 1996.

Dey A, Choudhury H, Mazumder A, Thaosen S, Sarma D: *Psilorhynchus nahlongthai*, a new fish species (Teleostei: Psilorhynchidae) from the Brahmaputra drainage, northeast India, *J Fish Biol* 96(3):642−650, 2020.

Diogo R, Chardon M, Vandewalle P: Osteology and mycology of the cephalic region and pectoral girdle of *Glyptothorax fukiensis* (Rendahl, 1925), comparison with other sisorids, and comments on the synapomorphies of the Sisoridae (Teleostei: Siluriformes), *Belgium J Zool* 132(2):95−103, 2002.

Dishma M, Vishwanath W: *Barilius profundus*, a new cyprinid fish (Teleostei: Cyprinidae) from Koladyne basin, India, *J Threatened Taxa* 4(2):2363−2369, 2012.

Dishma M, Vishwanath W: A new species of the genus *Pethia* from Mizoram, northeastern India (Teleostei: Cyprinidae), *Zootaxa* 3736(1):82−88, 2013.

Dishma M, Vishwanath W: *Parambassis serrata*, a new species of glassperch (Teleostei: Ambassidae) from the Kaladan drainage, India, *Zootaxa* 4040(5):583−588, 2015.

Dudgeon D, Arthington AH, Gessner MO, Kawabata Z-I, Knowler DJ, L.v.que C, et al: Freshwater biodiversity: importance, threats, status and conservation challenges, *Biol Rev* 81:163−182, 2006.

Eshchmeyer WN, Fricke R, van der Laan R: Catalog of fishes: genera, species, References California Adademy of Science. 2020 <http://www.calacademy.org/catalogoffishes/ Accessed November 30, 2020.

Ezung S, Shangningam B, Pankaj PP: A new fish species of the genus *Garra* (Teleostei: Cyprinidae) from the Brahmaputra basin, Nagaland, India, *J Exp Zool India* 23(2):1333−1339, 2020.

Fang F: Introduction. In: Phylogeny and species diversity of the South and Southeast Asian cyprinid genus *Danio* Hamilton (Teleostei, Cyprinidae). 2001. Department of Zoology, Stockholm University, Sweden, pp. 7−26.

Ferraris CJ Jr.: *Rita sacerdotum*, a valid species of catfish from Myanmar (Pisces, Bagridae), *Bull Nat Hist Mus Lond (Zool)* 65(1):15−21, 1999.

Ferraris CJ Jr.: A new species of the Asian schilbid catfish genus *Clupisoma* from Myanmar, with a redescription of *Clupisoma prateri* Hora (Osteichthyes: Siluriformes: Schilbidae), *Zootaxa* 437:1−10, 2004.

Ferraris CJ Jr, Runge KE: Revision of the South Asian bagrid catfish genus *Sperata*, with the description of a new species from Myanmar, *Proc Calif Acad Sci* 51(10):397−424, 1999.

Fowler HW: Some fishes from Borneo, *Proc Acad Nat Sci Phila* 57:455−523, 1905.

Fowler HW: Notes on clupeoid fishes, *Proc Acad Nat Sci Phila* 63:204−221, 1911.

Fowler HW: Descriptions of new fishes obtained 1907 to 1910, chiefly in the Philippine Islands and adjacent seas, *Proc Acad Nat Sci Phila* 85:233−267, 1934a.

Fowler HW: Zoological results of the Third de Schauensee Siamese Expedition, I. Fishes, *Proc Acad Nat Sci, Phila* 86:67−163, 1934b. Pl. 12.

Fowler HW: A collection of fishes from Rangoon, Burma, *Not Naturae (Phila)* 17:1−12, 1939.

Fujita K: *The caudal skeleton of teleostean fishes*, Tokyo, 1990, Tokyo Univ. Press, p 899.

Garsault FAPD: Les figures des plantes et animaux d'usage en medecine, décrits dans la matiere medicale de Mr. Geoffroy medecin, dessinés d'après nature. Garsault, Paris, 1764; 5: 1−3, Pls. 644−729, pp. 1−20.

Gauba RK: The endoskeleton of *Bagarius bagarius* (Hamilton), part I- The skull, *Agra Univ J Res* 11(1):75−90, 1962.

Gauba RK: Studies on the osteology of Indian sisorid catfishes II, *The skull of Glyptothorax cavia, Copeia* 4:802−810, 1966.

Gauba RK: The head skeleton of *Glyptosternum reticulatum* McClelland & Grifith, *Monitore Zoologico Italiano* 3:1−17, 1969.

Geetakumari K: *Parambassis bistigmata*, a new species of glassperch from north-eastern India (Teleostei: Ambassidae), *Zootaxa* 3317:59−64, 2012.

Geetakumari K, Basudha C: *Parambassis waikhomi*, a new species of glassfish (Teleostei: Ambassidae) from Loktak Lake, northeastern India, *J Threatened Taxa* 4(14):3327−3332, 2012.

Geetakumari K, Kadu K: *Badis singenensis*, a new fish species (Teleostei: Badidae) from Singen River, Arunachal Pradesh, northeastern India, *J Threatened Taxa* 3(9):2085−2089, 2011.

Geetakumari K, Vishwanath W: *Badis dibruensis*, a new species (Teleostei: Badidae) from northeastern India, *J Threatened Taxa* 2(1):644−647, 2010.

Geetakumari K, Vishwanath W: *Channa melanostigma*, a new species of freshwater snakehead from north−east India (Teleostei: Channidae), *J Bombay Nat Hist Soc* 107(3):231−235, 2011.

Gibert J, Deharveng L: Subterranean ecosystems: a truncated functional biodiversity, *BioScience* 52:473−481, 2002.

Gill TN: Description of a new generic form of Gobinae from the Amazon River, *Ann Lyceum Nat Hist N Y* 7:45−48, 1859. (nos 1−3, art. 10).

Gill TN: Descriptive enumeration of a collection of fishes from the western coast of Central America, presented to the Smithsonian Institution by Captain John M. Dow. *Proc Acad Nat Sci, Phila* 1863; 15, pp. 162−174.

Gleick PH: Water resources. In Schneider SH, editor: *Encyclopedia of climate and weather*, New York, 1996, Oxford University Press, pp 817−823.

Grant S: A new subgenus of *Acanthocobitis* Peters, 1861 (Teleostei: Nemacheilidae). *Ichthyofile* 2:1−9, 2007.

Gray JE: Description of twelve new genera of fish, discovered by Gen. Hardwicke, in India, the greater part in the British Museum, *Zool Misc*:7−9, 1831.

Gray JE: Illustrations of Indian Zoology, chiefly selected from the collection of Major−General Hardwicke FRS. 1830−1835; 20 parts in 2 vols. Pls. 1−202.

Gregory JW: The evolution of the river system of south-eastern Asia, *Scott Geogr J* 41:129−141, 1925.

Gregory WK: Fish skulls. A study of the evolution of natural mechanisms, *Trans Am Philos Soc* 23:75−481, 1933.

Günther A: Catalogue of the fishes in the British Museum. Catalogue of the Physostomi, containing the families Siluridae, Characinidae, Haplochitonidae, Sternoptychidae, Scopelidae, Stomiatidae in the collection of the British Museum. 1864; 5: i−xxii + 1−455.

Günther A: Catalogue of the fishes in the British Museum. Catalogue of the Physostomi, containing the families Heteropygii, Cyprinidae, Gonorhynchidae, Hyodontidae, Osteoglossidae, Clupeidae, [thru]. Halosauridae, in the collection of the British Museum 1868; 7: i−xx + 1−512.

Günther A: On some fishes from Kilima-Njaro District. Proceedings of the Zoological Society of London 1889; (1): 70−72, Pl. 8.

Guo X, He S, Zhang Y: Phylogeny and biogeography of Chinese sisorid catfishes using mitochondrial cytochrome b and 16S rRNA gene sequences, *Mol Phyolog Evol* 35(2):344−362, 2005.

Hamilton, F: A journey from Madras through the countries of Mysore, Canara, and Malabar, performed under the orders of the most noble the Marquis Wellesley, Governor General of India. T. Cadell and W. Davies etc., London. 1807; 3: i−iv + 1−479 + i−xxxi + index.

Hamilton F: An account of the fishes found in the river Ganges and its branches. Edinburgh/London: Archibald Constable/Hurst, Robinson: 1822.

Havird JC, Page LM: A revision of *Lepidocephalichthys* (Teleostei: Cobitidae) with descriptions of two new species from Thailand, Laos, Vietnam, and Myanmar, *Copeia* 1:137−159, 2010.

Hawksworth DJ, Kalin-Arroyo MT: Magnitude and distribution of biodiversity. In Heywood VH, editor: *Global biodiversity assessment*, Cambridge, 1995, Cambridge University Press, pp 107−191.

He S-P: A new species of the genus *Gagata* (Pisces: Sisoridae), *Acta Zootaxonom Sin* 21(3):380−382, 1996.

He S, Wenxuan CAO, Chen Y: The uplift of Qinghai-Ziang (Tibet) plateau and the vicarian speciation of glyptosternoid fishes (Siluriformes: Sisoridae), *Sci China (Ser C)* 44(6):644−651, 2001.

Heckel JJ: Fische aus Caschmir gesammelt und herausgegeben von Carl Freiherrn von Hügel, beschrieben von J. J. Heckel. Wien. 1838; 1−86, Pls. 1−12.

Heckel JJ: Ichthyologie. Stuttgart. Ichthyologie. In Russegger; 1 (2) In von Russegger J, editor: *Reisen in Europa, Asien und Afrika, mit besonderer Rücksicht auf die naturwissenschaftlichen Verhältnisse der betreffenden Länder unternommen in den Jahren 1835 bis 1841, etc*, 1843, E. Schweizerbart'sche Verlagshandlung, pp 991−1099.

Hollister G: Clearing and dying fishes for bone study, *Zoologica* 12:89−101, 1934.

Holly M: Zur Nomenklatur der Siluridengattung *Macrones* C. Duméril, *Zoologischer Anz* 125(5/6):143, 1939.

Hora SL: Fish and fisheries of Manipur with some observations on those of the Naga Hills, *Rec Indian Mus* 22(3):165−214, 1921a. Pls. 9−12.

Hora SL: Indian cyprinoid fishes belonging to the genus *Garra*, with notes on related species from other countries, *Rec Indian Mus* 22(5):633−687, 1921b. Pls. 24−26.

Hora SL: On some new or rare species of fish from the eastern Himalayas, *Rec Indian Mus* 22(5):731−744, 1921c. Pl. 29.

Hora SL: Notes on fishes in the Indian Museum, V. On the composite genus *Glyptosternon* McClelland, *Rec Indian Mus* 25(1):1−44, 1923. Pls. 1−4.

Hora SL: Notes on fishes in the Indian Museum. VIII. On the loaches of the genus *Aborichthys* Chaudhuri, *Rec Indian Mus* 27:231−236, 1925.

Hora SL: Notes on fishes in the Indian Museum. XVII. Loaches of the genus *Nemachilus* from Burma, *Rec Indian Mus* 31(4):311−334, 1929. Pls. 14−15.

Hora SL: Classification, bionomics and evolution of homalopterid fishes, *Mem Indian Mus* 12(2):263−330, 1932. Pls. 10−12.

Hora SL: Notes on fishes in the Indian Museum. XXIV. Loaches of the genus *Nemachilus* from eastern Himalayas, with the description of a new species from Burma and Siam, *Rec Indian Mus* 37(1):49−67, 1935. Pl. 3.

Hora SL: On a further collection of fish from the Naga Hills, *Rec Indian Mus* 38(3):317−331, 1936a.

Hora SL: Siluroid fishes of India, Burman and Ceylon, *Rec Indian Mus* 38:109−209, 1936b.

Hora SL: Notes on fishes in the Indian Museum. XXXI. On a small collection of fish from Sandoway, *Lower Burma, Rec Indian Mus* 39(4):323−331, 1937a.

Hora SL: The game fishes of India. Part III, *J Bombay Nat Hist Soc* 39(4):659−678, 1937b. Pl. 1.

Hora SL: Geographical distribution of Indian freshwater fishes and its bearing on the probable land connections between India and the adjacent countries, *Curr Sci* 5(7):351−356, 1937c.

Hora SL: On the Malayan affinities of the freshwater fish fauna of Peninsular India, and its bearing on the probable age of the Garo-Rajmahal gap, *Proc Natl Inst Sci India* 10:425−493, 1944.

Hora SL: Satpura hypothesis of the distribution of the Malayan fauna and flora to Peninsular India, *Proc Natl Inst Sci India* 15:309−314, 1949.

Hora SL: Siluroid fishes of India, Burma and Ceylon. XIII. Fishes of the genera *Erethistes* Müller and Troschel, *Hara* Blyth and of two new allied genera, *Rec Indian Mus* 47(2):183−202, 1950. Pls. 1−2.

Hora SL, Misra KS: Fishes collected by the Vernay-Hopwood Upper Chindwin Expedition, 1935, *J Bombay Nat Hist Soc* 42(3):478−482, 1941. Pl. 1.

Hora SL, Mukerji DD: Notes on fishes in the Indian Museum. XXII. On a collection of fish from the S. Shan States and the Pegu Yomas, Burma, *Rec Indian Mus* 36(1):123−138, 1934a.

Hora SL, Mukerji DD: Notes on fishes in the Indian Museum. XXIII. On a collection of fish from the S. Shan States, Burma, *Rec Indian Mus* 36(3):353−370, 1934b.

Hora SL, Mukerji DD: Fishes of the Naga Hills, Assam, *Rec Indian Mus* 37(3):381−404, 1935. Pl. 7.

Hora SL, Silas EG: Notes on fishes in the Indian Museum. XLVII.—Revision of the glyptosternoid fishes of the family Sisoridae, with descriptions of new genera and species, *Rec Indian Mus* 49:5−29, 1952. Pl. 1.

Howes GJ: The anatomy, phylogeny, and classification of Bariline cyprinid fishes, *Bull Br Mus Nat Hist (Zool)* 37(3):129−198, 1980.

Howes GJ: Additional notes on bariliine cyprinid fishes, *Bull Br Mus Nat Hist (Zool)* 45:95−101, 1983.

Hubbs CL, Lagler CF: Fishes of the Great Lakes Region. Cranbrook Instt, *Science* 26:1−186, 1946.

ICZN. *International code of zoological nomenclature*, ed 4, International Trust for Zoological Nomenclature. 1999; pp. 1−306.

ICZN: Ammendment of Articles 8, 9, 10, 21 and 78 of the International Code of Zoological Nomenclature to expand and refine methods of publication, *Bull Zool Nomencl* 69(2):161−169, 2012.

Jayaram KC: A new species of sisorid fish from the Kameng Frontier Division, Nefa, *J Zool Soc India* 15(1):85−87, 1966.

Jordan DS: The genera of fishes, Pt 2, from Agassiz to Bleeker, 1833−1858, twenty-six years, with the accepted type of each. A contribution to the stability of scientific nomenclature. Leland Stanford Jr. University Publications, University Ser, 1919a; 36: i−ix + 163−284 + i−xiii.

Jordan DS: New genera of fishes, *Proc Acad Nat Sci Phila* 70:341−344, 1919b.

Kaup JJ: Uebersicht der Lophobranchier, *Arch für Naturgeschichte* 19(1):226−234, 1853.

Khynriam D, Sen N: On a new species *Badis triocellus* (Pisces: Perciformes: Badidae) from North East India, *Rec Zool Surv India* 111(4):65−72, 2013.

Khanna SS: The hyobranchial skeleton of some fishes, *Ind J Zoot* 2(1):1−5, 1961.

Knight JDM: *Pethia aurea* (Teleostei: Cyprinidae), a new species of barb from West Bengal, India, with redescription of *P. gelius* and *P. canius*, *Zootaxa* 3700(1):173−184, 2013.

Knight JDM: *Oreichthys andrewi* (Teleostei: Cyprinidae) a new species from Assam, northeastern India, *J Threatened Taxa* 6(1):5357−5361, 2014.

Knight JDM: *Channa pardalis*, a new species of snakehead (Teleostei: Channidae) from Meghalaya, northeastern India, *J Threatened Taxa* 8(3):8583−8589, 2016.

Kosygin L: *Aborichthys waikhomi*, a new species of fish (Teleostei: Nemacheilidae) from Arunachal Pradesh, India, *Rec Zool Sur India* 112(1):49−55, 2012.

Kosygin L, Vishwanath W: A new cyrpinid fish *Garra compressus* from Manipur, India, *J Freshw Biol* 10(1−2):45−48, 1998.

Kosygin L, Shangningam N, Gopi KC: *Olyra parviocula*, a new species of bagrid catfish (Actinopterygii: Siluriformes), from northeastern India, *Env Biol Fishes* 101(4):589–593, 2018.

Kosygin L, Gurumayum SD, Singh P, Chowdhury BR: *Aborichthys iphipaniensis*, a new species of loach (Cypriniformes: Nemacheilidae) from Arunachal Pradesh, India, *J Zool* 39(2):69–75, 2019.

Kottelat M: A new noemacheiline loach from Thailand and Burma, *Japanese J Ichthyol* 29(2):169–172, 1982.

Kottelat M: Indian and Indochinese species of *Balitora* (Osteichthyes: Cypriniformes) with descriptions of two new species and comments on the family–group names Balitoridae and Homalopteridae, *Rev Suisse de Zoologie* 95(2):487–504, 1988.

Kottelat M: Zoogeography of the fishes from Indochinese inland waters with an annotated checklist, *Bull Zoölogisch Mus* 12(1):1–54, 1989.

Kottelat M: Indochinese nemacheilines. A revision of nemacheiline loaches (Pisces: Cypriniformes) of Thailand, Burma, Laos, Cambodia and southern Viet Nam. Verlag Dr. Friedrich Pfeil, München 1990, pp. 1–262.

Kottelat M: On the valid generic names for the Indian fishes usually referred to *Salmostoma* and *Somileptes* (Teleostei: Cyprinidae and Cobitidae), *J South Asian Nat History* 3(1):117–119, 1998.

Kottelat M: *Fishes of Laos*, Colombo, 2001, WHT Publications (Pte) Ltd.

Kottelat M: The fishes of the inland waters of southeast Asia: a catalogue and core bibliography of the fishes known to occur in freshwaters, mangroves and estuaries, *Raffles Bull Zool* 27:1–663, 2013. Suppl.

Kottelat M: *Mustura celata*, a new genus and species of loaches from northern Myanmar, and an overview of *Physoschistura* and related taxa (Teleostei: Nemacheilidae), *Ichthyol Explor Freshw* 28(4):289–314, 2018.

Kottelat M: *Rhyacoschistura larreci*, a new genus and species of loach from Laos and redescription of *R. suber* (Teleostei: Nemacheilidae), *Zootaxa* 4612(2):151–170, 2019.

Kottelat M, Whitten T: Freshwater biodiversity in Asia with special reference to fish. World Bank Technical Paper 343. 1996.

Kottelat M, Whitten AJ, Kartikasari SN, Wirjoatmodjo S: Freshwater fishes of Western Indonesia and Sulawesi. Periplus, Hong Kong, 1993.

Kottelat M, Harries DR, Proudlove GS: *Schistura papulifera*, a new species of cave loach from Meghalaya, India (Teleostei: Balitoridae), *Zootaxa* 1393:34–44, 2007.

Kullander SO: Taxonomy of chain *Danio*, an Indo–Myanmar species assemblage, with descriptions of four new species (Teleostei: Cyprinidae), *Ichthyol Explor Freshw* 25(4):357–380, 2015.

Kullander SO, Britz R: Revision of the family Badidae (Teleostei: Perciformes), with description of a new genus and ten new species, *Ichthyol Explor Freshw* 13(4):295–372, 2002.

Kullander SO, Fang F: Seven new species of *Garra* (Cyprinidae: Cyprininae) from the Rakhine Yoma, southern Myanmar, *Ichthyol Explor Freshw* 15(3):257–278, 2004.

Kullander SO, Fang F: *Danio aesculapii*, a new species of danio from south–western Myanmar (Teleostei: Cyprinidae), *Zootaxa* 2164:41–48, 2009.

Kullander SO, Liao T–Y, Fang F: *Danio quagga*, a new species of striped danio from western Myanmar (Teleostei: Cyprinidae), *Ichthyol Explor Freshw* 20(3):193–199, 2009.

Kullander SO, Rahman MM, Norén M, Mollah AR: *Devario* in Bangladesh: Species diversity, sibling species, and introgression within danionin cyprinids (Teleostei: Cyprinidae: Danioninae), *PLoS One* 12(11):1–37, 2017. e0186895.

Kullander SO, Norén M, Rahman MM, Mollah AR: Chameleonfishes in Bangladesh: hipshot taxonomy, sibling species, elusive species, and limits of species delimitation (Teleostei: Badidae), *Zootaxa* 4586(2):301–337, 2019.

Kumar P, Yuan X, Kumar MR, Kind R, Li X, Chadha RK: The rapid drift of the Indian tectonic plate, *Nature* 449L:894–897, 2007. Available from: https://doi.org/10.1038/nature06214.

Lacepède BGE: Histoire naturelle des poissons 1800; 2: i–lxiv + 1–632, Pls. 1–20.

Lacepède BGE: Histoire naturelle des poissons. 1803; 5: i–lxviii + 1–803 + index, Pls. 1–21.

Lalhlimpuia DV, Lalronunga S, Lalramliana: *Channa aurantipectoralis*, a new species of snakehead from Mizoram, north−eastern India (Teleostei: Channidae), *Zootaxa* 4147(3):343−350, 2016.

Lalramliana: *Schistura aizawlensis*, a new species of loach from Mizoram, northeastern India (Cypriniformes: Balitoridae), *Ichthyol Explor Freshw* 23(2):97−104, 2012.

Lalramliana, Knight JDM, Laltlanhlua Z: *Pethia rutila* (Teleostei: Cyprinidae), a new species from Mizoram, northeast India, *Zootaxa* 3827(3):366−374, 2014a.

Lalramliana, Lalronunga S, Vanramliana, Lalthanzara H: *Schistura mizoramensis*, a new species of loach from Mizoram, northeastern India (Teleostei: Nemacheilidae), *Ichthyol Explor Freshw* 25(3):205−212, 2014b.

Lalramliana, Solo B, Lalronunga S, Lalnuntluanga: *Psilorhynchus khopai*, a new fish species (Teleostei: Psilorhynchidae) from Mizoram, northeastern India, *Zootaxa* 3793(2):265−272, 2014c.

Lalramliana, Lalnuntluanga, Lalronunga S: *Psilorhynchus kaladanensis*, a new species (Teleostei: Psilorhynchidae) from Mizoram, northeastern India, *Zootaxa* 3962(1):171−178, 2015a.

Lalramliana, Lalronunga S, Lalnuntlunga, Ng HH: *Exostoma sawmteai*, a new sisorid catfish from northeast India (Teleostei: Sisoridae), *Ichthyol Explor Freshw* 26(1):59−64, 2015b.

Lalramliana, Vanlalhlimpuia D, Singh M: *Laubuka parafasciata*, a new cyprinid fish species (Teleostei: Cyprinidae) from Mizoram northeastern India, *Zootaxa* 4244(2):269−276, 2017.

Lalramliana, Knight JDM, Lalhlimpuia DV, Singh M: Integrative taxonomy reveals a new species of snakehead fish, *Channa stiktos* (Teleostei: Channidae), from Mizoram, North Eastern India, *Vert Zool* 68(2):165−175, 2018a.

Lalramliana, Lalronunga S, Kumar S, Singh M: DNA barcoding revealed a new species of *Neolissochilus* Rainboth, 1985 from the Kaladan River of Mizoram, North East India, *Mitochond DNA A* 30(1):52−59, 2018b.

Lalramliana, Solo B, Lalronunga S, Lalnuntluanga: *Hemimyzon indicus*, a new species of balitorid fish from the Kaladan basin, Mizoram, northeast India (Teleostei: Balitoridae), *Ichthyol Explor Freshw* 28(2):107−113, 2018c.

Lalramliana, Lalronunga S, Singh M: *Cabdio crassus*, a new species of cyprinid fish (Teleostei: Cyprinidae) from the Kaladan River of Mizoram, India, *Zootaxa* 4657(1):159−169, 2019.

Lalronunga S, Lalnuntluanga, Lalramliana: *Schistura maculosa*, a new species of loach (Teleostei: Nemacheilidae) from Mizoram, northeastern India, *Zootaxa* 3718(6):583−590, 2013.

Linnaeus C: Systema naturae, ed 10. 1758; 1: i−ii + 1−824.

Linthoingambi I, Vishwanath W: Two new fish species of the genus *Puntius* Hamilton (Cyprinidae) from Manipur, India, with notes on *P. ticto* (Hamilton) and *P. stoliczkanus* (Day), *Zootaxa* 1450:45−56, 2007.

Linthoingambi I, Vishwanath W: Two new catfish species of the genus *Amblyceps* from Manipur, India (Teleostei: Amblycipitidae), *Ichthyol Explor Freshw* 19(2):167−174, 2008.

Linthoingambi I, Vishwanath W: *Oreoglanis majusculus*, a new glyptosternine catfish from Arunachal Pradish, India (Teleostei: Sisoridae), *Zootaxa* 2754:60−66, 2011.

Lokeshwor Y, Vishwanath W: *Schistura fasciata*, a new nemacheiline species (Cypriniformes: Balitoridae) from Manipur, India, *J Threatened Taxa* 3(2):1514−1519, 2011.

Lokeshwor Y, Vishwanath W: *Physoschistura chindwinensis*, a new balitorid loach from Chindwin basin, Manipur, India, *Ichthyol Res* 59(3):230−234, 2012a.

Lokeshwor Y, Vishwanath W: *Physoschistura dikrongensis*, a new loach from Arunachal Pradesh, India (Teleostei: Nemacheilidae), *Zootaxa* 3586:249−254, 2012b.

Lokeshwor Y, Vishwanath W: A new loach of the genus *Physoschistura* Bănărescu & Nalbant (Teleostei: Nemacheilidae) from Chindwin basin, Manipur, India, *Zootaxa* 3586:95−102, 2012c.

Lokeshwor Y, Vishwanath W: *Schistura koladynensis*, a new species of loach from the Koladyne basin, Mizoram, India (Teleostei: Nemacheilidae), *Ichthyol Explor Freshw* 23(2):139−145, 2012d.

Lokeshwor Y, Vishwanath W: *Schistura ferruginea*, a new species of loach from northeast India (Teleostei: Nemacheilidae), *Ichthyol Explor Freshw* 24(1):49−56, 2013a.

Lokeshwor Y, Vishwanath W: *Schistura porocephala*, a new nemacheilid loach from Koladyne basin, Mizoram, India (Teleostei: Nemacheilidae), *Ichthyol Res* 60(2):159−164, 2013b.

Lokeshwor Y, Vishwanath W, Shanta K: *Physoschistura tuivaiensis*, a new species of loach (Teleostei: Nemacheilidae) from the Tuivai River, Manipur, India, *Taprobanica* 4(1):5−11, 2012.

Lokeshwor Y, Vishwanath W, Kosygin L: *Schistura paucireticulata*, a new loach from Tuirial River, Mizoram, India (Teleostei: Nemacheilidae), *Zootaxa* 3682(5):581−588, 2013.

Lundberg G, Kottelat M, Smith GR, Stiassny MLJ, Gill AC: So many fishes, so little time: an overview of recent ichthyological discovery in continental waters, *Ann Mo Botanical Gard* 87:26−62, 2000.

Matthews ED: Climate and evolution, *Ann N Y Acad Sci* 24:171−318, 1915.

McClelland J: Observations on six new species of Cyprinidae, with an outline of a new classification of the family, *J Asiat Soc Bengal* 7(2):941−948, 1838. Pls. 55−56.

McClelland J: Indian Cyprinidae, *Asiatic Res* 19(2):217−471, 1839. Pls. 37−61.

McClelland J: On the fresh−water fishes collected by William Griffith, Esq., F. L. S. Madras Medical Service, during his travels under the orders of the Supreme Government of India, from 1835 to 1842, *Calcutta J Nat Hist* 2(8):560−589, 1842. Pls. 6, 15, 18, 21.

McClelland J: Description of four species of fishes from the rivers at the foot of the Boutan Mountains, *Calcutta J Nat Hist* 5(18):274−282, 1845.

Menon AGK: Further observations on the fish fauna of the Manipur State, *Rec Indian Mus* 52(1):21−26, 1955.

Menon AGK: Monograph of the cyprinid fishes of the genis Garra Hamilton, *Mem Indian Mus* 14:173−260, 1964.

Menon AGK:The fauna of India and the adjacent countries. Pisces. Vol. IV. Teleostei − Cobitoidea. Part 1. Homalopteridae. Zool Surv India, Calcutta 1987; i−x + 1−259, Pls. 1−16.

Menon AGK, Datta AK: Zoological results of the Indian Cho−Oyu Expedition (1958) in Nepal. Part 7. −Pisces (concluded). *Psilorhynchus pseudecheneis*, a new cyprinid fish from Nepal, *Rec Indian Mus* 59(3):253−255, 1964. Pls. 16−17.

Menon AGK, Rema Devi K, Vishwanath W: A new species of *Puntius* (Cyprinidae: Cyprininae) from Manipur, India, *J Bombay Nat Hist Soc* 97(2):263−268, 2000. Pls. 1−2.

Mirza MR, Saboohi N: A note on the freshwater fishes of the river Dasht with the description of *Tariqilabeo* new subgenus (Pisces, Cyprinidae), *Pak J Zool* 22(4):405−406, 1990.

Misra KS: *Teleostomi: Cypriniformes; Siluri*, ed 2, The fauna of India and the adjacent countries. Pisces, 3, 1976, Government of India Press, pp 1−15. i−xxi + 1−367, Pls.

Mohiuddin SA, Mustafa AM: 2-D- modelling of the anticlinal structures and structural development of the eastern fold belt of the Bengal basin, Bangladesh, *Sediment Geol* 155(3−4):20−26, 2003.

Mukerji DD: Report on the Burmese fishes collected by Lt Col RW Burton from the tributary streams of Myitkyina District (Upper Burma) 2, *J Bombay Nat Host Soc* 37(1):38−80, 1934.

Mukhopadhyay M, Sujit D: Deep structure and tectonics of tge Burmese Arc: constraints from earthquake and gravity data, *Tectonophysics* 149(3):299−322, 1988.

Müller J. Ueber Nebenkiemen und Wundernetze. Archiv für Anatomie, Physiologie und Wissenschaftliche Medicin, in Verbindung mit mehreren Gelehrten Herausgebern 1840; 7, pp. 101−142.

Müller J, Troschel FH: Horae Ichthyologicae. Beschreibung und Abbildung neuer Fische. Berlin. 1849; 3: 1−27 + additional p. 24, Pls. 1−5.

Musikasinthorn P: *Channa aurantimaculata*, a new channid fish from Assam (Brahmaputra River basin), India, with designation of a neotype for *C. amphibeus* (McClelland, 1845), *Ichthyol Res* 47:27−37, 2000.

Myers GS: On a small collection of fishes from upper Burma, *Am Mus* 150:1−7, 1924.

Nath P, Dey SC: Two new fish species of the genus *Amblyceps* Blyth from Arunachal Pradesh, India, *J Assam Sci Soc* 32(1):1−6, 1989.

Nath P, Dam D, Bhutia PT, Dey SC, Das DN: A new fish species of the genus *Bhavania* (Homalopteridae: Homalopterinae) from river Noadhing drainage, Arunachal Pradesh, India, *Rec Zool Surv India* 107(3):71−78, 2007.

Nebeshwar K, Vishwanath W: Three new species of *Garra* (Pisces: Cyprinidae) from north−eastern India and redescription of *G. gotyla*, *Ichthyol Explor Freshw* 24(2):97−120, 2013.

Nebeshwar K, Vishwanath W: Two new species of *Garra* (Pisces: Cyprinidae) from the Chindwin River basin in Manipur, India, with notes on some nominal *Garra* species of the Himalayan foothills, *Ichthyol Explor Freshw* 25(4):305−321, 2015.

Nebeshwar K, Vishwanath W: On the snout and oromandibular morphology of genus *Garra*, description of two new species from the Koladyne River basin in Mizoram, India and redescription of *G. manipurensis* (Teleostei: Cyprinidae), *Ichthyol Explor Freshw* 28(1):17−53, 2017.

Nebeshwar K, Bagra K, Das DN: A new species of the cyprinoid genus *Psilorhynchoides* Yazdani et al. (Cypriniformes: Psilorhynchidae) from Arunachal Pradish, India, *Zoos' Print J* 22(3): 2632−2636 + 1 web page, 2007. Pl. [Image] 1.

Nebeshwar K, Vishwanath W, Das DN: *Garra arupi*, a new cyprinid fish species (Cypriniformes: Cyprinidae) from upper Brahmaputra basin in Arunachal Pradish, India, *J Threatened Taxa* 1(4):197−202, 2009.

Nebeshwar K, Bagra K, Das DN: *Garra kalpangi*, a new cyprinid fish species (Pisces: Teleostei) from upper Brahmaputra basin in Arunachal Pradesh, India, *J Threatened Taxa* 4(2):2353−2362, 2012.

Nelson JS: *Fishes of the World*, 4th Edn, Hoboken, NJ, 2006, John Wiley & Sons, 601 PP.

Nelson JS, Grande TC, Wilson MVH: *Fishes of the world*, 5th Edn, Hoboken, NJ, 2016, John Wiley & Sons, 752 PP.

Ng HH: A revision of the south Asian sisorid catfish genus *Sisor* (Teleostei: Siluriformes), *J Nat Hist* 37(23):2871−2883, 2003.

Ng HH: *Amblyceps carinatum*, a new species of hillstream catfish from Myanmar (Teleostei: Amblycipitidae), *Raffles Bull Zool* 53(2):243−249, 2005a.

Ng HH: *Conta pectinata*, a new erethistid catfish (Teleostei: Erethistidae) from northeast India, *Ichthyol Explor Freshw* 16(1):23−28, 2005b.

Ng HH: *Erethistoides sicula*, a new catfish (Teleostei: Erethistidae) from India, *Zootaxa* 1021:1−12, 2005c.

Ng HH: *Glyptothorax botius* (Hamilton, 1822), a valid species of catfish (Teleostei: Sisoridae) from northeast India, with notes on the identity of *G. telchitta* (Hamilton, 1822), *Zootaxa* 930:1−19, 2005d.

Ng HH: *Gogangra laevis*, a new species of riverine catfish from Bangladesh (Teleostei: Sisoridae), *Ichthyol Explor Freshw* 16(3):279−286, 2005e.

Ng HH: *Pseudolaguvia foveolata*, a new catfish (Teleostei: Erethistidae) from northeast India, *Ichthyol Explor Freshw* 16(2):173−178, 2005f.

Ng HH: Two new species of *Pseudolaguvia* (Teleostei: Erethistidae) from Bangladesh, *Zootaxa* 1044:35−47, 2005g.

Ng HH: *Erethistoides infuscatus*, a new species of catfish (Teleostei: Erethistidae) from South Asia, *Ichthyol Explor Freshw* 17(3):281−287, 2006a.

Ng HH: *Pseudolaguvia ferula*, a new species of sisoroid catfish (Teleostei: Erethistidae) from India, *Zootaxa* 1229:59−68, 2006b.

Ng HH: The identity of *Batasio tengana* (Hamilton, 1822), with the description of two new species of *Batasio* from north-eastern India (Teleosei: Bagridae), *J Fish Biol* 68(A):101−118, 2006c.

Ng HH: The identity of *Pseudecheneis sulcata* (M'Clelland, 1842), with descriptions of two new species of rheophilic catfish (Teleostei: Sisoridae) from Nepal and China, *Zootaxa* 1254:45−68, 2006d.

Ng HH: *Akysis vespertinus*, a new species of catfish from Myanmar (Siluriformes: Akysidae), *Ichthyol Explor Freshw* 19(3):255−262, 2008a.

Ng HH: *Batasio procerus*, a new species of catfish from northern Myanmar (Siluriformes: Bagridae), *Ichthyol Explor Freshw* 19(1):1−6, 2008b.

Ng HH: Two new species of *Pseudolaguvia*, sisorid catfishes (Teleostei: Siluriformes) from northeastern India, *Ichthyol Explor Freshw* 20(3):277–288, 2009a.

Ng HH: Redescription of *Batasio merianiensis*, a catfish (Teleostei: Bagridae) from northeastern India, *J Threatened Taxa* 1(5):253–256, 2009b.

Ng HH, Conway KW: *Pseudolaguvia assula*, a new species of crypto–benthic sisorid catfish from central Nepal (Teleostei: Sisoridae), *Ichthyol Explor Freshw* 24(2):179–185, 2013.

Ng HH, Edds DR: *Batasio macronotus*, a new species of bagrid catfish from Nepal (Teleostei: Bagridae), *Ichthyol Explor Freshw* 15(4):295–300, 2004.

Ng HH, Edds DR: Two new species of *Erethistoides* (Teleostei: Erethistidae) from Nepal, *Ichthyol Explor Freshw* 16(3):239–248, 2005a.

Ng HH, Edds DR: Two new species of *Pseudecheneis*, rheophilic catfishes (Teleostei: Sisoridae) from Nepal, *Zootaxa* 1047:1–19, 2005b.

Ng HH, Ferraris CJ Jr.: A new species of anguilliform catfish (Actinopterygii: Siluriformes: Bagridae) from Bangladesh and northeastern India, *Zootaxa* 4079(3):381–387, 2016.

Ng HH, Jaafar Z: A new species of leaf fish, *Nandus andrewi* (Teleostei: Perciformes: Nandidae) from notheastern India, *Zootaxa* 1731:24–32, 2008.

Ng HH, Kottelat M: A review of the catfish genus *Hara*, with the description of four new speceis (Siluriformes: Erethistidae), *Rev Suisse de Zoologie* 114(3):471–505, 2007.

Ng HH, Kottelat M: *Glyptothorax rugimentum*, a new species of catfish from Myanmar and western Thailand (Teleostei: Sisoridae), *Raffles Bull Zool* 56(1):129–134, 2008a.

Ng HH, Kottelat M: *Batasio feruminatus*, a new species of bagrid catfish from Myanmar (Siluriformes: Bagridae), with notes on the identity of *B. affinis* and *B. fluviatilis*, *Ichthyol Explor Freshw* 18(4):289–300, 2008b. [for 2007].

Ng HH, Kottelat M: A new species of *Mystus* from Myanmar (Siluriformes: Bagridae), *Copeia* 2:245–250, 2009.

Ng HH, Kottelat M: Revision of the Asian catfish genus *Hemibagrus* Bleeker, 1862 (Teleostei: Siluriformes: Bagridae), *Raffles Bull Zool* 61(1):205–291, 2013.

Ng HH, Kullander SO: *Glyptothorax igniculus*, a new species of sisorid catfish (Teleostei: Siluriformes) from Myanmar, *Zootaxa* 3681(5):552–562, 2013.

Ng HH, Lalramliana: *Pseudolaguvia spicula*, a new sisorid catfish (Teleostei: Sisoridae) from Bangladesh and northeastern India, *Zootaxa* 2558:61–68, 2010.

Ng HH, Lalramliana: *Glyptothorax maceriatus*, a new species of sisorid catfish (Actinopterygii: Siluriformes) from north–eastern India, *Zootaxa* 3416:44–52, 2012a.

Ng HH, Lalramliana: *Glyptothorax scrobiculus*, a new species of sisorid catfish (Osteichthyes: Siluriformes) from northeastern India, *Ichthyol Explor Freshw* 23(1):1–9, 2012b.

Ng HH, Lalramliana: *Glyptothorax radiolus*, a new species of sisorid catfish (Osteichthyes: Siluriformes) from northeastern India, with a redescription of *G. striatus* McClelland 1842, *Zootaxa* 3682(4):501–512, 2013.

Ng HH, Wright JJ: *Amblyceps cerinum*, a new catfish (Teleostei: Amblycipitidae) from northeastern India, *Zootaxa* 2672:50–60, 2010.

Ng HH, Lalramliana, Lalronunga S, Lalnuntluanga: *Eutropiichthys cetosus*, a new riverine catfish (Teleostei: Schilbeidae) from northeastern India, *J Threatened Taxa* 6(8):6073–6081, 2014a.

Ng HH, Lalramliana, Lalthanzara H: *Olyra saginata*, a new species of bagrid catfish (Actinopterygii: Siluriformes) from northeastern India, *Zootaxa* 3821(2):265–272, 2014b.

Ng HH, Lalramliana, Lalronunga S: A new diminutive sisorid catfish (Actinopterygii: Siluriformes) from northeastern India, *Zootaxa* 4105(6):546–556, 2016.

Ng HH, Lalramliana, Lalronunga S: *Pterocryptis subrisa*, a new silurid catfish (Teleostei: Siluridae) from northeastern India, *Zootaxa* 4500(1):126–134, 2018.

Niebuhr C: Descriptiones animalium avium, amphibiorum, piscium, insectorum, vermium; quae in itinere orientali observavit Petrus Forskål. Post mortem auctoris edidit Carsten Niebuhr, *Hauniae*: 1−20 + i−xxxiv + 1−164, map.

Oken LVK: Fische. Isis (Oken) 1817; 8 (148): 1779−1782.

Pallas PS: Spicilegia Zoologica quibus novae imprimis et obscurae animalium species iconibus, descriptionibus atque commentariis illustrantur, *Berolini, Gottl August Lange* 1(7):1−42, 1769. Pls. 1−6.

Peng Z, Simon YWH, Zhang Y, He S: Uplift of the Tibetal plateau: evidence from divergence times of glyptosternine catfishes, *Mol Phylogenetics Evolution* 39:568−572, 2006.

Peters W (CH): Diagnosen von neuen Flussfischen aus Mossambique. Monatsberichte der Königlichen Preussischen Akademie der Wissenschaften zu Berlin 1852; 275−276, 681−685.

Peters, W (C.H.): Eine neue vom Herrn Jagor im atlantischen Meere gefangene Art der Gattung Leptocephalus, und über einige andere neue Fische des Zoologischen Museums. Monatsberichte der Königlichen Preussischen Akademie der Wissenschaften zu Berlin: 1859, pp. 411−413.

Peters W (CH): Über zwei neue Gattungen von Fischen aus dem Ganges. Monatsberichte der Königlichen Preussischen Akademie der Wissenschaften zu Berlin 1861; pp. 712−713.

Pethiyagoda R, Kottelat M, Silva A, Maduwage K, Meegaskumbura M: A review of the genus *Laubuca* in Sri Lanka, with description of three new species (Teleostei: Cyprinidae), *Ichthyol Explor Freshw* 19(1):7−26, 2008.

Pethiyagoda R, Meegaskumbura M, Maduwage K: A synopsis of the South Asian fishes referred to *Puntius* (Pisces: Cyprinidae), *Ichthyol Explor Freshw* 23(1):69−95, 2012.

Pillai RS, Yazdani GM: Two new species and two records of *Lepidocephalichthys* Bleeker [Pisces: Cobitidae] from Assam and Meghalaya, India, with a key to the known species, *J Zool Soc India* 26(1−2):11−17, 1976.

Playfair RL: On the fishes of Cachar, *Proc Zool Soc Lond* 1(3):14−17, 1867. Pl. 3.

Potthoff T: Clearing and staining techniques. Lawrence KS Spl Publ 1, American Society of Ichthyologists and Herpetologists In Moser HG, Richards WJ, Cohen DM, Fahay MP, Kendall AW Jr, Richardson SL, editors: *Ontogeny and systematics of fishes*, 1984, Allen Press, pp 35−37.

Prashad B, Mukerji DD: The fish of the Indawgyi Lake and the streams of the Myitkyina District (Upper Burma), *Rec Indian Mus* 31(3):161−223, 1929. Pls. 7−10.

Premananda N, Kosygin L, Saidullah B: *Glyptothorax senapatiensis*, a new species of catfish (Teleostei: Sisoridae) from Manipur, India, *Ichthyol Explor Freshw* 25(4):323−329, 2015.

Qin T, Chen Z-Y, Xu L-L, Zaw P, Kyaw YMM, Maung KW, et al: Five newly recorded Cyprinid fish (Teleostei: Cypriniformes) Myanmar, *Zool Res* 38(5):300−309, 2017.

Rahman MM, Mollah AR, Norén M, Kullander SO: *Garra mini*, a new small species of rheophilic cyprinid fish (Teleostei: Cyprinidae) from southeastern hilly areas of Bangladesh, *Ichthyol Explor Freshw* 27(2):173−181, 2016.

Rainboth WJ: *Neolissochilus*, a new genus of South Asian cyprinid fishes, *Beaufortia* 35(3):25−35, 1985.

Rainboth WJ: Fishes of the Asian cyprinid genus *Chagunius*, *Occ Pap Mus Zool* 712:1−17, 1986. University of Michigan.

Rainboth WJ: *Fishes of the Cambodian Mekong*, Rome, 1996a, FAO, United Nations.

Rainboth WJ: The taxonomy, systematics, and zoogeography of *Hypsibarbus*, a new genus of large barbs (Pisces, Cyprinidae) from the rivers of southeastern Asia, *Univ Calif Publ Zool* 129: i−xiii + 1−199.

Rameshori Y, Vishwanath W: A new catfish of the genus *Glyptothorax* from the Kaladan basin, Northeast India (Teleostei: Sisoridae), *Zootaxa* 3538:79−87, 2012a.

Rameshori Y, Vishwanath W: *Glyptothorax jayarami*, a new species of catfish (Teleostei: Sisoridae) from Mizoram, northeastern India, *Zootaxa* 3304:54−62, 2012b.

Rameshori Y, Vishwanath W: *Glyptothorax verrucosus*, a new sisorid catfish species from the Koladyne basin, Mizoram, India (Teleostei: Sisoridae), *Ichthyol Explor Freshw* 23(2):147−154, 2012c.

Ramsay EP, Ogilby JD: A contribution to the knowledge of the fish-fauna of New Guinea, *Proc Linn Soci N South Wales (Ser 2)* 1(1):8−20, 1886.

Raveenraj J, Uma A, Moulitharan N, Singh SG: A new species of dwarf *Channa* (Teleostei: Channidae) from Meghalaya, northeast India, *Copeia* 107(1):61–70, 2019.

Regan CT: The classification of the teleostean fishes of the order Synentognathi, *Ann Mag Nat Hist (Ser 8)* 7(40):327–335, 1911a. Pl. 9.

Regan CT: The classification of the teleostean fishes of the order Ostariophysi.–I. Cyprinoidea, *Ann Mag Nat Hist* 8(43):13–32, 1911b. Pl. 2.

Regan CT: A revision of the clupeoid fishes of the genera *Pomolobus*, *Brevoortia* and *Dorosoma* and their allies, *Ann Mag Nat Hist (Ser 8)* 19(112):297–316, 1917.

Richardson J: The Fish. In: Fauna Boreali-Americana; or the zoology of the northern parts of British America: containing descriptions of the objects of natural history collected on the late northern land expeditions, under the command of Sir John Franklin, R.N. J. Bentley, London. 1836; 3: i–xv +1–327, Pls. 74–97.

Richardson J: Ichthyology.–Part 3. In Hinds RB, editor: *The zoology of the voyage of H. M. S. Sulphur, under the command of Captain Sir Edward Belcher, R. N., C. B., F. R. G. S., etc., during the years 1836–42, No. 10*, London, 1845, Smith, Elder & Co, pp 99–150. Pls. 55–64.

Richardson J: Report on the ichthyology of the seas of China and Japan. Report of the British Association for the Advancement of Science 15th meeting [1845]. 1846, pp. 187–320

Roberts TR: A revision of the south and southeastern Asian angler-catfishes (Chacidae), *Copeia* 4:895–901, 1982.

Roberts TR: Revision of the south and southeast Asian sisorid catfish genus *Bagarius*, with description of a new species from the Mekong, *Copeia* 2:435–445, 1983.

Roberts TR: *Danionella translucida*, a new genus and species of cyprinid fish from Burma, one of the smallest living vertebrates, *Environ Biol Fishes* 16(4):231–241, 1986a.

Roberts TR: Systematic review of the Mastacembelidae or spiny eels of Burma and Thailand, with description of two new species of *Macrognathus*, *Jpn J Ichthyol* 33(2):95–109, 1986b.

Roberts TR: Review of the tropical Asian cyprinid fish genus *Poropuntius*, with descriptions of new species and trophic morphs, *Nat Hist Bull Siam Soc* 46:105–135, 1998.

Roberts TR: *Ayarnangra estuarius*, a new genus and species of sisorid catfish from the Ayeyarwaddy basin, Myanmar, *Nat Hist Bull Siam Soc* 49(1):81–87, 2001.

Roberts TR: The "Celestial Pearl Danio," a new genus and species of colourful minute cyprinid fish from Myanmar (Pisces: Cypriniformes), *Raffles Bull Zool* 55(1):131–140, 2007.

Roberts TR, Ferraris CJ Jr.: Review of South Asian sisorid catfish genera *Gagata* and *Nangra*, with descriptions of a new genus and five new species, *Proc Calif Acad Sci* 50(14):315–345, 1998.

Romer AS: Review of the Labyrinthodontia, *Bull Mus Comp Zool* 99(1):368, 1947. Havard.

Roni N, Vishwanath W: *Garra biloborostris*, a new labeonine species from north–eastern India (Teleostei: Cyprinidae), *Vert Zool* 67(2):133–137, 2017.

Roni N, Vishwanath W: A new species of the genus *Garra* (Teleostei: Cyprinidae) from the Barak River drainage, Manipur, India, *Zootaxa* 4374(2):263–272, 2018.

Roni N, Sarbojit T, Vishwanath W: *Garra clavirostris*, a new cyprinid fish (Teleostei: Cyprinidae: Labeoninae) from the Brahmaputra drainage, India, *Zootaxa* 4244(3):367–376, 2017.

Rüber L, Britz R, Kullander SO, Zardoya R: Evolutionary and biogeographic patterns of the Badidae (Teleostei: Perciformes) inferred from mitochondrial and nuclear DNA sequence data, *Mol Phyl Evol* 32:1010–1022, 2004.

Rüber L, Tan HH, Britz R: Snakehead (Teleostei: Channidae) diversity and the Eastern Himalaya biodiversity hotspot, *J Zool Syst Evol Res* 58:356–386, 2020.

Russell A: Natural history of Aleppo, ed 2, revised by P. Russell. 1794; 2: i–vii + 1–430 + i–xxxiv + 26 p. index, Pls. 1–16.

Sala OE, Chaplin FS, Armesto JJ, Berlow R, Bloomfield J, Dirzo R, et al: Global biodiversity scenarios for the year 2100, *Science* 87:1770–1774, 2000.

Saxena SC, Chandy M: The pelvic girdle and fin in certain Indian Hill stream fishes, *J Zool Lond* 148:167−190, 1965.

Schäfer F: *Oreichthys crenuchoides*, a new cyprinid from west Bengal, India, *Ichthyol Explor Freshw* 20(3):201−211, 2009.

Scopoli JA: Introductio ad historiam naturalem, sistens genera lapidum, plantarum et animalium hactenus detecta, caracteribus essentialibus donata, in tribus divisa, subinde ad leges naturae, *Prague*: i−x + 1−506, 1777.

Selim K, Vishwanath W: A new freshwater cyprinid fish *Aspidoparia* from the Chatrikhong River, Manipur, India, *J Bombay Nat Hist Soc* 98(2):245−257, 2001.

Selim K, Vishwanath W: A new cyprinid fish species of *Barilius* Hamilton from the Chatrickong River, Manipur, India, *J Bombay Nat Hist Soc* 99(2):267−270, 2002.

Sen N: Description of a new species of *Brachydanio* Weber and de Beaufort, 1916 (Pisces: Cypriniformes: Cyprinidae) from Meghalaya, north east India with a note on comparative studies of other known species, *Rec Zool Surv India* 107(4):27−31, 2007.

Sen N: Description of a new species of *Aborichthys*, Chaudhuri from North−east India (Pisces: Cypriniformes: Balitoridae), *Rec Zool Surv India* 109(2):13−20, 2009.

Sen N, Biswas BK: On a new species of *Nangra* Day (Pisces: Siluriformes: Sisoridae) from Assam, north east India with a note on comparative studies of other known species, *Rec Zool Surv India* 94(2−4):441−446, 1994.

Sen N, Dey SC: Two new fish species of the genus *Danio* Hamilton (Pisces: Cyprinidae) from Meghalaya, India, *J Assam Sci Soc* 27(2):60−68, 1985.

Shangningam BD, Kosygin L: A new sisorid catfish of the genus *Exostoma* Blyth from the Chindwin-Irrawaddy drainage in northeastern India (Teleostei: Siluriformes), *Copeia* 108(3):545−550, 2020.

Shangningam B, Vishwanath W: Validation of *Garra namyaensis* Shangningam & Vishwanath, 2012 (Teleostei: Cyprinidae: Labeioninae), *Ichthyol Explor Freshw* 2(1):10, 2012.

Shangningam B, Vishwanath W: A new species of *Psilorhynchus* (Teleostei: Psilorhynchidae) from the Chindwin basin of Manipur, India, *Zootaxa* 3694(4):381−390, 2013a.

Shangningam B, Vishwanath W: *Psilorhynchus maculatus*, a new species of torrent minnow from the Chindwin basin, Manipur, India (Teleostei: Psilorhynchidae), *Ichthyol Explor Freshw* 24(1):57−62, 2013b.

Shangningam B, Vishwanath W: Redescription of *Psilorhynchus homaloptera*, a torrent minnow from north-east India (Teleostei: Psilorhynchidae), *Ichthyol Explor Freshw* 24(3):237−248, 2014a.

Shangningam B, Vishwanath W: *Psilorhynchus ngathanu*, a new torrent minnow species (Teleostei: Psilorhynchidae) from the Chindwin Basin, Manipur, India, *Ichthyol Res* 61:27−31, 2014b.

Shangningam B, Vishwanath W: Two new species of *Garra* from the Chindwin basin, India (Teleostei: Cyprinidae), *Ichthyol Explor Freshw* 26(3):263−272, 2015.

Shangningam B, Vishwanath W: *Psilorhynchus konemi*, a new species of torrent minnow from northeast India (Teleostei: Psilorhynchidae), *Ichthyol Explor Freshw* 27(4):289−296, 2016.

Shangningam BD, Kosygin L, Vishwanath W: Redescription of *Psilorhynchus rowleyi* Hora and Misra 1941 (Cypriniformes: Psilorhynchidae), *Ichthyol Res* 60(3):249−255, 2013.

Shangningam B, Kosygin L, Gopi KC: *Psilorhynchus bichomensis*, a new species of torrent minnow from Arunachal Pradesh, northeast India (Teleostei: Psilorhynchidae), *Fish Taxa* 4(3):130−139, 2019a.

Shangningam B, Kosygin L, Sinha B, Gurumayum SD: *Aborichthys kailashi* and *A. pangensis* (Cypriniformes: Nemacheilidae), two new species from Arunachal Pradesh, India, *Ichthyol Explor Freshw* 29(4):361−370, 2019b.

Shangningam BD, Kosygin L, Sinha B: A new species of rheophilic cyprinid fish (Teleostei: Cyprinidae) from the Brahmaputra Basin, northeast India, *Zootaxa* 4695(2):148−158, 2019c.

Shaw GE, Shebbeare EO: The fishes of northern Bengal, *J R Asiat Soc Bengal, Sci* 3(1):1−2, 1937. 137, pl. 1−6.

Shrestha AB, Devkota LP: Climate change in the Eastern Himalayas: observed trends and model projections. Climate change impact and vulnerability in the Eastern Himalays-Technical Report 1, ICIMOD (International Centre for Integrated Mountain development), 2010, pp. 1−14.

Singer RA, Page LM: Revision of the zipper loaches, *Acanthocobitis* and *Paracanthocobitis* (Teleostei: Nemacheilidae), with descriptions of five new species, *Copeia* 103(2):378−401, 2015.

Singer RA, Pfeiffer JM, Page LM: A revision of the *Paracanthocobitis zonalternans* (Cypriniformes: Nemacheilidae) species complex with descriptions of three new species, *Zootaxa* 4324(1):85−107, 2017.

Singh A, Sen N, Bănărescu PM, Nalbant TT: New noemacheiline loaches from India (Pisces, Cobitidae), *Trav du Muséum d'Histoire Naturelle "Grigore Antipa"* 23:201−212, 1982.

Sinha B, Tamang L: *Creteuchiloglanis arunachalensis*, a new species of glyptosternine catfish (Teleostei: Sisoridae) from Arunachal Pradesh, northeastern India, *Ichthyol Res* 62(2):189−196, 2014.

Smith HM: Descriptions of new genera and species of Siamese fishes, *Proc United States Nat Mus* 79(2873):1−48, 1931. Pl. 1.

Smith HM: Contributions to the ichthyology of Siam. II−VI, *J Siam Soc Nat Hist Suppl* 9(1):53−87, 1933. Pls. 1−3.

Smith HM: *Chagunius*, a new genus of Asiatic cyprinoid fishes, *Proc Biol Soc, Wash* 51:157−158, 1938.

Smith HM: The freshwater fishes of Siam, or Thailand, *Bull United States Nat Mus* 1881: i−xi + 1−622, Pls. 1−9.

Soibam I, Hemanta Singh RK: Transtensional basin in oblique subduction margin: Imphal valley, an example. In Saklani PS, editor: *Himalaya (Geographical aspects)*, New Delhi, 2007, Satish Serial Publ House, pp 273−297.

Soibam I, Khuman MCH, Subhamenon SS: Ophiolitic rockes of the Indo-Myanmar ranges, NE India: relicts of an inverted and tectonically imbricated hyper-extended continental margin basin?. In *Sedimentary basins and crustal processes at continental margins: from modern hyper-extended margins of deformed ancient analogues*, Gibson GM, Roure F, Manatschal (Eds.) The Geological Society, London. 2015, pp. 301−331.

Starks EC: Synonymy (SEC) of the fish skeleton, *Proc Wash Acad Sci* 3:507−539, 1901. 43−45 pls.

Steindachner F: Ichthyologische Mittheilungen, IX. Verhandlungen der K.-K. zoologisch-botanischen Gesellschaft in Wien. 1866; 16: 761−796, Pls. 131−18.

Stiassny MLJ: The medium is the message: freshwaterbiodiversity in peril. In Cracraft J, Grifo FT, editors: *The living planet in crisis: biodiversity science and policy*, New York, 1999, Columbia University Press, pp 53−71.

Stiassny MLJ: Conservation of freshwater fish biodiversity: the knowledge impediment, *Verhandlungen der Ges fuÄNr Ichthyologie* 3:7−18, 2002.

Suckley G: Notices of certain new species of North American Salmonidae, chiefly in the collection of the N. W. Boundary Commission, in charge of Archibald Campbell, Esq., Commissioner of the United States, collected by Doctor C. B. R. Kennerly, naturalist to the... Ann Lyceum Nat Hist New York 1861; 7 (30), pp. 306−313.

Sufi SMK: Revision of the Oriental fishes of the family Mastacembelidae, *Bull Raffles Mus* 27:93−146, 1956. Pls. 13−26.

Swainson W: *On the natural history and classification of fishes, amphibians, & reptiles, or monocardian animals*, 1, London, 1838, A. Spottiswoode, i−vi + 1−368.

Swainson W: *On the natural history and classification of fishes, amphibians, & reptiles, or monocardian animals*, 2, London, 1839, Spottiswoode & Co., i−vi + 1−452.

Sykes WH: On the fishes of the Deccan, *Proc Zool Soc Lond* 6:157−165, 1839.

Taki Y, Katsuyama A, Urushido T: Comparative Morphology and interspecific relationships of the Cyprinid Genus *Puntius*, *Jpn J Ichth* 25(1):1−8, 1978.

Talwar PK, Jhingran AG: *Inland fishes of India and adjacent countries*, 1, New Delhi, 1991a, Oxford & IBH Publ Co, p 205.

Talwar PK, Jhingran AG: *Inland fishes of India and adjacent countries*, 2 vols, New Delhi, 1991b, Oxford & IBH Publishing Co., pp 1−2. i−xvii + 36 unnumbered + 1−1158, 1 pl, 1 map.

Talwar PK, Yazdani GM, Kundu DK: On a new eel-like fish of the genus *Pillaia* Yazdani (Pisces: Mastacembeloidei) from India, *Proc Indian Acad Sci B (Anim Sci)* 85(2):53−56, 1977.

Tamang L: *Garra magnidiscus*, a new species of cyprinid fish (Teleostei: Cypriniformes) from Arunachal Pradesh, northeastern India, *Ichthyol Explor Freshw* 24(1):31−40, 2013.

Tamang L, Chaudhry S: *Glyptothorax dikrongensis*, a new species of catfish (Teleostei: Sisoridae) from Arunchal Pradesh, northeastern India, *Ichthyol Res* 58(1):1−9, 2011.

Tamang L, Sinha B: Two new species of the South Asian catfish genus *Pseudolaguvia* from northeastern India (Teleostei: Sisoridae), *Zootaxa* 3887(1):37−54, 2014.

Tamang L, Sinha B: *Physoschistura walongensis*, a new species of loach (Teleostei: Nemacheilidae) from Arunachal Pradesh, northeastern India, *Zootaxa* 4173(3):280−288, 2016.

Tamang L, Chaudhry S, Choudhury D: *Erethistoides senkhiensis*, a new catfish (Teleostei: Erethistidae) from India, *Ichthyol Explor Freshw* 19(2):185−191, 2008.

Tamang L, Sinha B, Gurumayum SD: *Exostoma tenuicaudata*, a new species of glyptosternine catfish (Siluriformes: Sisoridae) from the upper Brahmaputra drainage, northeastern India, *Zootaxa* 4048(3):441−445 & 4111(5): 598−600, 2015.

Tang KL, Agnew MK, Hirt MV, Sado T, Schneider LM, Freyhof J, et al: Systematics of the subfamily Danioninae (Teleostei: Cypriniformes: Cyprinidae), *Mol Phylog Evol* 57:189−214, 2010.

Tejavej A: Taxonomic review of the cyprinid fish genus *Barilius* Hamilton, 1822 from Indochina (Cypriniformes: Cyprinidae). Master of Science (Fishery Science), Kasetsart University, Bangkok, 2010, pp. 1−148.

Thoni RJ, Gurung DB: *Parachiloglanis bhutanensis*, a new species of torrent catfish (Siluriformes: Sisoridae) from Bhutan, *Zootaxa* 3869(3):306−312, 2014.

Thoni RJ, Gurung DB: Morphological and molecular study of the torrent catfishes (Sisoridae: Glyptosterninae) of Bhutan including the description of five new species, *Zootaxa* 4476(1):40−68, 2018.

Thoni RJ, Hart R: Repatriating a lost name: notes on McClelland and Griffith's *Cobitis boutanensis* (Cypriniformes: Nemacheilidae), *Zootaxa* 3999(2):291−294, 2015.

Thoni RJ, Gurung DB, Mayden RL: A review of the genus *Garra* Hamilton 1822 of Bhutan, including the descriptions of two new species and three additional records (Cypriniformes: Cyprinidae), *Zootaxa* 4169(1):115−132, 2016.

Tilak R: The osteocranium and Weberian apparatus of *Eutropichthyes murius* (Ham): A study of interrelationship, *Zool Anz* 167:413−430, 1963.

Tilak R: A new sisorid catfish of the genus *Gagata* Bleeker from India, *Zoologische Mededelingen* 44(14):207−215, 1970.

Tilak R, Husain A: Description of a new psilorhynchid, *Psilorhynchus sucatio nudithoracicus* (Psilorhynchidae: Cypriniformes) from Uttar Pradesh, with notes on zoogeography, *Mitteilungen aus dem Zoologischen Mus Berl* 56(1):35−40, 1980.

Tilak R, Husain A: Description of a new loach, *Nemacheilus chindwinicus* sp. nov. (Homalopteridae, Cypriniformes) from Manipur, India, with notes on the systematic status of the genus *Nemacheilus* and the subfamily Nemacheilinae, *Mitteilungen aus dem Zoologischen Mus Berl* 66(1):51−58, 1990.

Van der Voo R: A plate-tectonic model for the Paleozoic assembly of Pangea based on paleomagnetic data. IN Contributions to the tectonics and geophysics of mountain chains, Hatcher RD Jr, Williams H, Zietz I. 1983, pp.19−23.

van Hasselt JC: Uittreksel uit een' brief van den Heer J. C. van Hasselt, aan den Heer C. J. Temminck, geschreven uit Tjecande, Residentie Bantam, den 28sten December 1822. Algemeene Konst− en Letter−bode voor het Jaar II Deel 1823; 35, pp.130−133.

Vierke J: Ein farbenfroher neuer Schlangenkopffisch aus Assam: *Channa bleheri* spec. nov, *Das Aquar* 25(259):20−24, 1991. Magazin für zeitgemässe Vivaristik.

Vinciguerra D: Viaggio di Leonardo Fea in Birmania e regioni vicine. XXIV. Pesci. Annali del Museo Civico di Storia Naturale di Genova, 1890; 29 [Ser. 2, vol. 9]: 129–362, pls. 7–11.

Vishwanath W, Kosygin L: Species status of *Poropuntius burtoni* (Mukerji 1934), (Cypriniformes: Cyprinidae) with a systematic note on *Poropuntius clavatus* (McClelland 1845), *J Bombay Nat Hist Soc* 98(1):31–37, 2001.

Vishwanath W: On a collection of fishes of the genus *Garra* Hamilton from Manipur, India, with a description of a new species, *J Freshw Biol* 5(1):59–68, 1993.

Vishwanath W, Darshan A: A new catfish species of the genus *Sisor* Hamilton (Teleostei: Siluriformes) from Manipur, *India Zoos' Print J* 20(8):1952–1954, 2005.

Vishwanath W, Darshan A: Two new catfish species of the genus *Pseudecheneis* Blyth (Teleostei: Siluriformes) from northeastern India, *Zoos' Print J* 2(3): 2627–2631 + 2 web pages, Pls. 1–2.

Vishwanath W, Joyshree H: A new species of genus *Garra* Hamilton–Buchanan (Teleostei: Cyprinidae) from Manipur, India, *Zoos' Print J* 20(4):1832–1834, 2005.

Vishwanath W, Joyshree H: A new sisorid catfish of the genus *Exostoma* Blyth from Manipur, India, *Zoos' Print J* 22(1):2531–2534, 2007.

Vishwanath W, Kosygin L: A new sisorid catfish of the genus *Myersglanis* Hora & Silas 1951, from Manipur, India, *J Bombay Nat Hist Soc* 96(2):291–296, 1999. Pl. 1.

Vishwanath W, Kosygin L: *Garra elongata*, a new species of the subfamily Garrinae from Manipur, India (Cyprinidae, Cypriniformes), *J Bombay Nat Hist Soc* 97(3):408–414, 2000a.

Vishwanath W, Kosygin L: Fishes of the cyprinid genus *Semiplotus* Bleeker 1859, with description of a new species from Manipur, India, *J Bombay Nat Hist Soc* 97(1):92–102, 2000b.

Vishwanath W, Laisram J: Two new species of *Puntius* Hamilton–Buchanan (Cypriniformes: Cyprinidae) from Manipur, India, with an account of *Puntius* species from the state, *J Bombay Nat Hist Soc* 101(1):130–137, 2004.

Vishwanath W, Laisram J: A new species of *Rasbora* Bleeker (Cypriniformes: Cyprinidae) from Manipur, India, *J Bombay Nat Hist Soc* 101(3):429–432, 2005.

Vishwanath W, Linthoingambi I: A new sisorid catfish of the genus *Glyptothorax* Blyth from Manipur, India, *J Bombay Nat Hist Soc* 102(2):201–203, 2006.

Vishwanath W, Linthoingambi I: Fishes of the genus *Glyptothorax* Blyth (Teleostei: Sisoridae) from Manipur, India, with description of three new species, *Zoos' Print J* 22(3): 2617–2626 + 3 web pages, 2007. Pls. [images] 1–3.

Vishwanath W, Linthoingambi I: Redescription of *Garra abhoyai* Hora (Teleostei: Cyprinidae" Garrinae) with a note on *Garra rupecula* from Manipur, India, *J Bombay Nat Hist Soc* 105(1):101–104, 2008.

Vishwanath W, Manojkumar W: Fishes of the cyprinoid genus *Psilorhynchus* McClelland from Manipur, India, with description of a new species, *Jpn J Ichthyol* 42(3/4):249–253, 1995.

Vishwanath W, Manojkumar W: A new bariline cyprinid fish of the genus *Barilius* Hamilton, from Manipur, India, *J Bombay Nat Hist Soc* 99(1):86–89, 2002.

Vishwanath W, Nebeshwar Sharma K: *Schistura reticulata*, a new species of balitorid loach from Manipur, India, with redescription of *S. chindwinica*, *Ichthyol Explor Freshw* 15(4):323–330, 2004.

Vishwanath W, Nebeshwar Sharma K: *Pterocryptis barakensis* Vishwanath & Nebeshwar Sharma, sp. nov. Pp. 99–100 + Pl. V (fig. 3) In Jayaram KC, editor: *Catfishes of India*, 2006, Narendra Publishing House.

Vishwanath W, Sarojnalini C: A new cyprinid fish *Garra manipurensis*, from Manipur, India, *Jpn J Ichthyol* 35(2):124–126, 1988.

Vishwanath W, Shanta K: *Schistura khugae*, a new replacement name for *S. macrocephalus* Vishwanath & Shanta, 2004 (Teleostei: Balitoridae), *Ichthyol Explor Freshw* 15(4):330, 2004a.

Vishwanath W, Shanta K: A new fish species of the Indo-Burmese genus *Badis* Bleeker (Teleostei: Perciformes) from Manipur, India, *Zoos' Print J* 19(9):1619–1621, 2004b.

Vishwanath W, Shanta Devi K: A new fish species of the genus *Garra* Hamilton–Buchanan (Cypriniformes: Cyprinidae) from Manipur, India, *J Bombay Nat Hist Soc* 102(1):86–88, 2005.

Vishwanath W, Shanta Kumar M: A new Nemacheiline fish of the genus *Schistura* McClelland (Cypriniformes: Balitoridae) from Manipur, India, *J Bombay Nat Hist Soc* 102(2):210–213, 2006.

Vishwanath W, Tombi Singh H: A new species of the genus *Puntius* Hamilton from Manipur, *Rec Zool Surv India* 83(1–2):129–133, 1986a.

Vishwanath W, Tombi Singh H: First record of the bagrid fish *Mystus microphthalmus* from India, *Japanese J Ichthyol* 33(2):197–199, 1986b.

Vishwanath W, Ng HH, Britz R, Kosygin Singh L, Chaudhury S, Conway KW: The status and distribution of freshwater fishes of the eastern Himalaya region. Chapter 3 In Allen DJ, Molur S, Daniel BA, editors: *The status and distribution of freshwater biodiversity in the Eastern Himalaya*, Cambridge and Gland, 2010, IUCN, and Coimbatore, India: Zoo Outreach Organisation, pp 22–41.

Walbaum, JJ: Petri Artedi sueci genera piscium. In quibus systema totum ichthyologiae proponitur cum classibus, ordinibus, generum characteribus, specierum differentiis, observationibus plurimis. Redactis speciebus 242 ad genera 52. Ichthyologiae pars III. Ant. Ferdin. Rose, Grypeswaldiae [Greifswald]. Part 3: [i–viii] + 1–723, Pls. 1–3, 1792. [Reprint 1966 by J. Cramer].

Wang Y, Sieh K, Tun ST, Lai K-Y, Than M: Active tectonics and earthquake potential of the Myanmar region, *J Geophys Res Solid Earth* 119:3767–3822, 2014.

Wegener A: Die Herausbildung der Grossformen der Erdrinde (Kontinente und Ozeane), auf geophysikalischer Grundlage." Petermanns Geographische Mitteilungen (in German). Annual meeting, German Geol Soc, Frankfurt am Main 1912; 63: 185–195, 253–256, 305–309.

Weitzman SH: The osteology of *Brycon meeki*, a generalized Characid fish, with an osteological definition of the family, *Stanf Ichthol Bull* 8(1):1–77, 1962.

Whitley GP: Ichthyological notes and illustrations, *Australian Zool* 10(1):1–50, 1941. Pls. 1–2.

Williams MW: Mountain Geography, University of Colarado at Boulder Week 11: Class notes. Accessed Dec 31, 2019. Snobear.colorado.edu. > Marckw > Mountains > week11.

Wu X−W, He MJ, Chu S−L: On the fishes of Sisoridae from the region of Xizang, *Oceanologia et Limnologia Sin* 12(1):74–79, 1981.

Yazdani GM: A new genus and species of fish from India, *J Bombay Nat Hist Soc* 69(1):134–135, 1972.

Yazdani GM, Talukdar SK: A new species of *Puntius* (Cypriniformes: Cyprinidae) from Khasi and Jaintia Hills (Meghalaya), India, *J Bombay Nat Hist Soc* 72(1):218–221, 1975.

Yazdani GM, Talwar PK: On the generic relationship of the eel-like fish, *Pillaia khajuriai* Talwar, Yazdani & Kundu (Perciformes, Mastacembeloidei), *Bull Zool Surv India* 4(3):287–288, 1981.

Zhang E, Chen Y-Y: *Garra tengchoingensis*, a new cyprinid species from the upper Irrawaddy River basin in Yunnan, China (Pisces: Teleostei), *Raff Bull Zool* 50(2):459–464, 2002.

Zhang E, Kottelat M: *Akrokolioplax*, a new genus of Southeast Asian labeonine fishes (Teleostei: Cyprinidae), *Zootaxa* 1225:21–30, 2006.

Zhou W, Li X, Thomson AW: A new genus of glyptosternine catfish (Siluriformes: Sisoridae) with descriptions of two new species from Yunnan, China, *Copeia* 2:226–241, 2011.

Zhu S-Q, Guo Q-Z: Descriptions of a new genus and a new species of noemacheiline loaches from Yunnan Province, China (Cypriniformes: Cobitidae), *Acta Zootaxonom Sin* 10(3):321–325, 1985.

Index

Note: Page numbers followed by "*f*" refer to figures.

A

Aborichthys, 33, 174–175
  A. boutanensis, 33, 175–176, 176*f*
  A. cataracta, 175, 373
  A. elongatus, 33, 176, 176*f*
  A. garoensis, 33, 177
  A. iphipaniensis, 33, 177, 177*f*
  A. kailashi, 33, 178, 178*f*
  A. kempi, 33, 175, 178, 179*f*, 373
  A. pangensis, 33, 178
  A. rosammae, 202
  A. tikaderi, 33, 179, 179*f*
  A. verticauda, 175, 373
  A. waikhomi, 33, 180, 180*f*
Acanthocobitis, 33, 174, 180, 188
  A. botia, 189*f*
  A. longipinnis, 180
  A. mackenziei, 190*f*
  A. marmorata, 191*f*, 374
  A. pavonacea, 33, 180, 181*f*
Acantopsis, 32, 162
  A. spectabilis, 32, 162, 162*f*
Adipose fin, 18
Ailia, 285
  A. coila, 38, 285, 286*f*
Ailiidae, 38–39, 285–289, 285*f*
  Asian schilbeids, 285–289
Air breathing catfishes, 280–283
Akysidae, 35, 210–212, 210*f*
  stream catfishes, 210–212
Akysis, 35, 211
  A. manipurensis, 35, 211, 211*f*
  A. prashadi, 35, 211, 212*f*
  A. variegatus variegatus, 211
Aluminum foil fish tag, 15, 15*f*
Ambassidae, 41, 329–333, 329*f*
  Asiatic glassfishes, 329–333
Amblyceps, 35, 206
  A. apangi, 35, 206, 206*f*
  A. arunachalensis, 207, 207*f*
    caudal complex, 207*f*
  A. arunchalensis, 35
  A. caecutiens, 206
  A. cerinum, 207
  A. mangois, 35, 208
  A. torrentis, 35, 209, 209*f*
  A. tuberculatum, 35, 209, 209*f*
  A. waikhomi, 35, 210, 210*f*

Amblyceps laticeps, 35, 208, 208*f*
Amblycipitidae, 35, 205–210, 206*f*
  torrent catfishes, 205–210
Amblypharyngodon, 27, 51, 53
  A. microlepis, 27, 54
  A. mola, 27, 53, 54*f*
Anabantidae, 42, 348–349, 348*f*
  climbing perches, 348–349
Anabas, 42, 348
  A. testudineus, 42, 348, 349*f*
Anchovies, 46–47
Angler fishes, 278–280
Anguilla, 27, 45
  A. bengalensis, 27, 45, 46*f*
Anguillidae, 27, 45, 45*f*
  freshwater eels, 45
Aplocheilidae, 40, 318–319
  Rivulines, 318–319
Aplocheilus, 40, 319
  A. panchax, 40, 319, 319*f*
Ariidae, 38, 283–284, 283*f*
  sea catfishes, 283–284
Arius
  A. burmanicus, 284
  A. spatula, 283
Asian schilbeids, 285–289
Asiatic glassfishes, 329–333
Awaous, 41, 346
  A. grammepomus, 42, 346, 347*f*

B

Badidae, 41, 337–343, 337*f*
  chameleon fishes, 337–343
Badis, 41, 337
  B. assamensis, 41, 337
  B. badis, 41, 338, 338*f*
    B. badis assamensis, 338
  B. blosyrus, 41, 338, 339*f*
  B. chittagongis, 41, 339, 339*f*
  B. dibruensis, 41, 339, 340*f*
  B. ferrarisi, 41, 340, 340*f*
  B. kanabos, 41, 341, 341*f*
  B. rhabdotus, 41, 341
  B. singenensis, 41, 342, 342*f*
  B. triocellus, 342
  B. tuivaiei, 41, 342, 342*f*
Bagarius, 35, 213–214
  B. bagarius, 35, 214, 214*f*

Bagarius (*Continued*)
    B. yarrelli, 35, 215, 215*f*
Bagrid catfishes, 292−315
Bagridae, 39−40, 292−315, 293*f*
    bagrid catfishes, 292−315
Bagrus
    B. aorellus, 313
    B. halepensis, 301
    B. lamarrii, 311, 313
Balitora, 15, 33, 170
    B. brucei, 33, 170−171, 171*f*
    B. burmanica, 33, 171, 171*f*
    B. eddsi, 33, 172
Balitoridae, 33, 170−174, 170*f*
    stream loaches, 170−174
Bangana, 27, 53−54, 365
    B. dero, 27, 55, 55*f*, 365
    B. devdevi, 27, 56, 56*f*, 365−366
    oromandibular structures, 365−366
Barbus
    B. hexagonolepis, 115
    B. progeneius, 145
Barilius, 27−28, 52, 56, 366−368
    B. barila, 27, 57, 57*f*, 368
    B. barna, 27, 57
    B. barnoides, 27, 58, 58*f*
    B. bendelisis, 27, 58, 59*f*
    B. cosca, 27, 59, 59*f*
    B. dogarsinghi, 27, 59, 60*f*
    B. gatensis, 368
    B. lairokensis, 28, 60, 60*f*
    B. profundus, 28, 61, 61*f*
    B. shacra, 28, 61, 61*f*
    B. tileo, 28, 62, 62*f*
    B. vagra, 28, 62, 62*f*
Batasio, 293−294
    B. affinis, 39, 294*f*, 295, 295*f*
    B. batasio, 39, 296, 296*f*
    B. buchanani, 294
    B. convexirostrum, 39, 296, 296*f*
    B. fasciolatus, 39, 297
    B. macronotus, 39, 297
    B. merianiensis, 39, 297, 297*f*
    B. procerus, 39
    B. spilurus, 39, 298, 298*f*
    B. tengana, 39, 298, 298*f*
Belone cancila, 318
Belonidae, 40, 317−318, 318*f*
    needle fishes, 317−318
Bengala, 28, 52, 62
    B. elanga, 28, 63
Bhavania arunachalensis, 172
Body depth (BD), 16
Bola coitor, 335

Botia, 32, 157
    B. dario, 32, 158, 158*f*
    B. histrionica, 32, 158, 158*f*
    B. lohachata, 32, 159, 159*f*
    B. rostrata, 32, 159, 160*f*
Botiid loaches, 157−161
Botiidae, 32, 157−161, 157*f*
    botiid loaches, 157−161
Branched rays, 20

C

Cabdio, 28, 51, 63
    C. crassus, 28, 63, 63*f*
    C. jaya, 28, 63, 64*f*
    C. morar, 28, 64, 64*f*
    C. ukhrulensis, 28, 64, 65*f*
Canthophrys, 33, 162
    C. gongota, 33, 162, 163*f*
Carps and minnows, 50−147
Caudal fins, 18
Caudal peduncle depth (CpD), 18
Caudal peduncle length (CpL), 18
Center of dispersal, 3−4
Chaca, 38, 279
    C. burmensis, 38, 279
    C. chaca, 38, 279, 280*f*
    C. hamiltonii, 279
Chacidae, 38, 278−280, 279*f*
    angler or frogmouth fishes, 278−280
Chagunius, 28, 53, 65
    C. chagunio, 28, 65
    C. nicholsi, 28, 66, 66*f*
Chameleon fishes, 337−343
Chanda, 41, 329
    C. baculis, 331
    C. lala, 332
    C. nama, 41, 329, 330*f*
Chandramara, 309, 374−375
Channa, 11, 42, 353
    C. amari, 42, 353, 353*f*
    C. amphibeus, 42, 353, 354*f*
    C. andrao, 42, 354
    C. aurantimaculata, 42, 354, 354*f*
    C. aurantipectoralis, 42, 355, 355*f*
    C. aurolineata, 42, 355, 356*f*, 376−377
    C. barca, 42, 356, 356*f*
    C. bleheri, 42, 356, 357*f*
    C. gachua, 42, 357, 357*f*
    C. lipor, 42, 358
    C. marulius, 42, 358, 358*f*, 376−377
    C. melanostigma, 42, 359, 359*f*
    C. orientalis, 353
    C. pardalis, 42, 359, 359*f*
    C. pomanensis, 354

C. punctata, 42, 360, 360*f*

C. stewartii, 42, 360

C. stiktos, 42, 361, 361*f*

C. striata, 42, 361, 361*f*

C. torsaensis, 42, 362, 362*f*

Channidae, 42, 352–362, 353*f*

Chaudhuriidae, 40, 324–325, 324*f*

earthworm eels, 324–325

Chela, 28, 51, 66

C. cachius, 28, 66

C. khujairokensis, 112

Chitala, 27, 43

C. chitala, 27, 44, 44*f*

Chopraia rupicola, 173

Chromis (Tilapia) mossambicus, 345

Cichlidae, 344–345, 344*f*

Cichlids, 344–345

Ciclidae, 41

Cirrhinus, 28, 52, 67

C. cirrhosus, 28

C. reba, 28, 68, 68*f*

Clarias, 38, 281

C. gariepinus, 38, 281, 281*f*

C. magur, 38, 281, 282*f*

Clariidae, 38, 280–283, 280*f*

air breathing catfishes, 280–283

Climbing perches, 348–349

Clupeidae, 27, 47–50, 47*f*

Clupisoma, 285–286

C. garua, 38, 286, 286*f*

C. montanum, 38, 287, 287*f*

C. pateri, 38

C. prateri, 287

Cobitidae, 32–33, 161–170, 161*f*

loaches, 161–170

Cobitis, 202

C. botia, 190

C. boutanensis, 175

C. corica, 186

C. fasciata, 186

C. pavonacea, 180

C. zonalternans, 188, 374

Cochlefelis, 283

C. burmanicus, 38, 283, 284*f*

Coius nandus, 336

Coloration, 23

Conta, 35, 213, 215

C. conta, 35, 215

C. pectinata, 35, 216, 216*f*

Continental drift, 5

Corica, 27, 48

C. soborna, 27, 48

Creteuchiloglanis, 35, 214, 216

C. arunachalensis, 35, 217, 217*f*

C. bumdelingensis, 35, 218

C. kamengensis, 35, 218, 218*f*

C. longipectoralis, 216

C. payjab, 35, 219, 219*f*

C. tawangensis, 35, 219*f*, 220, 220*f*

ventral surface of mouth and thorax of, 217*f*

Ctenopharyngodon, 28, 52, 68

C. idella, 28, 68, 68*f*

Cyprinidae, 27–32, 50–147, 50*f*

carps and minnows, 50–147

Cyprinion, 372

C. semiplotum, 372

Cyprinus, 28, 52, 69

C. cachius, 67

C. carpio, 28, 69–70, 69*f*

C. catla, 105

C. cirrhosus, 67, 67*f*

C. cosuatis, 118

C. dangila, 71

C. daniconius, 135

C. dyocheilus, 110

C. gonius, 111

C. richardsonii, 140

C. sarana, 143

C. ticto, 129

D

Danio, 28, 52

D. assamila, 28, 70, 71*f*

D. dangila, 28, 71

D. jaintianensis, 28, 71

D. meghalayensis, 28, 71, 71*f*

D. quagga, 28, 72, 72*f*

D. rerio, 28, 72–73, 72*f*

D. yuensis, 79

Danionella, 28, 51, 73

D. priapus, 28, 73, 73*f*

Danionin notch and mandibular symphyseal knob, 368

Dario, 41, 343

D. dario, 41, 343

D. kajal, 41, 343

Devario, 28, 52, 73

D. acuticephala, 28, 74, 74*f*

D. aequipinnatus, 28, 74, 75*f*, 369

D. coxi, 28, 74, 75*f*

D. deruptotalea, 28, 76, 76*f*

D. devario, 28, 76, 77*f*

D. horai, 28, 77, 369

D. manipurensis, 28, 77, 77*f*

D. naganensis, 28, 78, 78*f*

D. yuensis, 28, 78, 78*f*

Doryichthys, 321

Drainage basin evolution, 5–6

Drums, croakers, 334–335

E

Earthworm eels, 324–325
Eastern Himalaya (EH), 1–3, 2f
Engraulididae, 27, 46–47, 46f
Erethistes, 15, 35, 213, 220
    E. hara, 35, 220
    E. horai, 35, 221
    E. jerdoni, 35, 221, 222f
    E. koladynensis, 35, 221, 222f
    E. pusillus, 35, 220, 222, 223f
Erethistoides, 35–36, 213, 223
    E. ascita, 35, 223
    E. cavatura, 35, 224
    E. infuscatus, 35, 224
    E. montana, 36, 223–224, 224f
    E. senkhiensis, 36, 225
    E. sicula, 36, 225
    pectoral spine of, 223f
Esomus, 29, 52, 79
    E. danrica, 79
    E. danricus, 29
Esox
    E. cancila, 318
    E. panchax, 319
Euchiloglanis
    E. kamengensis, 219
    E. kishinouvei, 374
Euchiloglanis, 8–9
Eutropiichthys, 285, 287
    E. burmannicus, 38, 288, 288f
    E. cetosus, 38, 288
    E. murius, 39, 288, 288f
    E. vacha, 39, 289, 289f
Eutropius macropthalmos, 290–291
Exostoma, 36, 214, 225
    E. barakensis, 36, 225, 226f
    E. berdmorei, 225
    E. blythii, 252–253
    E. dujangensis, 36, 226
    E. kottelati, 36, 227, 227f
    E. labiatum, 36, 227, 227f
    E. mangdechhuensis, 36, 227
    E. sawmteai, 36, 228, 228f
    E. tenuicaudatum, 36, 228, 228f

F

Featherfin knifefishes, 43–45
Fins, 18
    ray counts, 19–20
Fish
    biogeography, 3–11
        center of origin of freshwater fishes, 3–4
        drainage basin evolution, 5–6

        plate tectonics, 5
    collection, 15
    morphological features, 16–18
        measurements, 16–18
    photography, 16
    preservation, 15–16
Freshwater
    biodiversity, 1
    center of origin of freshwater fishes, 3–4
    ecosystems, 1
    eels, 45
Frogmouth fishes, 278–280

G

Gagata, 36, 213, 229, 229f
    G. cenia, 36, 229, 230f
    G. dolichonema, 36, 230, 230f
    G. gagata, 36, 230
    G. gasawyuh, 230
    G. sexualis, 36, 231
Garo, 40, 324
    G. khajuriai, 40, 324
Garra, 29–30, 50, 79, 369
    G. abhoyai, 29, 80, 80f
    G. annandalei, 29, 80
    G. arunachalensis, 29, 80, 80f
    G. arupi, 29, 81, 81f
    G. biloborostris, 29, 82, 82f
    G. bimaculacauda, 29, 83
    G. birostris, 29, 83, 83f
    G. chakpiensis, 29, 84, 84f
    G. chathensis, 29, 85
    G. chindwinensis, 29, 85
    G. chivaensis, 29
    G. clavirostris, 29, 85, 85f, 86f
    G. compressa, 29, 86–87, 86f
    G. cornigera, 29, 87, 87f
    G. elongata, 29, 88, 88f
    G. gotyla, 29, 88, 88f, 89f
    G. kalpangi, 29, 89
    G. kempi, 29, 90, 90f
    G. khawbungi, 29, 89
    G. koladynensis, 29, 91, 91f
    G. lamta, 29, 92, 92f
    G. lissorhynchus, 29, 92, 92f
    G. litanensis, 29, 92, 93f
    G. magnacavus, 29, 93
    G. magnidiscus, 29, 93, 93f, 94f
    G. manipurensis, 29, 94, 94f, 95f
    G. matensis, 29, 95, 95f
    G. mini, 29, 96
    G. naganensis, 29, 96, 96f
    G. nambulica, 29, 97, 97f
    G. namyaensis, 29, 97, 97f

G. nasuta, 29, 98, 98*f*, 371–372
G. nepalensis, 29, 98
G. paralissorhynchus, 29, 99, 99*f*
G. parastenorhynchus, 29, 99, 99*f*
G. paratrilobata, 29, 100, 100*f*
G. quadratirostris, 29, 100, 101*f*
G. rakhinica, 29, 101, 102*f*
G. ranganensis, 83
G. rupicola, 29, 102
G. substrictorostris, 29, 102, 103*f*
G. trilobata, 30, 103, 103*f*
G. tyao, 101
G. ukhrulensis, 30, 104, 104*f*, 105*f*
snout
    morphology, 369–371
    with proboscis group, 371
    with smooth dorsal surface group, 369–371
    with transverse lobe group, 371
Gibelion, 30, 105
    G. catla, 30, 105, 105*f*
Gill raker, 23
Glossogobius, 42, 347
    G. giuris, 42, 347, 347*f*
Glyptosternoids, 8–9
Glyptosternon
    G. labiatus, 227
    G. striatus, 231, 248, 248*f*, 249*f*
    G. sulcatus, 261, 263
Glyptosternum, 374
    G. hodgarti, 260
Glyptothorax, 15, 36–37, 213, 231
    G. ater, 36, 232, 232*f*
    G. botius, 36, 232, 232*f*
    G. burmanicus, 36, 233, 233*f*, 374
    G. caudimaculatus, 36, 233, 234*f*
    G. cavia, 36, 234, 234*f*, 374
    G. chavomensis, 245
    G. chimtuipuiensis, 36, 235, 235*f*
      thoracic adhesive apparatus, 235*f*
    G. churamanii, 36, 236, 236*f*
    G. dikrongensis, 36, 237, 237*f*
    G. giudikyensis, 36, 237
    G. gopii, 36, 238
    G. gracilis, 36, 238, 238*f*
    G. granulus, 36, 238, 238*f*
    G. igniculus, 36, 239
    G. indicus, 36, 239, 239*f*, 240*f*, 241*f*
    G. jayarami, 36, 240, 241*f*, 242*f*
    G. kailashi, 36, 240
    G. maceriatus, 36, 241, 243*f*
    G. manipurensis, 36, 244, 244*f*
    G. mibangi, 36, 245, 245*f*
    G. ngapang, 36, 245, 245*f*
    G. pantherinus, 36, 246, 246*f*

    G. radiolus, 36, 246
    G. rugimentum, 36, 246*f*, 247, 247*f*
    G. scrobiculus, 36, 247, 248*f*
    G. senapatiensis, 36, 248
    G. striatus, 36
    G. telchitta, 36, 232, 249, 249*f*
    G. tuberculatus, 264
    G. ventrolineatus, 37, 250, 250*f*
    G. verrucosus, 37, 250, 250*f*, 251*f*
    shapes of thoracic adhesive apparatus, 231*f*
Gobies, 345–348
Gobiidae, 41–42, 345–348
    gobies, 345–348
Gobius
    G. grammepomus, 346
    G. ocellaris, 346
    G. platycephalus, 347
Gogangra, 37, 213, 229*f*, 251
    G. laevis, 37, 251
    G. viridescens, 37, 252, 252*f*
Gonialosa, 27, 48
    G. manmina, 27, 48
Gonorhynchus rupicolus, 102
Gouramies, 349–352
Gudusia, 27, 48
    G. chapra, 27, 49, 49*f*

H

Hara
    H. hara, 221
    H. horai, 221
    H. jerdoni, 221
    H. koladynensis, 222
Head length (HL), 16, 18
Helicophagus hypophthalmus, 292
Hemibagrus, 293, 299
    H. menoda, 39, 299, 299*f*
    H. microphthalmus, 39, 299, 300*f*
    H. peguensis, 39, 300, 300*f*
Hemimyzon, 33, 172
    H. arunachalensis, 33, 172, 172*f*
    H. indicus, 33, 173, 173*f*
Herrings, 47–50
Heteropneustes, 38, 281–282
    H. fossilis, 38, 282, 282*f*
Homaloptera formosana, 172
Homalopteroides, 33, 173
    H. rupicola, 33, 173, 174*f*
Horobagridae, 39, 289–291, 290*f*
    imperial or sun catfishes, 289–291
Hypophthalmichthys, 30, 51, 106
    H. molitrix, 30, 106, 106*f*
    H. nobilis, 30, 106, 106*f*
Hypsibarbus, 30, 53, 107

Hypsibarbus (*Continued*)
  H. myitkyinae, 30, 107, 107*f*

I

Ichthyofaunal diversity, 11–12
Imperial catfishes, 289–291
Indo-Burman range. *See* Indo-Myanmar range
Indo-Myanmar range, 10–11

J

Johnius, 41, 334
  J. coitor, 41, 334, 335*f*

K

Kryptopterus indicus, 277

L

Labeo, 30, 53, 107–108
  L. angra, 30, 108
  L. bata, 30, 108, 108*f*, 366
  L. boga, 30, 109, 109*f*
  L. calbasu, 30, 109, 109*f*
  L. dyocheilus, 30, 110, 110*f*
  L. gonius, 30, 110, 110*f*
  L. nandina, 30, 111
  L. pangusia, 30, 111, 111*f*
  L. rohita, 30, 111, 112*f*
Labrus
  L. badis, 337–338
  L. dario, 343
Laguvia
  L. manipurensis, 211
  L. ribeiroi, 269
  L. shawi, 269
Last ray of dorsal and anal fins, 20
Lateral line scales, 21–22
Laubuka, 30, 51, 112
  L. khujairokensis, 30, 112, 112*f*
  L. laubuca, 30, 113, 113*f*
  L. parafasciata, 30, 113, 113*f*
Leiodon, 42, 363
  L. cutcutia, 42, 363, 363*f*
Lepidocephalichthys, 33, 163
  L. alkaia, 33, 163, 163*f*
  L. annandalei, 33, 164, 164*f*
  L. arunachalensis, 33, 164
  L. berdmorei, 33, 165, 165*f*
  L. goalparensis, 33, 165, 166*f*
  L. guntea, 33, 165, 167*f*
  L. irrorata, 33, 166, 168*f*, 372–373
    type locality, 372–373
  L. micropogon, 166, 168*f*
Leuciscus
  L. belangeri, 120

L. nobilis, 107
Livebearers, 320–321
Loaches, 161–170

M

Macrognathus, 40–41, 326
  M. aral, 40, 326, 326*f*
  M. armatus, 328
  M. lineatomaculatus, 41, 327
  M. morehensis, 41, 327, 327*f*
  M. pancalus, 41, 327*f*, 328
  M. siangensis, 328
Macrones
  M. bleekeri, 301
  M. merianiensis, 297
  M. microphthalmus, 300
  M. montanus var. *dibrugarensis*, 303
  M. peguensis, 300
  M. pulcher, 305
Macropteronotus magur, 281
Malapterurus coila, 286
Malapterus bengalensis, 285
Mastacembelidae, 40–41, 325–329, 325*f*
  spiny eels, 325–329
Mastacembelus, 41, 326, 328
  M. armatus, 41, 328, 328*f*
  M. tinwini, 41, 329, 329*f*
Mesonoemacheilus reticulofasciatus, 202
Meyersglanis, 214
  M. jayarami, 253, 253*f*, 255*f*
Micro glass fish, 73
Microphis, 40, 323
  M. deocata, 40, 321, 321*f*
Monopterus, 40, 322
  M. cuchia, 40, 322, 322*f*
  M. ichthyophoides, 40, 323
  M. javanensis, 40, 322–323, 323*f*
  M. rongsaw, 40, 323
Mountain carps, 147–157
Mugil
  M. corsula, 316
  M. hamiltoni, 317
Mugilidae, 40, 316–317, 316*f*
  mullets, 316–317
Mullets, 316–317
Mustura, 33–34, 175, 181
  M. celata, 181
  M. chhimtuipuiensis, 33, 182
  M. chindwinensis, 33, 182, 182*f*
  M. dikrongensis, 182, 183*f*
  M. harkishorei, 33, 183, 183*f*
  M. prashadi, 34, 183
  M. tigrina, 34, 184, 184*f*
  M. tuivaiensis, 34, 185, 185*f*

M. walongensis, 34, 185, 186*f*
pectoral fin of male, 181*f*
Myersglanis, 37, 252
   M. blythii, 37, 252, 255*f*
   M. jayarami, 37
Mystus, 293, 301
   M. bleekeri, 39, 301, 301*f*
   M. carcio, 39, 301−302, 301*f*, 375
   M. cavasius, 39, 302, 302*f*
   M. cineraceus, 39, 302, 303*f*
   M. dibrugarensis, 39, 303, 303*f*
   M. falcarius, 39, 303, 304*f*
   M. microphthalmus, 300
   M. ngasep, 39, 304, 304*f*
   M. prabini, 39, 305, 305*f*
   M. pulcher, 39, 305, 305*f*
   M. rufescens, 39, 306, 306*f*
   M. tengara, 39, 306, 306*f*
   M. vittatus, 375

N

Nandidae, 41, 335−336, 335*f*
   Asian leaffishes, 335−336
Nandus, 41, 336
   N. andrewi, 41, 336
   N. marmoratus, 336
   N. nandus, 41, 336, 336*f*
Nangra, 37, 213, 254
   N. assamensis, 37, 254, 255*f*
   N. bucculenta, 37, 256
   N. nangra, 37, 256, 256*f*
   N. ornata, 37, 256
Needle fishes, 317−318
Nemacheilidae, 33−35, 174−205, 174*f*
   river loaches, 174−205
Nemacheilus, 34, 175, 186
   N. brunneanus, 192
   N. chindwinicus, 196
   N. corica, 34, 186, 186*f*
   N. multifasciatus, 199
   N. prashadi, 183, 184*f*
   N. zonalternans, 374
Nemachilus
   N. beavani, 195
   N. kangjupkhulensis, 198
   N. mackenziei, 190
   N. manipurensis, 194
   N. sikmaiensis, 204
   N. zonalternans, 191
Neoeucirrhichthys, 33, 168
   N. maydelli, 33, 169, 169*f*
Neolissochilus, 30, 53, 114
   N. dukai, 30, 114, 114*f*
   N. heterostomus, 30, 114, 115*f*

N. hexagonolepis, 30, 115, 115*f*
N. kaladanensis, 30, 115, 116*f*
N. spinulosus, 30, 116
N. stevensonii, 30, 116, 117*f*
N. stracheyi, 30, 117, 117*f*
Neonoemacheilus, 34, 175, 187, 187*f*
   N. assamensis, 34, 187, 187*f*
   N. morehensis, 188
   N. peguensis, 34, 188, 188*f*
Noemacheilus
   N. arunachalensis, 164, 164*f*, 205
   N. devdevi, 196
   N. nagaensis, 200
   N. sijuensis, 203
   N. singhi, 204
Notopteridae, 27, 43−45, 43*f*
   featherfin knifefishes, 43−45
Notopterus, 27, 44
   N. notopterus, 27, 44, 44*f*

O

Occipital process (OP), 220
Ocellus, 23
Olyra, 293, 307
   O. kempi, 39, 307, 307*f*, 375−376
   O. laticeps, 208, 376
   O. longicaudata, 39, 307, 307*f*, 376
   O. parviocula, 39, 308
   O. praestigiosa, 39, 308, 308*f*
   O. saginata, 39, 309, 309*f*
Ompok, 38, 274
   O. bimaculatus, 38, 274, 274*f*
   O. pabda, 38, 274, 275*f*
   O. pabo, 38, 275, 275*f*
   O. siluroides, 274
Oncorhynchus, 40, 315
   O. mykiss, 40, 315, 316*f*
Ophicephalus
   O. amphibeus, 354
   O. aurolineatus, 355
   O. gachua, 357
   O. marulius, 358
   O. punctatus, 360
   O. stewartii, 360, 360*f*
   O. striatus, 361
Ophidium aculeatum, 326
Opsarius, 366−368
   O. guttatus, 135
   O. tileo, 368
Oreichthys, 30, 52, 117
   O. andrewi, 30, 118, 118*f*
   O. cosuatis, 30, 118
   O. crenuchoides, 30, 118, 118*f*
Oreinus, 30, 50, 119

Oreinus (Continued)
  O. molesworthi, 30, 119, 119*f*
  O. progastus, 139
Oreochromis, 41, 344
  O. hunteri, 344
  O. mossambica, 41, 344, 344*f*
  O. niloticus, 41, 345, 345*f*
Oreoglanis, 37, 214, 257
  O. majuscula, 37
  O. majusculus, 257*f*, 258, 258*f*
  O. pangenensis, 37, 259, 259*f*
  O. siamensis, 257
Osphronemidae, 42, 349—352, 349*f*
Osteobrama, 30, 51, 119
  O. belangeri, 30, 120, 120*f*
  O. cotio, 30, 120, 120*f*
  O. cunma, 30, 121, 121*f*
  O. feae, 30, 121, 121*f*

P

Pachypterus, 285, 289
  P. atherinoides, 39, 290, 290*f*
Pangaea, 5
Pangasianodon, 292
  P. gigas, 292
  P. hypophthalmus, 39, 292, 292*f*
Pangasiidae, 39, 291—292, 291*f*
  shark catfishes, 291—292
Pangio, 33, 169
  P. apoda, 33, 169
  P. pangia, 33, 169, 169*f*
Paracanthocobitis, 34, 188, 189*f*
  P. abutwebi, 34, 189
  P. botia, 34, 189
  P. linypha, 34, 190, 190*f*
  P. mackenziei, 34, 190
  P. marmorata, 34, 191, 373—374
  P. triangula, 34, 192
  P. tumitensis, 191
Parachiloglanis, 37, 213, 259
  P. benjii, 37, 259
  P. bhutanensis, 37, 259
  P. dangmechhuensis, 37, 260
  P. drukyulensis, 37, 260
  P. hodgarti, 37, 260, 260*f*
Parambassis, 41, 330
  P. baculis, 41, 330, 330*f*
  P. bistigmata, 41, 331, 331*f*
  P. lala, 41, 331, 332*f*
  P. ranga, 41, 332, 332*f*
  P. serrata, 41, 333, 333*f*
  P. waikhomi, 41, 333, 334*f*
Perca
  P. nilotica, 345

P. scandens, 348
Pethia, 30—31, 53, 122
  P. ater, 30, 122, 122*f*
  P. aurea, 31, 123
  P. canius, 31, 123
  P. conchonius, 31, 123, 123*f*
  P. expletiforis, 31, 124, 124*f*
  P. gelius, 31, 124, 124*f*
  P. guganio, 31, 125, 125*f*
  P. khugae, 31, 125, 125*f*
  P. manipurensis, 31, 126, 126*f*
  P. meingangbii, 31, 126
  P. phutunio, 31, 127, 127*f*
  P. rutila, 31, 127
  P. shalynius, 31, 127, 128*f*
  P. stoliczkana, 31, 128, 128*f*
  P. ticto, 31, 129, 129*f*
  P. yuensis, 31, 129, 129*f*
Physoschistura, 34, 175, 184, 192
  P. chhimtuipuiensis, 182
  P. chindwinensis, 182
  P. dikrongensis, 182
  P. elongata, 34, 192, 192*f*
  P. harkishorei, 183
  P. tigrinum, 184
  P. tuivaiensis, 185
  P. walongensis, 185
Pillaia, 40, 325
  P. indica, 40, 325, 325*f*
  P. khajuriai, 324
Pimelodus
  P. aor, 312
  P. bagarius, 214
  P. batasio, 296
  P. batasius, 302
  P. botius, 232
  P. carcio, 302
  P. cavasius, 302
  P. cavia, 234
  P. cenia, 229
  P. chandramara, 309—310, 374—375
  P. conta, 215
  P. gagata, 229—230
  P. hara, 221
  P. mangois, 208, 208*f*
  P. menoda, 299
  P. murius, 288
  P. nangra, 254, 256
  P. rama, 309—310, 374—375
  P. rita, 310
  P. telchitta, 249
  P. tengana, 299, 306
  P. vacha, 289
  P. variegatus, 211

P. viridescens, 252
Pipe fishes, 321
Plate tectonics, 5
Platycara nasuta, 98, 371
Platystacus chaca, 279
Poecilia, 40, 320
    P. reticulata, 40, 320, 320*f*
    P. vivipara, 320
Poeciliidae, 40, 320–321, 320*f*
    livebearers, 320–321
Poropuntius, 31, 53, 130
    P. burtoni, 31, 131, 131*f*
    P. clavatus, 31, 131, 131*f*
    P. margarianus, 31, 132
    P. normani, 130
    P. shanensis, 31, 132, 132*f*, 372
Post-Himalayas orogenic movement, 9
Predorsal length (PDL), 18
Principal rays of caudal fin, 20
Proboscis, 371
Procurrent rays, 20
Proeutropiichthys, 39, 285, 290
    P. macropthalmos, 290, 291*f*
Pseudecheneis, 37, 213, 261, 261*f*
    P. crassicauda, 37, 261
    P. eddsi, 37, 262
    P. koladynae, 37, 262, 262*f*
    P. serracula, 37, 263
    P. sirenica, 37, 263
    P. sulcata, 37, 263, 264*f*
    P. ukhrulensis, 37, 264, 264*f*
Pseudolaguvia, 37–38, 213, 264, 265*f*
    P. assula, 37, 265
    P. ferruginea, 37, 265, 266*f*
    P. ferula, 37, 266, 266*f*
    P. flavida, 37, 266
    P. foveolata, 37, 267, 267*f*
    P. fucosa, 37, 267
    P. inornata, 37, 267
    P. jiyaensis, 37, 268, 268*f*
    P. magna, 37, 268
    P. muricata, 37, 269, 269*f*
    P. nubila, 37, 269
    P. ribeiroi, 37, 269
    P. shawi, 37, 269, 270*f*
    P. spicula, 38, 270
    P. viriosa, 38, 270, 270*f*
Psidium mastacembelus, 328
Psilorhynchidae, 32, 147–157, 147*f*
    mountain carps, 147–157
Psilorhynchus, 15, 32, 147
    P. amplicephalus, 32, 147, 148*f*
    P. arunachalensis, 32, 148–149, 148*f*
    P. balitora, 32, 149, 149*f*

    P. bichomensis, 32, 149, 149*f*
    P. chakpiensis, 32, 150, 150*f*
    P. hamiltoni, 32, 150, 150*f*
    P. homaloptera, 32, 151, 151*f*
    P. kaladanensis, 32, 151, 151*f*
    P. khopai, 32, 152, 152*f*
    P. konemi, 32, 152–153, 152*f*
    P. maculatus, 32, 153, 153*f*
    P. microphthalmus, 32, 153, 153*f*
    P. nahlongthai, 32, 154, 154*f*
    P. nepalensis, 32, 154
    P. ngathanu, 32, 154
    P. nudithoracicus, 32, 155
    P. pseudecheneis, 32, 156
    P. rowleyi, 32, 156, 156*f*
    P. sucatio, 32, 156, 157*f*
Pterocryptis, 38, 275
    P. barakensis, 38, 276, 276*f*
    P. berdmorei, 38, 276, 276*f*
    P. gangelica, 38, 275, 277
    P. indica, 38, 277, 277*f*
    P. subrisa, 38, 277
Puffers, globefishes, 362–363
Puntius, 25, 31, 53, 132
    P. bizonatus, 127
    P. chola, 31, 132, 133*f*
    P. khugae, 126
    P. meingangbii, 126*f*, 127
    P. sophore, 31, 133, 133*f*
    P. stoliczkanus, 129
    P. terio, 31, 133, 134*f*

R

Raiamas, 31, 51, 133–134
    R. bola, 134, 134*f*
    R. guttatus, 134, 135*f*
Rama, 293, 309, 374–375
    R. chandramara, 40, 309, 309*f*
Rasbora, 31, 51, 135
    R. bola, 31
    R. daniconius, 31, 135, 135*f*
    R. guttatus, 31
    R. ornata, 31, 136, 136*f*
    R. rasbora, 31, 136, 136*f*
Rays in paired fins, 20
Rhinomugil, 40, 316
    R. corsula, 40, 316, 317*f*
Rhyacoschistura, 34, 175, 193
    R. ferruginea, 34, 193, 193*f*
    R. larreci, 193
    R. maculosa, 34, 193
    R. manipurensis, 34, 194, 194*f*, 374
    R. porocephala, 34, 194, 195*f*
Rhynchobdella aral, 327

Rita, 293, 295*f*, 310
  R. rita, 40, 310, 310*f*
  R. sacerdotum, 40, 310−311, 311*f*
River loaches, 174−205
Rivulines, 318−319
Rostral flap, 370
Rostral lobe, 370−371

S

Salmo
  S. mykiss, 315
  S. scouleri, 315
Salmonidae, 18, 40, 315, 315*f*
  trouts, 315
Salmostoma, 31, 51, 137
  S. bacaila, 31, 137, 137*f*
  S. phulo, 31, 137, 137*f*
  S. sladoni, 31, 138, 138*f*
Scale counts, 21−26
  lateral line scales, 21−22
  transverse scales, 22−26
Schistura, 34−35, 175, 195
  S. aizawlensis, 34, 195
  S. beavani, 34, 195
  S. chindwinica, 34, 196, 196*f*, 373
  S. devdevi, 34, 196, 197*f*
  S. fasciata, 34, 197, 197*f*
  S. ferruginea, 193
  S. kangjupkhulensis, 34, 197, 197*f*, 373−374
  S. khugae, 34, 198, 198*f*
  S. koladynensis, 34, 198, 198*f*
  S. maculosa, 193*f*, 194
  S. minuta, 34
  S. minutus, 199
  S. mizoramensis, 34, 199, 199*f*
  S. multifasciata, 34, 199
  S. nagaensis, 34, 199, 200*f*
  S. papulifera, 34, 200, 200*f*
  S. paucireticulata, 34, 201, 201*f*
  S. porocephala, 195
  S. rebuw, 34, 201, 201*f*
  S. reticulata, 34, 202, 202*f*
  S. reticulofasciata, 34, 202
  S. rosammae, 34, 202
  S. rupecula, 195
  S. savona, 34, 202, 203*f*
  S. scaturigina, 34, 203
  S. sijuensis, 35, 203, 203*f*
  S. sikmaiensis, 35, 204, 204*f*
  S. singhi, 35, 204
  S. syngkai, 35, 204, 204*f*
  S. tirapensis, 35, 205
  S. zonata, 35, 205, 205*f*
Schizothorax, 31, 50, 138

  S. chivae, 31, 138
  S. esocinus, 138
  S. progastus, 31, 139, 139*f*
  S. richardsonii, 31, 139, 139*f*
Sciaenidae, 41, 334−335, 334*f*
  drums, croakers, 334−335
Sea catfishes, 283−284
Sea horses, 321
Securicula, 31, 51, 140
  S. gora, 31, 140, 140*f*
Semiplotus, 31−32, 52, 140, 372
  S. cirrhosus, 31, 141
  S. modestus, 32, 141, 142*f*
  S. semiplotus, 32, 142, 142*f*, 372
Setipinna, 27, 46
  S. phasa, 27, 46, 47*f*
Shark catfishes, 291−292
Sheat fishes, 273−278
Sicamugil, 40, 317
  S. cascasia, 40, 317, 317*f*
Silurichthys berdmorei, 276
Siluridae, 38, 273−278, 273*f*
  sheat fishes, 273−278
Silurus
  S. anguillaris, 281
  S. atherinoides, 289−290
  S. attu, 278
  S. bimaculatus, 274
  S. fossilis, 282
  S. gariepinus, 281
  S. garua, 286
  S. muelleri, 278
  S. pabda, 274
  S. pabo, 275
Simple rays, 19
Sisor, 38, 213, 271
  S. barakensis, 38, 271*f*, 272
  S. chennuah, 38, 272, 272*f*
  S. rabdophorus, 38, 271−272, 273*f*
  S. rheophilus, 38, 272
  S. torosus, 38, 273
Sisorid catfishes, 212−273
Sisoridae, 35−38, 212−273, 212*f*
Snakeheads, 352−362
Sperata, 293, 311
  S. acicularis, 40, 311, 312*f*, 314*f*
  S. aor, 40, 312, 312*f*, 314*f*
  S. aorella, 40, 313, 313*f*, 314*f*
  S. lamarii, 40, 313, 313*f*, 314*f*
  S. seenghala, 313
Spines, 19
Spiny eels, 325−329
Standard length (SL), 16
Stream catfishes, 210−212

Stream loaches, 170−174
Sun catfishes, 289−291
Swamp eels, 322−324
Synbranchidae, 40, 322−324, 322*f*
    swamp eels, 322−324
Syncrossus, 32, 160
    S. berdmorei, 32, 160, 160*f*
Syngnathidae, 40, 321, 321*f*
    pipe fishes and sea horses, 321
Syngnathus deocata, 321
Systematic index, 27−42
Systomus, 32, 53, 143
    S. sarana, 32, 143, 143*f*

T

Tariqilabeo, 32, 51, 143
    T. burmanicus, 32, 143, 144*f*
    T. latius, 32, 144, 144*f*
Tectonics, 5
Tenualosa, 27, 49
    T. ilisha, 27, 49, 49*f*
Tetraodon
    T. cutcutia, 363
    T. marmoratus, 363
Tetraodontidae, 42, 362−363, 363*f*
Tor, 32, 145
    T. mosal, 32, 145, 145*f*
    T. putitora, 32, 145, 145*f*
    T. tor, 32, 146, 146*f*
    T. yingjiangensis, 32, 146, 146*f*, 372
Torrent catfishes, 205−210
Total length (TL), 16
Transverse lobe, 371
Transverse scales, 22−26
    color marks, 23

diagnosis, 26
distribution, 26
figures, 25
fish coloration, 23
gill raker, 23
mouth position, 22
osteology, 23
preparation, 23−25
recording, 25
scales, 22
terminology, 25
Trichogaster, 42, 350
    T. fasciata, 42, 350, 350*f*
    T. labiosa, 42, 350
    T. lalius, 42, 351, 351*f*
    T. sota, 42, 352, 352*f*
Trouts, 315

U

Unibranchiapertura cuchia, 323

V

Vicariance, evolution by, 7−11

W

Wallago, 38, 278
    W. attu, 38, 278, 278*f*
Wetlands, 1

X

Xenentodon, 40, 318
    X. cancila, 40, 318, 318*f*

Z

Zoogeography, 7−8